PHYSICS OF ATOMIC COLLISIONS

FIZIKA ATOMNYKH STOLKNOVENII

ФИЗИКА АТОМНЫХ СТОЛКНОВЕНИЙ

# The Lebedev Physics Institute Series

## Editor: Academician D. V. Skobel'tsyn
### Director, P. N. Lebedev Physics Institute, Academy of Sciences of the USSR

*Proceedings (Trudy) of the P. N. Lebedev Physics Institute*

Volume 51

# PHYSICS OF ATOMIC COLLISIONS

### Edited by
## Academician D. V. Skobel'tsyn
*Director, P. N. Lebedev Physics Institute*
*Academy of Sciences of the USSR, Moscow*

Translated from Russian by Paul Robeson, Jr.

CONSULTANTS BUREAU
NEW YORK–LONDON
1971

ISBN 978-1-4684-1592-6          ISBN 978-1-4684-1590-2 (eBook)
DOI 10.1007/978-1-4684-1590-2

The Russian text was published by Nauka Press in Moscow in 1970 for the
Academy of Sciences of the USSR as Volume 51 of the Proceedings (Trudy)
of the P. N. Lebedev Physics Institute.   The present translation is published under an agreement with Mezhdunarodnaya Kniga, the Soviet book ex-
port agency.

Library of Congress Catalog Card Number 79-157934

© 1971 Consultants Bureau, New York
A Division of Plenum Publishing Corporation
227 West 17th Street, New York, N. Y. 10011

United Kingdom edition published by Consultants Bureau, London
A Division of Plenum Publishing Company, Ltd.
Davis House (4th Floor), 8 Scrubs Lane, Harlesden, NW10 6SE, London, England

# CONTENTS

# ON APPROXIMATE METHODS OF CALCULATING
# THE EXCITATION CROSS SECTIONS OF ATOMS AND IONS

## L. A. Vainshtein

At present the theory of atomic collisions has developed into a broad field of physics, and it is practically impossible to give a survey of all the existing approximate methods of calculating effective cross sections in one paper. Papers on collision theory can, sometimes somewhat arbitrarily, be subdivided into two groups. The first includes investigations of various singularities in the scattering amplitudes, in particular various kinds of resonances, threshold effects, etc. As a rule these singularities are concentrated in a fairly narrow energy range. The other group consists of papers on the creation of approximate methods of calculating cross sections in a comparatively broad energy range without consideration of the singularities mentioned above. For applications (physics of a gas discharge, astrophysics, etc.) the second trend is especially useful, since in practical problems it is necessary to carry out averaging over the energy distribution of the electrons in a fairly wide range. Papers on collision theory which are presented in the present collection belong basically to this group.

Collisions with slow electrons, for which the simplest method of calculating the cross sections — the Born method — is, strictly speaking, inapplicable, are of fundamental interest for applications. However, a comparison of the results of specific calculations with experiments shows that the error of the Born method in the region of the cross section maximum usually does not exceed a factor of the order of 1.5-2.0. Such accuracy is fully acceptable for many applications.

The essential merit of the Born method is its universality. For a very broad class of transitions in atoms the Born method yields a cross section which is too high by a factor of 1.5-2.0 (in the region of the maximum; i.e., for $E \sim 2\Delta E$, where $\Delta E$ is the threshold energy). Hereafter it is convenient to call an error of this kind the standard error of the Born method. Heretofore, notwithstanding numerous attempts, success has not been achieved in constructing a substantially more exact method which would be just as universal as the Born method. For individual transitions the accuracy of the calculation can be raised very considerably, however, for transitions of different types this is achieved by different means.

In light of what has been said, we should evidently acknowledge the fact that at present the calculations of cross sections required for the solution of numerous applied problems must be carried out on the basis of the Born method. In this connection the question arises of the possible broadening of the class of transitions in which it might be possible to guarantee an error within the limits of the above-mentioned standard error of the Born method.

1

One of the first attempts at refining the Born method was the so-called method of distorted waves. In this method an outer electron was described by the distorted average field $U_{00}(r)$ or $U_{11}(r)$ of the atom, rather than by a plane wave (as in the Born method). Here $U_{nn}$ is the diagonal matrix element of the potential; the subscripts 0 and 1 refer to the initial and final states. Systematic calculations carried out by the method of distorted waves have shown that this method not only fails to provide improved accuracy compared with the Born method, but leads as a rule to a deterioration of the results [1, 2]. The cause of the failure can be understood from the following intuitive (semiclassical) consideration. The potential $U_{nn}$ describes the attraction of an outer electron toward the center of the atom. If under these conditions the distortion of the atomic wave functions by the outer electron is neglected, then the consideration of $U_{nn}$ leads to a reduction of the average distance $\bar{r}_{12}$ between the outer and atomic electrons. However, in reality the distance between them is not decreased but increased due to the repulsion of these electrons. Physically, it is obvious that in problems involving inelastic collisions it is precisely $\bar{r}_{12}$ which is the principal factor. In the Born method $\bar{r}_{12}$ is assumed to be unperturbed, which leads to an overestimated result. In the method of distorted waves $\bar{r}_{12}$ turns out to be still smaller, and the cross section becomes larger.

The factor indicated must also be considered in constructing other approximate methods. In the well-known method of strong coupling a system of equations is solved for several channels. The equations include $U_{nn}$ from all the channels. Therefore, the role of the additional channels reduces to a considerable degree to compensation of the attraction caused by $U_{nn}$. Of course, for an infinite number of channels (with allowance for a continuous spectrum) we would have obtained an exact result, but the convergence of the method turns out to be extremely slow. For this reason the methods of strong coupling, which are very effective for describing the fine threshold effects mentioned at the beginning of the paper, turned out to be poorly suited for the calculation of absolute values of $\sigma(E)$ over a wider range of energies.

The method of strong coupling is based on the direct expansion of an approximate system of equations describing the collisions. The other group is made up of methods which are based on some model representations. The momentum approximation [3] and the model advanced in [4] are typical. The consideration of the repulsion effects directly in the basis wave functions of the system is a common feature of these methods. The attraction of the nucleus, which partially compensates this repulsion, is practically neglected in the momentum approximation. As might be expected, this leads to highly underestimated values of the cross section. In the model [4] the attraction is partially considered. As a result a very good agreement with experiment was obtained for the cross sections of the excitation of 1s − 2p and 1s − 2s transitions in H and resonance transitions in the atoms of alkali metals. The model yields a highly underestimated cross section for transitions between excited levels. At present we cannot say what is the cause of this — the basic scheme of the "model" or the additional approximations which had to be made in order to complete the calculations.

Both the method of strong coupling and the model method, which in a number of cases yields more exact results than the Born method, fail to allow any essential broadening of the class of transitions for which results can be obtained at least within the limits of the standard error of the Born method.

Let us return to the problem of broadening the range of applicability of the Born method. This broadening can be achieved by eliminating three physically obvious shortcomings of this method.

1. In calculating the excitation or ionization cross sections of ions it is necessary to consider the long-range Coulomb field which leads to a nonzero value of the cross section at threshold. This effect is basically connected with distant transits and consequently is funda-

mentally different from the distortion of the incident and scattered waves by the field of a neutral atom. In light of the above it is expedient to consider only the Coulomb part of the atomic field* $U = -2Z/r$. We obtain the so-called Born – Coulomb approximation which differs in principle from the Born method solely in the substitution (in the case of ions) of Coulomb wave functions for plane waves. An important property of this approximation lies in the fact that with increasing Z its accuracy increases (for transitions with a change of the principal quantum number).

2. For transitions between close levels the Born cross section can turn out to be larger than the theoretical limit which is determined by the condition of conservation of the number of particles. Actually, in the Born approximation the probability of excitation $\sim |U_{01}|^2$ and increases without limit with an intensification of the interaction. The elimination of this defect shall be called "normalization" of the cross section. In the quasi-classical approximation normalization is achieved simply. It is sufficient to place the probability of the transition for a stipulated impact parameter $\rho$ equal to

$$w_{01} = \frac{w_{01}^B}{1 + \sum_n w_{0n}^B},$$ (1)

where $w_{0n}^B$ is the transition probability in the Born approximation. The quantum-mechanical analogy of this procedure, which was advanced by Seaton [5], is based on using an R-matrix:

$$S = \frac{1 + i\mathbf{R}}{1 - i\mathbf{R}}.$$ (2)

If $\mathbf{R}$ is a Hermite matrix, then S is a matrix calculated by means of (2), which is unitary with respect to the channels considered regardless of the approximation for $\mathbf{R}$. In order to accomplish normalization of the $0-1$ transition it is sufficient to assume that the elements $R_{0n}$ are nonvanishing. Calculating the latter in the Born approximation, we find

$$|S_{01}|^2 = \frac{|S_{01}^B|^2}{(1 + R^2)^2}, \qquad R^2 = \frac{1}{4} \sum_n |S_{0n}^B|^2.$$ (3)

Finally, the third and evidently most consistent method of normalization is based on the solution of the simplified system of strong-coupling equations

$$(L_0 + k_0^2) F_0 = \sum_n U_{0n} F_n, \quad (L_n + k_n) F_n = U_{0n} F_0,$$
$$L_n = \frac{d^2}{dr^2} + \frac{l_n(l_n + 1)}{r^2}.$$ (4)

In order to consider just renormalization effects, the diagonal potentials $U_{nn}$ have been omitted in these systems (see what was said above concerning the method of distorted waves), as well as the potential $U_{nm}$ (n, m ≠ 0) which describes transitions through the intermediate level in the higher orders of perturbation theory. The solution of the system (4) ensures the unitarity of the S-matrix within the limits of the levels considered.

---

*Atomic units with the Rydberg energy unit are used throughout; the factor 2 is connected with the use of Rydberg units.

In the case of transitions between neighboring levels it is sufficient to limit the analysis to one term with n = 1 in the sums over n in (1), (3), (4). This is the one-channel approximation (i.e., normalization of the given transition only). For transitions to higher levels the situation is more complicated. Very often $R_{01} \ll 1$ under these conditions; i.e., the normalization effect due to the natural transition is small. Nevertheless, the normalization due to a strong 0−2 transition (usually through a neighboring level) can be substantial. Under these conditions $R = R_{02} \gg 1$, and the one-channel approximation is inapplicable in principle.

It should also be noted that the normalization effect in (3) is stronger than in (1): $S \rightarrow 0$ for $S^B \rightarrow \infty$, whereas $w \rightarrow 1$ for $w^B \rightarrow \infty$. Calculations by the method (4) substantiate Eq. (3).

3. Transitions with change of atomic spin (intercombinational transitions) are wholly caused by exchange interaction. The corresponding generalization of the Born method turns out to be very difficult, since the first and most natural method (the Born − Oppenheimer method) is clearly unsatisfactory. Ochkur [6] indicated the possible cause of this failure and advanced a modified method. For intercombinational transitions in neutral atoms the Ochkur method yields the same accuracy as the Born method for ordinary transitions. However, the consideration of the Coulomb field and normalization within the framework of this method encounter serious difficulties. In this sense the method [7], which is based on orthogonalization of the wave functions of the electron−atom system in the initial and final states, is more universal. This method introduces consideration of exchange within the framework of the first order of perturbation theory. For transitions in neutral atoms the Ochkur method and the method of orthogonalized functions give closely similar results if normalization is nonessential.

The Born method, augmented by the consideration of the effect of normalization, the Coulomb field (for ions), and exchange (for intercombinational transitions), can be called the generalized Born method. Although the normalization procedure is nonlinear, this nonlinearity reduces to a simple algebraic equation during the last stage of the calculation. Therefore, the generalized Born method extends only negligibly beyond the framework of the first order of perturbation theory.

In the computational sense the generalized Born method is somewhat more complex than the conventional Born method. The point is that the latter does not require expansion in partial waves. In using one-electron atomic wave functions, the Born cross section is represented in the form

$$\sigma_{01}^B = \sum_{\varkappa} Q_{\varkappa} \sigma_{\varkappa}, \qquad \varkappa = |l_1 - l_0|, \; |l_1 - l_0| + 2, \ldots, |l_0 + l_1|, \tag{5}$$

$$\sigma_{\varkappa} = \frac{8}{k_0^2}(2\varkappa + 1)(2l_1 + 1)\begin{pmatrix} l_0 & l_1 & \varkappa \\ 0 & 0 & 0 \end{pmatrix}^2 \int_{k_0-k_1}^{k_0+k_1} \frac{dq}{q^3} |R_{01}^{\varkappa}(q)|^2, \tag{6}$$

$$R_{01}^{\varkappa}(q) = \int_0^{\infty} P_0(r) P_1(r) [j_{\varkappa}(qr) - \delta_{\varkappa 0}] \, dr, \tag{7}$$

where $P_0$, $P_1$ are the radial functions of the optical electron of the atom; $j_{\varkappa}$ is the spherical Bessel function. For the resultant transition between two electron configurations $l_0^N \rightarrow l_0^{N-1} l_1$, summed over the terms, the factor $Q_{\varkappa} - N$. For transitions between individual terms $Q_{\varkappa}$ depends on the quantum numbers of the angular momenta. The equations for calculating $Q_{\varkappa}$ in the cases which are of practical interest have been presented in [8].

Calculations within the framework of the generalized Born method are so far possible only in the partial-wave representation. Under these conditions the integral with respect to $q$ in (6) is replaced by an infinite sum over the orbital momenta of the outer electrons $\lambda_0$, $\lambda_1$ (the computation equations are presented in the subsequent papers of the present collection). The slow convergence of the sums noticeably complicates the calculations. Nevertheless, the use of electronic computers eliminates any serious difficulties from calculations carried out within the framework of the generalized Born method, and the calculations can be carried out according to a unified program for a very broad class of transitions.

The Born method yields the standard error for optically allowed transitions from the ground state in atoms of the H, He type in which all of the excited states differ from the ground state by the value of the principal quantum numbers. In the case of transitions without change of the principal number, for example, for excitation of the resonance levels of alkali elements (the $3s-3p$ transition in Na, etc.), the cross section turns out to be overestimated by a factor of 3-4. The consideration of normalization brings the error within the limits of the standard error of the Born method.

Experimental data on the excitation of ions are almost unavailable. The results of measuring the cross section for the $1s-2s$ transition in $He^+$ [9], as well as a number of indirect data, show that the Born – Coulomb approximation with allowance for normalization (if this is necessary) similarly yields an error which is within the limits of the standard error.

As has already been mentioned above, the same applies to intercombinational transitions when exchange by the Ockhur method or by the method of orthogonalized functions is considered.

Thus, the generalized Born method ensures a standard error for a broad class of transitions from the ground state. This primarily includes the majority of optically allowed transitions, as well as transitions of the $s-s$, $d-s$ type.

For quadrupole transitions of the $s-d$, $p-f$ type the cross section of a direct transition of the first order of perturbation theory is comparatively small. Under these conditions a transition via an intermediate level in the second order of perturbation theory turns out to be more probable. Such transitions are discussed in greater detail in [10]. At present it is difficult to predict exactly when a transition to an intermediate level will be essential. We merely note that in ions the role of such transitions decreases rapidly with a growth of the charge Z. Evidently, their probability is small in transitions with $\Delta l = 0, -2$.

The problem of transitions between excited levels is similarly far from clear. Data on the width of the spectral line show that the generlized Born method ensures a standard error for optically allowed transitions between neighboring levels. For transitions with a change of the principal quantum number it is evident that the second order plays a large role (as a rule, consideration of one intermediate level is sufficient) [11].

In conclusion we note that the problem of transitions between levels having $n \gg 1$, which are of special interest in a number of astrophysical applications, remains essentially open.

References

1. L. A. Vainshtein and G. G. Dolgov, Opt. i Spektr., 7:3 (1959).
2. M. Blaha, Bull. Astron. Inst. Czechoslov., 9:160 (1958); 13:81 (1961).
3. R. Akerib and S. Borowitz, Phys. Rev., 122:1177 (1961).
4. L. A. Vainshtein, L. P. Presnyakov, and I. I. Sobel'man, Zh. Éksperim. i Teor. Phys., 45:2015 (1963).
5. M. J. Seaton, Proc. Phys. Soc., 77:184 (1961).
6. V. I. Ochkur, Zh. Éksperim. i Teor. Fiz., 45:734 (1963).

7.   I. L. Beigman and L. A. Vainshtein, Zh. Éksperim. i Teor. Fiz., 52:185 (1967).

8.   L. A. Vainshtein and I. I. Sobel'man, Preprint of the Physics Institute, Academy of Sciences of the USSR, No. 66 (1967). J. Quant. Spectr. Radiative Transfer, 8:1491 (1968).

9.   D. F. Dance, M. F. Harrison, and A. C. H. Smith, Proc. Roy. Soc., A290:74 (1966).

10.   L. A. Vainshtein and L. P. Presnyakov, this volume, p. 37.

11.   A. S. Khaikin, Zh. Éksperim. i Teor. Fiz., 54:52 (1968).

# EFFECTIVE EXCITATION AND
# IONIZATION CROSS SECTIONS OF IONS

## I. L. Beigman and L. A. Vainshtein

### 1. Introduction

In solving many problems in the physics of high-temperature plasma it is necessary to know the effective cross sections for the excitation and ionization of ions by electron impact. In the majority of cases the transition energy or ionization energy ($\Delta E$) is considerably higher than the plasma temperature (kT), and a fundamental role is played by the region near threshold where the scattered electron has a low energy. In calculating the cross sections in this region it is necessary to consider the long-range Coulomb field of the ions, which has a substantial effect on the behavior of the cross section near threshold. At the same time the presence of a strong Coulomb field in the case of a highly ionized atom allows the "small parameter" 1/Z, where Z−1 is the ion charge to be isolated in the problem. A consistent formulation of perturbation theory in this parameter not only provides the possibility of considering the effects of a long-range Coulomb field in the processes of ion excitation and ionization, but also yields considerably better results for small Z than the Born−Oppenheimer approximation for intercombinational transitions.

For transitions without spin change ($\Delta S = 0$) perturbation theory in the parameter 1/Z leads to the so-called Born−Coulomb approximation in which the outer electron is described by a Coulomb wave function. For neutral atoms (Z = 1) this function goes over into a plane wave, and we obtain the conventional Born method. It should be emphasized that in the Born−Coulomb approximation only the long-range Coulomb field, which leads to a qualitative effect in the threshold region, is considered. The short-range field, which is connected with incompleteness of the screening of the nucleus by atomic electrons, is not considered, so that for Z = 1 the method yields the same results as the Born approximation rather than the method of distorted waves (compare with [1]).

In the case of intercombinational transitions ($\Delta S \neq 0$) the wave function of the outer electron must in addition be orthogonalized to the atomic wave functions [2].

In the present paper we calculate the effective cross sections for excitation and ionization of ions for a number of transitions in ions of different multiplicities. Besides the cross section $\sigma$ itself, we have calculated the values of the excitation or ionization rate $\langle v\sigma \rangle$, averaged over the Maxwellian velocity distribution of the electrons.

For representing the results, and especially for using them in applications, it is convenient to approximate $\sigma$ and $\langle v\sigma \rangle$ by means of simple analytical equations. In the present paper we have used two-parameter equations; the parameters were chosen for each transition by the method of least squares from the results of a numerical calculation.

In the equations given below it is assumed that the atom is described by one-electron mutually orthogonal wave functions constructed according to the vector model scheme.

The radial functions of an optical electron are determined by the numerical solution of the radial Schrödinger equation with an atomic-core potential obtained by means of Slater functions. For ions with small Z it would be better to use a semiempirical method of the type [3]. However, the difference between the energy parameters of the two types of functions is of the order of the splitting into terms which, as a rule, has been neglected in the present paper. Moreover, for ions with large Z the experimental energies are often unknown.

The cross sections for the excitation of intercombinational transitions in He were calculated using semiempirical functions [3].

Below we use atomic units with the Ry unit for energy (Ry = $e^2/2a_0$ = 13.6 eV) throughout.

## 2. The Born — Coulomb Approximation for Transitions
## without a Change of the Atomic Spin

In the Born — Coulomb approximation an outer electron is described by Coulomb wave functions. At high incident-electron energies the Coulomb wave functions in this case differ negligibly from plane waves, and the Born — Coulomb approximation practically coincides with the Born approximation. At low energies the Coulomb wave function corresponding to the scattered electron contains an additional factor $k^{-1/2}$, and the cross section at the excitation threshold $(k \to 0)$ turns out to be nonvanishing.

The effective cross section in the Born — Coulomb approximation can be represented in the form (the subscripts 0, 1 refer to the initial and final states)

$$\sigma = \sum_{\varkappa} Q_{\varkappa} \sum_{\lambda_0 \lambda_1} \sigma_{\varkappa}(\lambda_0 \lambda_1), \tag{1}$$

$$\sigma_{\varkappa}(\lambda_0 \lambda_1) = \pi a_0^2 \frac{16}{k_0^2} \frac{(2l_1+1)(2\lambda_0+1)(2\lambda_1+1)}{2\varkappa+1} \begin{pmatrix} l_0 & \varkappa & l_1 \\ 0 & 0 & 0 \end{pmatrix}^2 \begin{pmatrix} \lambda_0 & \varkappa & \lambda_1 \\ 0 & 0 & 0 \end{pmatrix}^2 \times$$

$$\times \left| \iint dr_1 dr_2 \left( \frac{r_<^{\varkappa}}{r_>^{\varkappa+1}} - \frac{1}{r_1} \right) P_0(r_2) P_1(r_2) F_{\lambda_0}(r_1) F_{\lambda_1}(r_1) \right|^2, \tag{2}$$

where P (r) is the radial function of the optical electron of the ion; $F_{\lambda}$ is the Coulomb radial function, while

$$F_{\lambda}(r) \xrightarrow[r \to \infty]{} \frac{1}{\sqrt{k}} \sin\left( kr + \frac{Z}{k} \ln 2Zkr + \delta \right); \tag{3}$$

$\lambda$ and k are the orbital momentum and momentum of the outer electron.

The factor $Q_{\varkappa}$ which is included in (1) depends solely on the quantum numbers of the angular momenta of the ions. For the cross section summed over the terms for the transition between two electron configurations $l_0^N \to l_0^{N-1} l_1$ , the factor $Q_{\varkappa} = N$. In [4] equations for $Q_{\varkappa}$ are cited for various types of transitions between individual terms and components of the fine structure.

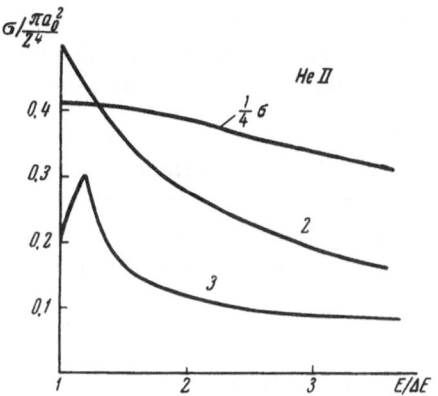

Fig. 1. Effective excitation cross sections of He II, calculated in the Born – Coulomb approximation for the 1s – 2p (1) and 1s – 2s (2) transitions, and the experimental data [6] for the 1s – 2s transition (3).

The sum over $\varkappa$ in (1) contains a small number of terms. The sum over $\lambda$ is infinite, while the convergence is fast for small k but extremely slow for large k. Therefore, in obtaining the total cross section it is convenient to rewrite (1) in the form

$$\sigma = \sigma^B + \sum_\varkappa Q_\varkappa \sum_{\lambda_0 \lambda_1} [\sigma_\varkappa(\lambda_0 \lambda_1) - \sigma_\varkappa^B(\lambda_0 \lambda_1)], \qquad (4)$$

where $\sigma^B$ is the Born cross section. For large $\lambda$ and k, $F_\lambda(r) \approx r j_\lambda(kr)$ (where $j_\lambda$ is a spherical Bessel function), and $\sigma_\varkappa(\lambda_0 \lambda_1)$ goes over into $\sigma^B(\lambda_0 \lambda_1)$. Therefore, the sum over $\lambda$ in (4) converges rapidly for all k. In practical calculations the restriction $\lambda_0, \lambda_1 \lesssim 10$ can be used. The total Born cross section can be calculated without resorting to expansion into partial waves [1, 4].

If the partial cross section exceeds the theoretical limit, it should be "normalized." For this purpose we make use of the R-matrix method [5]. By restricting the analysis to the one-channel approximation for the normalized* cross section $\sigma^N$, we obtain (compare with [1]):

$$\sigma_\varkappa^N(\lambda_0 \lambda_1) = (1 + R^2)^{-2} \sigma_\varkappa(\lambda_0 \lambda_1),$$
$$R^2 = \frac{k_0^2}{4(2\lambda_0 + 1)} \sum_{\varkappa' \lambda_1'} \sigma_{\varkappa'}(\lambda_0 \lambda_1'). \qquad (5)$$

The cross section for the excitation of the 1s – 2p, 1s – 2s transitions in the He II ion in the Born approximation are displayed in Fig. 1. As is evident, for 1s – 2s transition the calculated cross section near threshold exceeds the experimental value by a factor of approximately 1.5 [6]. Such an error is typical of the conventional Born method for neutral atoms. Since the Born – Coulomb approximation is the first order of perturbation theory in the parameter 1/Z, we can expect that for ions of higher multiplicity the magnitude of the error will be smaller. The growth of the experimental curve [6] from threshold is possibly connected with the presence of autoionization levels of the 3snl type in neutral helium [7, 8].

The dependence of the excitation cross section on the energy of the outer electron can be described by the following empirical equation† with sufficient accuracy for the majority of applications:

$$\sigma = \pi a_0^2 \left(\frac{Ry}{\Delta E}\right)^2 \left(\frac{E_1}{E_0}\right)^{3/2} \frac{Q}{2l_0 + 1} \frac{C}{u + \varphi}, \qquad u = \frac{E - \Delta E}{\Delta E}, \quad \Delta E = E_1 - E_0. \qquad (6)$$

---

*Strictly speaking, this normalization rule is not entirely consistent. The point is that the cross sections should be normalized separately for each value of the total angular momenta $L_T$ of the system, but in Eqs. (1), (2) summation over $L_T$ has already been carried out. However, it can be shown that such an error produces only an insignificant overestimation of the normalization effect.

†For optically allowed transitions it is necessary, generally speaking, to introduce a factor of the type ln u in Eq. (6). However, it begins to play a substantial role only for large u and therefore at electron temperatures $kT \gg \Delta E$. As a rule, this temperature range is not of practical interest.

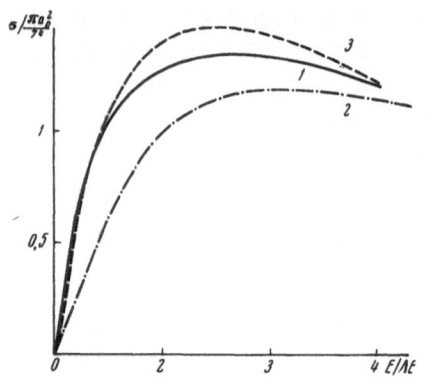

Fig. 2. Effective ionization cross sections of C VI in the Born – Coulomb (1), in the Born (2), and in the modified Born (3) approximations [Eqs. (9)].

Here $E_0$ and $E_1$ are the energies of the initial and final states, measured from the ionization boundary; the factor $Q = Q_{\varkappa_{min}}$ is the value of $Q_\varkappa$ for min $\varkappa = |l_0 - l_1|$.

The factor $(Ry/\Delta E)^2 (E_1/E_0)^{3/2}$ reflects the basic dependence of $\sigma$ on the parameters of the ions. Due to their introduction, the constant C depends weakly on the type of transition and the multiplicity of the ions.

The excitation rate of the zero – one transition for a Maxwellian velocity distribution of the electrons at the temperature T is equal to

$$\langle v\sigma \rangle = \frac{2\sqrt{\pi}}{m} a_0 \hbar \left( \frac{\Delta E}{Ry} \right)^{1/2} e^{-\beta} \beta^{3/2} \int_0^\infty (u+1) \frac{\sigma(u)}{\pi a_0^2} e^{-\beta u} \, du, \quad (7)$$

$$\beta = \Delta E/kT.$$

In order to describe $\langle v\sigma \rangle$ we make use of the equation

$$\langle v\sigma \rangle = 10^{-3} \left( \frac{Ry E_1}{\Delta E E_0} \right)^{3/2} \frac{Q}{2l_0 + 1} \frac{A\sqrt{\beta}(\beta+1)}{\beta+\chi} e^{-\beta} \, (\text{cm}^3 \cdot \text{sec}^{-1}). \quad (8)$$

The parameters C, $\varphi$ and A, $\chi$ were chosen for each transition from the results of a numerical calculation of $\sigma$ and $\langle v\sigma \rangle$ by the method of least squares.

## 3. The Ionization Cross Sections

The ionization cross section in the Born – Coulomb approximation (neglecting exchange) can be calculated by means of Eqs. (1) and (2) by adding summation over the angular momenta and integration over the momenta of the knocked-on electron. The effective cross sections for ionization from the 1s level of the C VI ion in the Born – Coulomb and Born approximations are displayed in Fig. 2. As is evident, the presence of the Coulomb field causes the ionization cross section maximum to increase and to be shifted into the range of lower energies. In the case of ionization from the 2p level the magnitude of the maximum remains approximately the same as in the Born approximation [9]. The behavior of the cross section near threshold changes very substantially: whereas in the Born approximation $\sigma \sim E^{3/2}$, we have $\sigma \sim E$ in the Born – Coulomb approximation.

Due to the presence of additional summation and integration, the calculation of the ionization cross sections in the Born – Coulomb approximation requires extremely large expenditures of machine time. However, from the calculations which have been carried out it is evident that the correction due to the Coulomb field in the case of ionization is considerably smaller than ' in the case of excitation. If the ionization cross section is assumed to equal

$$\sigma_i = \sqrt{\frac{u+1}{u}} \, \sigma_i^B, \qquad u = \frac{E - \Delta E}{\Delta E}, \quad (9)$$

where $\sigma_i^B$ is the cross section calculated in the conventional Born approximation, then we consider the qualitative effect of the Coulomb field and the error of the cross section obtained in

EFFECTIVE EXCITATION AND IONIZATION CROSS SECTIONS OF IONS

TABLE 1. Effective Cross Sections of Excitations of OVIII Ions

| Transition | C | φ | A | χ | Transition | C | φ | A | χ |
|---|---|---|---|---|---|---|---|---|---|
| 1s — 2s | 1 26 | 1 10 | 1 50 | 0 84 | 2s — 4f | 1 16 | 1 12 | 1 27 | 0 79 |
| 3s | 1 21 | 1 10 | 1 40 | 0 86 | 3s — 4f | 1 34 | 0 95 | 1 52 | 0 98 |
| 4s | 1 20 | 1 10 | 1 40 | 0 87 | 2p — 3s | 1 15 | 1 53 | 0 56 | 0 26 |
| 2s — 3s | 1 25 | 1 10 | 1 48 | 0 87 | 4s | 0 84 | 1 34 | 0 50 | 0 36 |
| 4s | 1 16 | 1 10 | 1 31 | 0 89 | 3p — 4s | 1 18 | 1 61 | 0 59 | 0 23 |
| 3s — 4s | 1 26 | 1 10 | 1 50 | 0 89 | 2p — 3p | 1 82 | 1 11 | 2 14 | 0 78 |
| 1s — 2p | 2 46 | 1 50 | 2 20 | 0 26 | 2p — 4p | 1 52 | 1 11 | 1 95 | 0 80 |
| 3p | 2 34 | 1 43 | 2 16 | 0 30 | 3p — 4p | 1 90 | 1 12 | 2 15 | 0 76 |
| 4p | 2 28 | 1 41 | 2 14 | 0 31 | 2p — 3d | 2 92 | 1 38 | 2 49 | 0 34 |
| 2s — 3p | 2 17 | 1 87 | 1 43 | 0 17 | 4d | 2 45 | 1 30 | 2 31 | 0 41 |
| 4p | 2 11 | 1 79 | 1 31 | 0 18 | 3p — 4d | 2 44 | 1 50 | 2 18 | 0 29 |
| 3s — 4p | 2 10 | 1 93 | 1 22 | 0 15 | 2p — 4f | 1 68 | 1 24 | 1 58 | 0 45 |
| 1s — 3d | 1 17 | 1 31 | 1 12 | 0 37 | 3p — 4f | 2 14 | 1 14 | 2 21 | 0 72 |
| 4d | 1 20 | 1 28 | 1 14 | 0 36 | 3d — 4s | 0 27 | 1 12 | 0 43 | 0 79 |
| 2s — 3d | 1 74 | 1 16 | 1 95 | 0 64 | 4p | 1 17 | 1 26 | 1 14 | 0 47 |
| 4d | 1 31 | 1 13 | 1 49 | 0 76 | 4d | 2 12 | 1 10 | 2 22 | 0 82 |
| 3s — 4d | 1 45 | 1 17 | 1 48 | 0 53 | 4f | 3 13 | 1 30 | 2 88 | 0 43 |
| 1s — 4f | —1 31 | 1 18 | —1 34 | 0 51 | | | | | |

The characteristic and mantissa of the number are given; for example, 1 26 denotes $0.26 \times 10^1$.

TABLE 2. The Parameters $A_\infty$, χ for Excitation and Ionization of H- and He-like Ions

| Transition | $A_\infty$ | χ | a Isoelectronic sequence | |
|---|---|---|---|---|
| | | | H | He |
| 1s — 2s | 5.3 | 0.83 | 0.82 | —1.5 |
| 2p | 21 | 0.21 | 5.0 | 7.0 |
| 3s | 4.4 | 0.83 | 0.43 | —1.3 |
| 3p | 17 | 0.25 | 4.0 | 4.5 |
| 3d | 1.3 | 0.34 | 0.18 | 0.70 |
| 1s — continuum | 8.1 | 0.62 | 0 | 0.20 |

this way will be small (see Fig. 2). Then the rate of ionization from the $l_0^N$ state can be described by the equation

$$\langle v\sigma_i \rangle = 10^{-8} \left( \frac{Ry}{\Delta E} \right)^{3/2} \frac{N}{2l_0+1} \frac{A \sqrt{\beta}}{\beta+\chi} e^{-\beta}, \quad (10)$$

where $\Delta E$ is the ionization energy.

## 4. The Results of the Calculations

The parameters C, φ, A, χ for all of the transitions between levels with $1 \leq n \leq 4$ in the OVIII ion are cited in Table 1. These same parameters can be used for any hydrogen-like ion having $Z \geq 3$ with an error not exceeding 25%.

In the general case these parameters depend on the type of transition, the isoelectronic sequence, and, within the limits of a given isoelectronic sequence, on the magnitude of Z. In the present paper only the parameters A, χ, which are most important in applications, are given.

For transitions from the ground state of H- and He-like ions it can be assumed with sufficiently good accuracy that

$$A = A_\infty - \frac{a}{Z-1}. \quad (11)$$

Here and hereafter Z coincides with the spectroscopic symbol of the ion; i.e., the ion charge is equal to Z-1. Table 2 cites the values of $A_\infty$, a, and χ for a series of transitions.

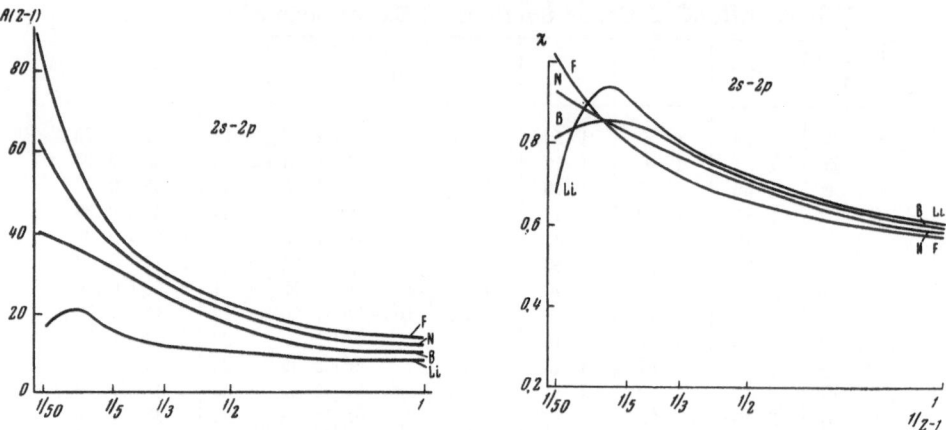

Fig. 3. The parameters A and χ for the 2s − 2p transition in the Li − F isoelectronic sequences.

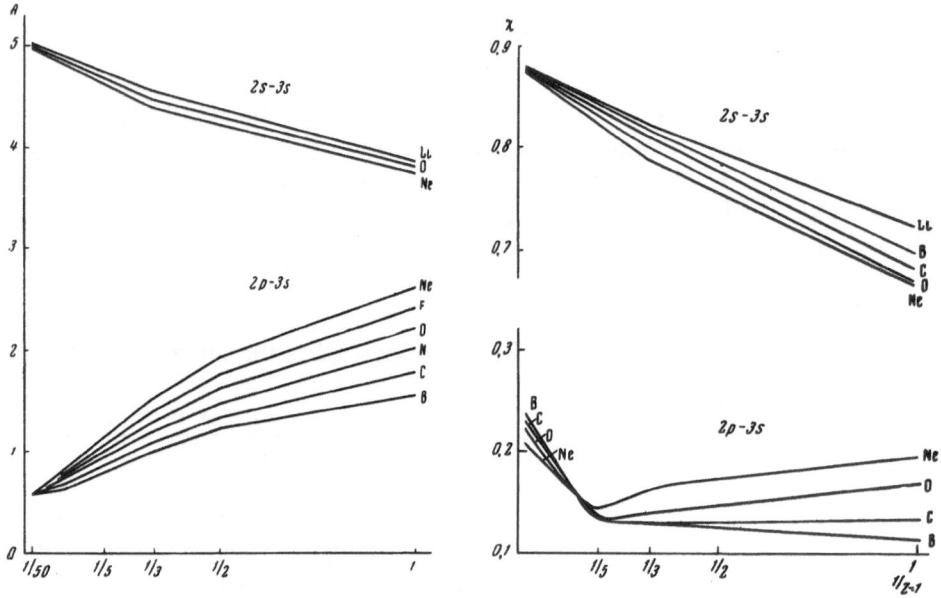

Fig. 4. The parameters A and χ for the 2s − 3s and 2p − 3s transitions in the
Li − Ne isoelectronic sequences.

For other isoelectronic sequences the dependence of A and χ on Z is more complicated. For a number of transitions in the isoelectronic sequences from Li to Ne the parameters A and χ are displayed as functions of Z in Figs. 3-8. In those cases in which the curves are closely spaced a portion of them is omitted.

## 5.    The Excitation Cross Sections of Intercombinational

## Transitions

The excitation of transitions with change of the atomic spin is possible in the LS-coupling approximation solely as a consequence of exchange interaction. The consideration of exchange

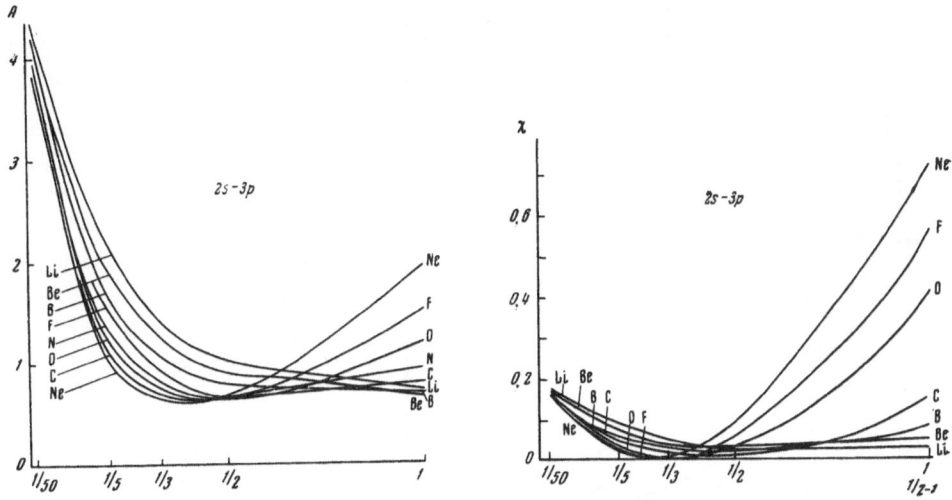

Fig. 5. The parameters A and χ for the 2s − 3p transition in the Li − Ne isoelectronic sequences.

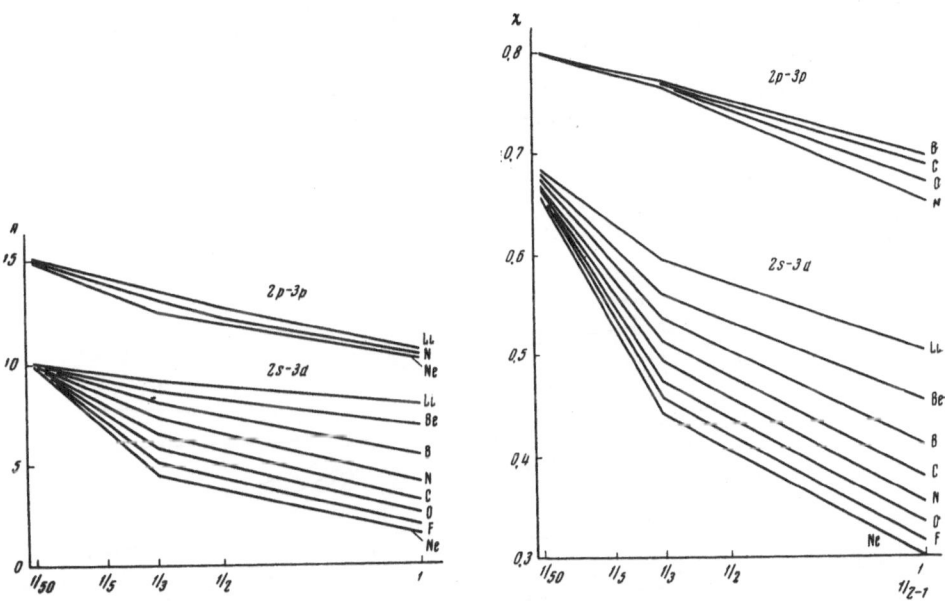

Fig. 6. The parameters A and χ for the 2p − 3p and 2s − 3d transitions in the
Li − Ne isoelectronic sequences.

within the framework of the well-known Born − Oppenheimer method as a rule leads to unsatisfactory results. The essential shortcoming of the Born − Oppenheimer method is the use of
nonorthogonal wave functions of the atom − electron system in the initial and final states.

In [2] a modified method was advanced which was based on the use of orthogonalized
wave functions. Without dwelling here on the substantiation of the method, we cite the results.

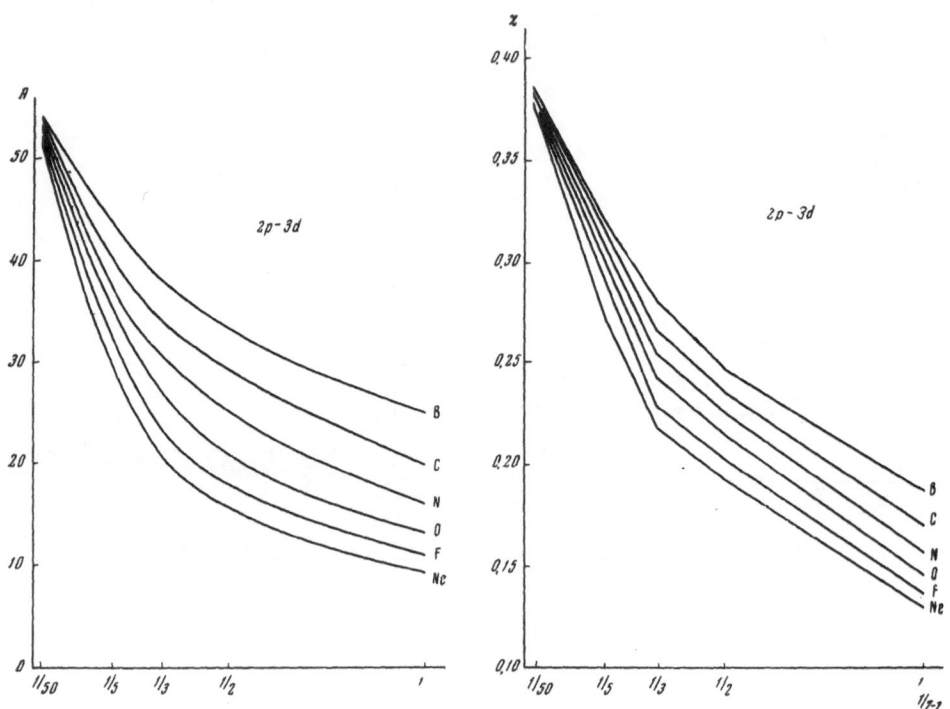

Fig. 7. The parameters A and χ for the 2p − 3d transition in the Li − Ne iso-electronic sequences.

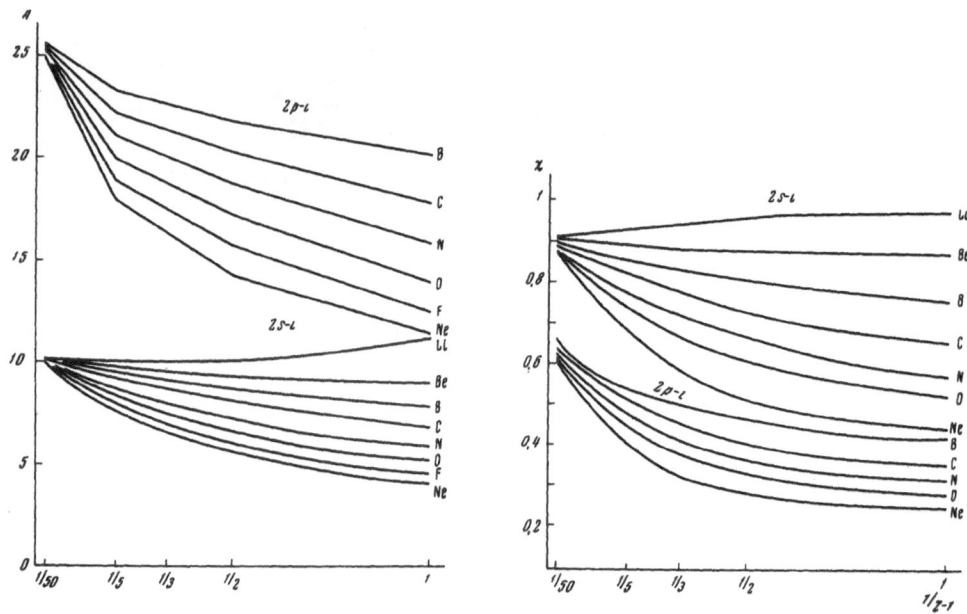

Fig. 8. The parameters A and χ for the rate of ionization from the 2s state in the Li − Ne isoelectronic sequences, and from the state 2p in the B − Ne isoelectronic sequences.

The exchange part of the transition amplitude is determined by the matrix element

$$\left\langle \Phi_1(\mathbf{r}_2, \mathbf{r}_1) \left| \frac{1}{|\mathbf{r}_1 - \mathbf{r}_2|} \right| \Phi_0(\mathbf{r}_1, \mathbf{r}_2) \right\rangle,$$

$$\Phi_0(\mathbf{r}_1, \mathbf{r}_2) = \varphi_0(\mathbf{r}_1)[F_0(\mathbf{r}_2) - \langle F_0 | \varphi_1 \rangle \varphi_1(\mathbf{r}_2)],$$
$$\Phi_1(\mathbf{r}_2, \mathbf{r}_1) = \varphi_1(\mathbf{r}_2)[F_1(\mathbf{r}_1) - \langle F_1 | \varphi_0 \rangle \varphi_0(\mathbf{r}_1)],$$

(12)

where $\varphi$ are the atomic wave functions; F is the Coulomb function (or the plane wave for a neutral atom).

Equations (12) are written for a hydrogen-like ion for the sake of simplicity, but their generalization for an arbitrary case does not present any substantial difficulties.

For the transitions $l_0^N L_0 S_0 \rightarrow l_0^{N-1} [L_p S_p] l_1 L_1 S_1$ ($L_p$, $S_p$ are the are the angular and spin momenta of the atomic core) the cross section has the form

$$\sigma = \pi a_0^2 \frac{16}{k_0^2} \sum_{\lambda_0 \lambda_1 L_T} \left| \sum_\varkappa b_\varkappa B_{\varkappa L_T} R_\varkappa \right|^2,$$

$$b_\varkappa = \left[ \frac{(2l_1+1)(2\lambda_0+1)(2\lambda_1+1)}{(2\varkappa+1)} \right]^{1/2} \begin{pmatrix} l_0 & \lambda_1 \\ 0 & 0 & 0 \end{pmatrix} \begin{pmatrix} l_1 & \varkappa & \lambda_0 \\ 0 & 0 & 0 \end{pmatrix},$$

(13)

$$B_{\varkappa L_T} = \sqrt{N}\, G_{L_p S_p}^{L_0 S_0} \left[ \frac{(2S_1+1)(2L_1+1)(2\varkappa+1)(2L_T+1)(2l_0+1)}{2(2S_p+1)} \right]^{1/2} \begin{Bmatrix} \lambda_0 & l_1 & \varkappa \\ L_0 & L_p & l_0 \\ L_T & L_1 & \lambda_1 \end{Bmatrix},$$

$$R_\varkappa = \iint dr_1\, dr_2\, P_0(r_1) g_{\lambda_0}(r_2) \frac{r_<^\varkappa}{r_>^{\varkappa+1}} P_1(r_2) g_{\lambda_1}(r_1),$$

where $G_{L_p S_p}^{L_0 S_0}$ is the genealogical coefficient; k and $\lambda$ denote the momentum and orbital momentum of the outer electron, and the functions $g_\lambda$ are equal to

$$g_{\lambda_0} = F_{\lambda_0} - \langle F_{\lambda_0} | P_1 \rangle P_1 \delta_{\lambda_0 l_1},$$
$$g_{\lambda_1} = F_{\lambda_1} - \langle F_{\lambda_1} | P_0 \rangle P_0 \delta_{\lambda_1 l_0}.$$

(14)

in accordance with Eqs. (12). Here $F_\lambda$, as previously, is the Coulomb radial function normalized according to (3).

In a number of particular cases Eqs. (13) allow considerable simplifications and can be written in the form

$$\sigma = Q \sum_{\lambda_0 \lambda_1} \sigma(\lambda_0 \lambda_1),$$

$$\sigma(\lambda_0 \lambda_1) = \pi a_0^2 \frac{16}{k_0^2} \sum_\varkappa | b_\varkappa R_\varkappa |^2.$$

(15)

Below we enumerate the most important of these cases (the principal quantum numbers n are omitted for simplicity).

TABLE 3.  The Parameters A, $\chi$ for Intercombinational Transitions
in He- and Ne-like Ions

| z | Transition | Ne | | | | Transition | He | | | |
|---|---|---|---|---|---|---|---|---|---|---|
| | | A | | $\chi$ | | | A | | $\chi$ | |
| 2 | 2p — 3s | 0 | 79 | 0 | 36 | 1s — 2s | 0 | 27 | —1 | 52 |
| 4 | | 1 | 11 | 0 | 29 | | 1 | 12 | 0 | 22 |
| 50 | | 0 | 73 | 0 | 38 | | 1 | 17 | 0 | 44 |
| 2 | 2p — 3p | 1 | 13 | 0 | 33 | 1s — 2p | 1 | 70 | 0 | 38 |
| 4 | | 0 | 43 | 0 | | | 2 | 10 | 0 | 58 |
| 50 | | 1 | 41 | 0 | 46 | | 1 | 97 | 0 | 65 |
| 2 | 2p — 3d | 1 | 104 | —1 | 37 | | | | | |
| 4 | | 1 | 71 | 0 | 57 | | | | | |
| 50 | | 2 | 11 | 0 | 59 | | | | | |

TABLE 4.  The Cross Sections for the Excitation
of Intercombinational Transitions in FeXVII

| $\sqrt{\dfrac{E-\Delta E}{\Delta \varepsilon}}$ | 2p — 3s | | 2p — 3p | | 2p — 3d | |
|---|---|---|---|---|---|---|
| 0.1 | 1 | 69 | 2 | 25 | 2 | 63 |
| 0.2 | 1 | 64 | 2 | 24 | 2 | 60 |
| 0.4 | 1 | 53 | 2 | 20 | 2 | 46 |
| 0.8 | 1 | 27 | 2 | 10 | 2 | 19 |
| 1.6 | 0 | 43 | 1 | 17 | 1 | 22 |
| 3.2 | —1 | 18 | —1 | 59 | —1 | 76 |
| A | 0 | 92 | 1 | 42 | 2 | 13 |
| $\chi$ | 0 | 39 | 0 | 37 | 0 | 61 |

1. Any of the angular momenta $L_p$, $l_0$, $l_1$ is equal to zero

$$Q (S_0 L_0, S_1 L_1) =$$

$$= N \, | \, G^{L_0 S_0}_{L_p S_p} |^2 \frac{(2S_1 + 1)\,(2L_1 + 1)}{2\,(2S_p + 1)\,(2L_p + 1)\,(2l_1 + 1)} . \quad (16)$$

2. The cross section summed over $L_1$ is

$$Q (l_0^N S_0 L_0, l_0^{N-1} [L_p S_p] \, l_1 S_1) =$$

$$= N \, | \, G^{L_0 S_0}_{L_p S_p} |^2 \frac{2S_1 + 1}{2\,(2S_p + 1)} . \quad (17)$$

3. In the case of a transition between states with a stipulated term of the atomic core $[L_p S_p]$ it is also possible to carry out averaging with respect to $L_0$.  Under these conditions

$$Q (l_0 S_0 L_0, l_1 S_1) = Q (l_0 S_0, l_1 S_1 L_1) = Q (l_0 S_0, l_1 S_1) = \frac{2S_1 + 1}{2\,(2S_p + 1)} . \quad (18)$$

For ions the cross section of an intercombinational transition is represented by the empirical equation

$$\sigma = \pi a_0^2 \left( \frac{Ry}{\Delta E} \right)^2 \left( \frac{E_1}{E_0} \right)^{2/3} \frac{Q}{(2l_0 + 1)} \frac{C}{(u+1)^2 \, (u + \varphi)} . \quad (19)$$

The factor Q is determined by Eqs. (17), (18).

For the excitation rate of intercombinational transitions we make use of the following empirical equation by analogy with Eqs. (7), (8):

$$\langle v\sigma \rangle = 10^{-8} \left( \frac{R_y}{\Delta E} \frac{E_1}{E_0} \right)^{2/3} \frac{Q}{2l_0 + 1} \frac{A\beta^{2/3}}{\beta + \chi} \, e^{-\beta} \; (\mathrm{cm^3 \cdot sec^{-1}}). \quad (20)$$

TABLE 5. Cross Sections for the Excitation of Intercombinational Transitions in the He Atom

| $\sqrt{\dfrac{E-\Delta E}{\Delta E}}$ | $1s^1S \to 2s^3S$ | | $1s^1S \to 3s^3S$ | | $1s^1S \to 2p^3P$ | | $1s^1S \to 3p^3P$ | | $1s^1S \to 4p^3P$ | | $1s^1S \to 4d^3D$ | |
|---|---|---|---|---|---|---|---|---|---|---|---|---|
| | 1 | 2 | 1 | 2 | 1 | 2 | 1 | 2 | 1 | 2 | 1 | 2 |
| 0.1 | —2 34 | —2 89 | —2 14 | —2 16 | —1 32 | —1 31 | —1 11 | —2 72 | —2 47 | —2 28 | —3 21 | —3 12 |
| 0.2 | —2 72 | —1 17 | —2 27 | —2 30 | —1 54 | —1 55 | —1 17 | —1 10 | —2 78 | —2 50 | —3 45 | —3 21 |
| 0.4 | —1 13 | —1 30 | —2 37 | —2 52 | —1 69 | —1 73 | —1 19 | —1 17 | —2 79 | —2 67 | —3 90 | —3 27 |
| 0.8 | —2 60 | —1 27 | —2 10 | —2 43 | —1 48 | —1 40 | —1 12 | —2 94 | —2 45 | —2 36 | —3 75 | —3 13 |
| 1.6 | —2 16 | —2 50 | —3 38 | —3 54 | —2 98 | —2 46 | —2 23 | —2 11 | —3 90 | —3 40 | —4 57 | —4 14 |
| 3.2 | —3 24 | —3 18 | —4 39 | —4 29 | —3 25 | —3 15 | —4 55 | —4 34 | —4 22 | —4 13 | —5 51 | —5 74 |
| $A$ | 0 93 | 1 29 | 1 15 | 1 28 | 2 11 | 2 14 | 2 13 | 2 13 | 2 14 | 2 13 | 1 17 | 0 58 |
| $\chi$ | 0 53 | 0 66 | 0 90 | 0 73 | 0 71 | 1 17 | 0 88 | 1 17 | 0 94 | 1 17 | 0 93 | 1 20 |

Note. 1. The method of orthogonalized functions; 2. The Ochkur method.

Table 3 gives the parameters for certain intercombinational transitions in the isoelectronic sequences of He and Ne. Tables 4 and 5 give the cross sections and parameters A, $\chi$ for a series of transitions in He and FeXVII, as well as the cross sections calculated by the Ochkur method for He.

The authors express their deep appreciation to A. V. Vinogradov for his help in the work and his valuable discussions.

# References

1. L. A. Vainshtein, this volume, p.1.
2. I. L. Beigman and L. A. Vainshtein, Zh. Éksperim. i Teor. Fiz., 52:185 (1967).
3. L. A. Vainshtein, Opt. i Spektr., 3:319 (1957).
4. L. A. Vainshtein and I. I. Sobel'man, Preprint of the Physics Institute of the Academy of Sciences of the USSR, No. 66 (1967); J. Quant. Spectr. Radiative Transfer, 8:1491 (1968).
5. M. J. Seaton, Proc. Phys. Soc., 77:184 (1961).
6. D. F. Dance, M. F. A. Harrison, and A. C. H. Smith, Proc. Roy. Soc. (London), A290:74 (1966).
7. S. Ormonde, W. Whitaker, and L. Lipsky, Phys. Rev. Letters, 19:1161 (1967).
8. N. R. Daly and R. E. Powell, Phys. Rev. Letters, 19:1165 (1967).
9. I. L. Beigman and L. A. Vainshtein, Astronom. Zh., 44:889 (1967).

# UNSEPARATED VARIABLES IN THE PROBLEM OF
# THE EXCITATION OF AN ATOM BY ELECTRON IMPACT

## L. P. Presnyakov

### §1. Introduction

Numerous papers on the calculation of the cross sections of inelastic collisions of electrons with atoms which have been published in recent years have shown that attempts at refining the Born method within the framework of conventional perturbation theory with separation of variables do not lead to success. The consideration of the distortion of the incident and scattered waves usually leads to an increase in the discrepancy between the calculated sections and the experimental data. Moreover, the exact solution of the system of equations for several levels in the case of electron – atom collisions improves the agreement with experiment so insignificantly that we can hardly count on obtaining reliable results in this way.

The cause of such failure, in our view, resides in the fact that in the original approximation the effect of the attraction of the incident electrons to the nucleus is considered; this effect bears no relation to the process under investigation. In fact, the main role in inelastic scattering is played by the distance between the optical and outer electrons. Because of the polarization of the atom, this distance practically does not decrease during the attraction of the outer electron to the nucleus.

Therefore, methods in which no resort is made to expansion in partial waves, which are connected with the separation of variables in the wave functions, are of definite interest. If we do not count the method of perturbed stationary states, which has not found any extensive application in the theory of electron – atom collisions, then the first attempt at introducing a function with unseparated variables was the paper by Akerib and Borowitz [1] which was carried out on the basis of the momentum approximation. However, a number of defects of the wave function derived in [1] in carrying over the standard formulation of the momentum approximation to the case of the Coulomb field led to a substantial discrepancy between theory and experiment and to a deterioration of the results compared with those obtained using the Born approximation.

With allowance for the above, Vainshtein, Sobel'man, and the author [2] advanced the model for calculating the cross sections of inelastic collisions of electrons with hydrogen atoms; in this model the repulsion of the outer and optical electrons is given first priority. The cross section calculated in this manner for the excitation [2] and ionization [3] of a hydrogen atom from the ground state turned out to be in very good agreement with experiments. While satisfying certain additional conditions, the method which has been developed also allows generalization for the case of an arbitrary atom [4].

19

The purpose of the present paper is the systematic exposition of the method indicated as it applies to the processes of excitation and ionization of atoms and ions by electron impact, and the discussion of already fairly numerous results on the calculation of cross sections from a unified point of view.

The atomic system of units with Rydberg energy units is used.

## §2.   Excitation of a Hydrogen Atom by Electron Impact

Let us consider the simplest case of excitation of a hydrogen atom. The problem consists in finding the solution of the Schrödinger equation

$$\left\{ \frac{1}{2}\Delta_1 + \frac{1}{2}\Delta_2 + \frac{1}{r_1} + \frac{1}{r_2} - \frac{1}{|\mathbf{r}_2 - \mathbf{r}_1|} + E \right\} \Psi_0(\mathbf{r}_1, \mathbf{r}_2) = 0, \tag{2.1}$$

which satisfies the boundary condition

$$\Psi_0(\mathbf{r}_1, \mathbf{r}_2) \underset{r_2 \to \infty}{\sim} \varphi_0(\mathbf{r}_1) e^{i\mathbf{k}_0 \mathbf{r}_2} + \sum_n f_n(\theta, \Phi) \varphi_n(\mathbf{r}_1) \frac{e^{ik_n r_2}}{r_2}, \tag{2.2}$$

where $\mathbf{r}_1$, $\mathbf{r}_2$ are the coordinates (position vector) of the atomic and outer electrons; $\mathbf{k}_0$, $\mathbf{k}_n$ are the wave vectors of the incident and scattered electrons; $\varphi_n$ are the unperturbed atomic functions.

Under these conditions, neglecting exchange, the expression for the effective cross section of a transition between two states 0 and 1 may be written in the form

$$\sigma_{01} = \frac{k_1 a_0^2}{4\pi^2 k_0} \int |\langle \varphi_1^*(\mathbf{r}_1) e^{-i\mathbf{k}_1\mathbf{r}_2} | V | \Psi_0(\mathbf{r}_1, \mathbf{r}_2) \rangle|^2 \, dO, \tag{2.3}$$

$$V = \frac{1}{|\mathbf{r}_2 - \mathbf{r}_1|} - \frac{1}{r_2}. \tag{2.4}$$

The consideration of exchange requires antisymmetrization of the functions $\Psi_0$ over all variables including the spin variables. In our case this requirement does not lead to any additional difficulties and can be carried out during the last stage after the function $\Psi_0(\mathbf{r}_1, \mathbf{r}_2)$ has been constructed (this will be discussed below).

We shall seek the function $\Psi_0$ in the form

$$\Psi_0(\mathbf{r}_1, \mathbf{r}_2) = \varphi_0(\mathbf{r}_1) g(\mathbf{r}_1, \mathbf{r}_2). \tag{2.5}$$

In the Born approximation $g(\mathbf{r}_1, \mathbf{r}_2) = \exp(i\mathbf{k}_0\mathbf{r}_2)$. In the general case the function g satisfies an equation which is conveniently written in the following form for later analysis:

$$\left\{ \Delta_1 + \Delta_2 + \frac{2}{|\mathbf{r}_2 + \mathbf{r}_1|} - \frac{2}{|\mathbf{r}_2 - \mathbf{r}_1|} + k_0^2 \right\} g(\mathbf{r}_1, \mathbf{r}_2) = \left\{ \frac{2}{|\mathbf{r}_2 + \mathbf{r}_1|} - \frac{2}{r_2} - 2(\nabla_1 \ln \varphi_0(\mathbf{r}_1)) \nabla_1 \right\} g(\mathbf{r}_1, \mathbf{r}_2). \tag{2.6}$$

Let us go over to the new variables $\rho = (r_2 - r_1)/2$, $R = (r_2 + r_1)/2$, which respectively describe the motion of the atomic and outer electrons and the motion of their center of inertia:

$$\left\{ \tfrac{1}{2} \Delta_R + \tfrac{1}{2} \Delta_\rho + \tfrac{1}{R} - \tfrac{1}{\rho} + k_0^2 \right\} g(R, \rho) = Q g(R, \rho), \tag{2.7}$$

where

$$Q = \left\{ \tfrac{1}{R} - \tfrac{2}{|R + \rho|} - (\nabla_1 \ln \varphi_0)(\nabla_R \ln g - \nabla_\rho \ln g) \right\}. \tag{2.8}$$

Equation (2.7) is exact. Let us now introduce the main assumption of the method considered, having taken the solution of (2.7) without its right side and the function $g(r_1, r_2)$. Under these conditions the function g will describe the scattering of free electrons by one another and the motion of their center of inertia in the field of the nucleus. In Eq. (2.7) without its right side the variables are separated, and the solution satisfying the necessary boundary conditions takes the form

$$g(R, \rho) = N \exp\left[ik_0 (\rho + R)\right] F(i\nu, 1, ik_0 R - i\mathbf{k}_0 R) F(-i\nu, 1, ik_0\rho - i\mathbf{k}_0\rho), \tag{2.9}$$

where F are degenerate hypergeometric functions,

$$\nu = \frac{1}{k_0}, \qquad N = \Gamma(1 - i\nu)\Gamma(1 + i\nu) = \frac{\pi\nu}{\sh \pi\nu}.$$

A shortcoming of the method which has been developed is the absence of the effect of screening of the interelectron repulsion, especially at small $k_0$. Let us consider one of the possible ways of refining g, having set out to reduce the magnitude of the discarded terms in the right side of (2.7) as much as possible and to introduce the dependence on the characteristics of the atomic state into g. It turns out that this can be done by introducing an appropriately chosen effective charge $\xi$ into the equation. Such a device does not pretend to be rigorous but is justified by its success. Let us rewrite (2.7) in the following form:

$$\left\{ \tfrac{1}{2} \Delta_R + \tfrac{1}{2} \Delta_\rho + \tfrac{\zeta}{R} - \tfrac{\zeta}{\rho} + k_0^2 \right\} g = \left\{ \tfrac{\zeta}{R} - \tfrac{2}{|R + \rho|} + \tfrac{1 - \zeta}{\rho} - (\nabla_1 \ln \varphi_0)(\nabla_R \ln \chi - \nabla_\rho \ln \chi) \right\} g, \tag{2.10}$$

where

$$\chi = \exp\left[-ik_0(R + \rho)\right] g. \tag{2.11}$$

The solution of this equation without its right side is determined by Eq. (2.9) with $\nu = \xi/k_0$. Substituting the function g defined in this manner into the right side of (2.10), it is not difficult to confirm that for large R and $\rho$,

$$\left| (\nabla_1 \ln \varphi_0) \nabla_R \ln \chi \right| \sim \sqrt{\varepsilon_0} \, \frac{\zeta}{k_0 R} + O\left(\frac{1}{R^2}\right),$$

$$\left| (\nabla_1 \ln \varphi_0) \nabla_\rho \ln \chi \right| \sim \sqrt{\varepsilon_0} \, \frac{\zeta}{k_0 \rho} + O\left(\frac{1}{\rho^2}\right), \tag{2.12}$$

where $\varepsilon_0$ is the ionization energy of the O state. (The factor $\sqrt{\varepsilon_0}$ arises in (2.12) for differentiation of the exponential included in $\varphi_0$.)

The first terms of the expansion in (2.12) increase ad infinitum for $\xi = \text{const}$ and $k_0 \to 0$. Let us choose $\xi$ in such a way as to eliminate this divergence, and to simultaneously ensure the identical order of magnitude of all of the discarded terms of the equation; for this purpose we require that

$$\frac{1-\zeta}{\rho} = \frac{\sqrt{\varepsilon_0}}{k_0} \frac{\zeta}{\rho}. \tag{2.13}$$

Hence,

$$\zeta = \frac{k_0}{k_0 + \sqrt{\varepsilon_0}}, \qquad \nu = \frac{1}{k_0 + \sqrt{\varepsilon_0}}. \tag{2.14}$$

For such a definition of $\xi$ the quantity $\nu^{-1}$ has the order of magnitude of the relative velocity of the electrons for any values of $k_0$, which corresponds to the physical meaning of this quantity. The method suggested for introducing the effective charge is obviously not the only possible one. Crothers and McCarroll [5] used Eqs. (2.12) with allowance for the signs and phases, rather than according to the norm, for the purpose of refining the definition of the effective charge $\xi$. As a result, they obtained the complex quantities

$$\zeta = \frac{k_0}{k_0 - i\sqrt{\varepsilon_0}}, \qquad \nu = \frac{1}{k_0 - i\sqrt{\varepsilon_0}}. \tag{2.15}$$

However, it is easy to confirm [6] that the function $\Psi_0$ with complex $\nu$ has an essentially incorrect asymptotic behavior; i.e., it does not satisfy the boundary condition of the scattering problem. It can also be shown [6] that the indicated violation of the asymptotic behavior leads to a substantial overestimation of the cross section in the range of intermediate and low collision energies. Therefore, in using other rules for choosing the effective charge it is necessary to see to it that the quantity $\nu$ is real.

The physical content of the approximation made in obtaining the wave function $\Psi_0(\mathbf{r}_1, \mathbf{r}_2)$ has been discussed above. Let us indicate a number of properties which can be established from an analysis of the structure of $\Psi_0$:

$$\Psi_0(\mathbf{r}_1, \mathbf{r}_2) = \varphi_0(\mathbf{r}_1) \exp[i\mathbf{k}_0(\rho + R)] N(\nu) F(-i\nu, 1, ik_0\rho - i\mathbf{k}_0\rho) F(i\nu, 1, ik_0R - i\mathbf{k}_0R), \tag{2.16}$$

$$\nu = \frac{1}{k_0 + \sqrt{\varepsilon_0}}, \qquad N(\nu) = \Gamma(1 - i\nu)\Gamma(1 + i\nu) = \frac{\pi\nu}{\sh \pi\nu}.$$

1. The wave function $\Psi_0$ is, generally speaking, nonorthogonal to none of the unperturbed atomic functions, and therefore in a specific approximation it considers the contribution to the polarization from an infinite number of perturbing levels. This is especially important in connection with the fact that within the framework of methods based on expansion in unperturbed atomic functions consideration of closed scattering channels creates mathematical difficulties of both a computational and fundamental character.

2. The wave function $\Psi_0$ preserves the basic analytical properties of the exact solution — it is an eigenfunction of a second order operator which is characterized by three poles in six-dimensional space. In the operator of the original Eq. (2.1) these are the points $r_1 = 0$, $\rho \equiv |\mathbf{r}_2 - \mathbf{r}_1|/2 = 0$, $r_2 = 0$. In the operator generated by (2.12) we respectively have $r_1 = 0$, $\rho \equiv |\mathbf{r}_2 - \mathbf{r}_1|/2 = 0$, $R = |\mathbf{r}_2 + \mathbf{r}_1|/2 = 0$.

3. The function (2.16) does not have additional singularities. The introduction of the effective charge liquidates the exponential singularity of the function (2.9) with $\nu = 1/k_0$ for $k_0 \to 0$.

4. The wave function $\Psi_0$ has a proper asymptotic behavior, while each separate degenerate hypergeometric function contains a logarithmic term in the phase of the forward wave.

The method expounded has certain features in common with the momentum approximation [1], but it permits a number of substantial shortcomings of the latter to be avoided. The momentum approximation describes mutual scattering of free electrons — an outer electron and an atomic electron which is distributed with the density of the unperturbed atomic function in the momentum representation. The interaction of the incident electron with the nucleus is not considered at all under these conditions; the motion of the center of inertia of the electrons is described by a plane wave. In the mathematical sense such an approach is more complex due to the presence of an additional integration in the momentum space.

The enumerated analytical properties 1-4 are what distinguishes the function (2.16) from the function with unseparated variables used by Akerib and Borowitz [1]. The latter contains a logarithmic term in the phase of the forward wave, has an additional singularity of an exponential nature for $k_0 \to 0$, and is an eigenfunction of an operator having just two poles in six-dimensional space: $r_1 = 0$, $\rho = 0$.

In the Appendix it is shown that the final result of [1] can be obtained only on the basis of a simplified version of the method being developed here.

Let us now calculate the matrix element of the transition. In the Born approximation only the first term in (2.4) yields a nonvanishing contribution. Within the framework of the method given the contribution from the second term can similarly be neglected. In order to calculate the matrix element

$$T_{01} = \iint d\mathbf{r}_1 \, d\mathbf{r}_2 \varphi_0(\mathbf{r}_1) \varphi_1^*(\mathbf{r}_1) e^{-i\mathbf{k}_1\mathbf{r}_2} |\mathbf{r}_2 - \mathbf{r}_1|^{-1} e^{i\mathbf{k}_0\mathbf{r}_2} N(\nu) F(-i\nu, 1, ik_0\rho - ik_0\rho) F(i\nu, 1, ik_0 R - ik_0\mathbf{R}) \quad (2.17)$$

we represent $\varphi_0(\mathbf{r}_1) \varphi_1^*(\mathbf{r}_1)$ in the form of a Fourier integral

$$\varphi_0(\mathbf{r}_1) \varphi_1^*(\mathbf{r}_1) = \int \widetilde{\varphi}(\varkappa) e^{-i\varkappa \mathbf{r}_1} d\varkappa. \quad (2.18)$$

Substituting (2.18) into (2.17), we obtain

$$T_{01} = 4N(\nu) \int d\mathbf{s} \varphi(\mathbf{q} - \mathbf{s}) \int d\mathbf{R} \, e^{i\mathbf{s}\mathbf{R}} F(i\nu, 1, ik_0 R - ik_0\mathbf{R}) \int d\rho \frac{\exp[i(2\mathbf{q} - \mathbf{s})\rho]}{\rho} F(-i\nu, 1, ik_0\rho - ik_0\rho), \quad (2.19)$$

where

$$\mathbf{q} = \mathbf{k}_0 - \mathbf{k}_1.$$

Since the integral with respect to $R$ increases ad infinitum for $\mathbf{s} \to 0$, let us replace the slowly varying function $\widetilde{\varphi}(\mathbf{q} - \mathbf{s})$ by $\widetilde{\varphi}(\mathbf{q})$. Under these conditions, in general, the contribution to (2.19) from the region $\mathbf{s} \sim \mathbf{q}$ is estimated incorrectly, since $\widetilde{\varphi}(0) = 0$. However, it can be shown that the error connected with this is compensated to a definite degree by the replacement of $\exp[i(2\mathbf{q} - \mathbf{s})\rho]$ by $\exp[i(-2\mathbf{q} - \mathbf{s})\rho]$. After such a substitution, we obtain the following result on the basis of using the convolution theorem which is well known in the theory of Fourier transformations:

$$T_{01} = 4N(\nu)\widetilde{\varphi}(\mathbf{q})\int dr \, \frac{e^{-i2\mathbf{q}\mathbf{r}}}{r} \, F(-i\nu, 1, ik_0 r - i\mathbf{k}_0\mathbf{r}) \, F(i\nu, 1, ik_0 r - i\mathbf{k}_0\mathbf{r}). \tag{2.20}$$

This integral can be calculated exactly by using the Nordsieck method [7].

Let us present the final result:

$$T_{01} = \frac{4\pi}{q^2} \, \widetilde{\varphi}(\mathbf{q}) f(\nu, \, x), \tag{2.21}$$

$$f(\nu, \, x) = \pi\nu \, [\text{sh} \, \pi\nu]^{-1} \, F(-i\nu, \, i\nu, \, 1, \, x), \tag{2.22}$$

$$x = \left[ \frac{qk_0}{q^2 + qk_0} \right]^2 = \left[ \frac{\Delta\varepsilon + q^2}{\Delta\varepsilon + 3q^2} \right]^2,$$

$$\nu = \frac{1}{k_0 + \sqrt{\varepsilon_0}},$$

where F is a hypergeometric function; $\Delta\varepsilon = k_0^2 - k_1^2$ is the excitation energy.

Equation (2.21) differs from the matrix element in the Born approximation solely by the real factor $f(\nu, \, x)$. For $k_0 \gg 1$, $f(\nu, \, x) \approx 1$; for $k_0 \lesssim 1$, $f(\nu, \, x)$ may differ considerably from unity. It is essential that $f(\nu, \, x) \leq 1$ for any values of the parameters. For specified $k_0$ (or $\Delta\varepsilon$) and $q \to 0$, x tends to unity. In this case $f(\nu, \, x)$ similarly tends to unity.

Let us now consider the exchange effect. The function g $(\mathbf{R}, \, \boldsymbol{\rho})$ in the approximation (2.9), (2.16) is symmetrical with respect to the coordinates $\mathbf{r}_1$ and $\mathbf{r}_2$; therefore, the transposition of the coordinates of the electrons in (2.16) leads merely to a replacement of $\varphi_0(\mathbf{r}_1)$ by $\varphi_0(\mathbf{r}_2)$ and $\boldsymbol{\rho}$ and $\boldsymbol{\rho}' = -\boldsymbol{\rho}$. As a result, the exchange integral has the form

$$T_{01}^{\text{exch}} = 4\iint d\mathbf{R} \, d\boldsymbol{\rho}'\varphi_1^*(\mathbf{r}_1) \, e^{-i\mathbf{k}_1\mathbf{r}_2} \frac{1}{\rho'} \, \varphi_0(\mathbf{r}_2) \, g(\mathbf{R}, \, \boldsymbol{\rho}') = 4N(\nu)\iint d\mathbf{s}_1 \, d\mathbf{s}\widetilde{\varphi}_1^*(\mathbf{s}_1)\widetilde{\varphi}_0(\mathbf{s} - \mathbf{q} - \mathbf{s}_1) \times$$

$$\times \int d\mathbf{R} e^{i\mathbf{s}\mathbf{R}} \, F(i\nu, 1, ik_0 R - i\mathbf{k}_0\mathbf{R}) \int d\boldsymbol{\rho}' \frac{\exp[i(2\mathbf{k}_0 - \mathbf{s} - 2\mathbf{s}_1)\boldsymbol{\rho}']}{\rho'} \, F(-i\nu, 1, ik_0\rho' - i\mathbf{k}_0\boldsymbol{\rho}'), \tag{2.23}$$

where $\widetilde{\varphi}_1^*$, $\widetilde{\varphi}_0$ are the Fourier transforms of the corresponding atomic functions. In the same approximation as that in which the integral (2.19) was calculated we can omit s in the argument of the function $\widetilde{\varphi}_0$. Moreover, in the integral with respect to $\rho'$ we omit the term $i2\mathbf{s}_1\boldsymbol{\rho}'$ in the exponent of the exponential. For large $k_0$ the possibility of such a simplification is obvious. In the range of small $k_0$ the role of this term, in general, increases, but the order of magnitude of the integral (2.23) is determined under these conditions by the small normalizing factor $N(\nu)$.

Using the convolution theorem and the Nordsieck method [7] again, we obtain

$$T_{01}^{\text{exch}} = 4N(\nu)\widetilde{\varphi}(\mathbf{q})\int dr \, \frac{e^{-i2\mathbf{k}_0\mathbf{r}}}{r} \, F(-i\nu, 1, ik_0 r - i\mathbf{k}_0\mathbf{r}) \, F(i\nu, 1, ik_0 r - i\mathbf{k}_0\mathbf{r}) =$$

$$= \frac{4N(\nu)\pi}{q^2} \, \widetilde{\varphi}(\mathbf{q}) F\left(-i\nu, \, i\nu, \, 1, \, \frac{1}{4}\right). \tag{2.24}$$

Ochkur [8] advanced the method which allows the well-known shortcomings of the Born – Oppenheimer method to be eliminated within the framework of the plane-wave approximation. In an analogous approximation $(g = \exp \, i\mathbf{k}_0\mathbf{r}_1)$ the method used above in calculating the integral (2.23) leads to the same results, although by a different way than the Ochkur method.

This fact is extremely essential. In verifying the Ochkur result the method discussed allows it to be refined. Note in this connection that the difference between the exact value of the integral (2.23) and the approximation (2.24) is smaller than the difference between the Born – Oppenheimer and Ochkur equations.

The method examined allows the effective cross section to be expressed in the form of integrals of analytic functions. Neglecting exchange,

$$\sigma_{01} = \frac{8\pi a_0^2}{k_0^2} \int_{k_0-k_1}^{k_0+k_1} \frac{dq}{q^3} |\langle 1|e^{i\mathbf{q}\mathbf{r}}|0\rangle|^2 [f(\nu, x)]^2. \tag{2.25}$$

Considering exchange, the equation for the effective cross section takes the form

$$\sigma_{01} = \sigma_{01}^+ + \sigma_{01}^-, \tag{2.26}$$

$$\sigma_{01}^{\pm} = c^{\pm} \frac{8\pi a_0^2}{k_0^2} \int_{k_0-k_1}^{[k_0+k_1]} \frac{dq}{q^3} |\langle 1|e^{i\mathbf{q}\mathbf{r}}|0\rangle|^2 \left\{ f(\nu, x) \pm \frac{q^2}{k_0^2} f\left(\nu, \frac{1}{4}\right)\right\}^2, \tag{2.27}$$

$$x = \left[\frac{\Delta\varepsilon + q^2}{\Delta\varepsilon + 3q^2}\right]^2, \quad f(\nu, x) = \pi\nu [\mathrm{sh}\,\pi\nu]^{-1} F(-i\nu, i\nu, 1, x).$$

In the case of the hydrogen atom, $c^+ = 1/4$, $c^- = 3/4$.

For $k_0 \gg 1$ Eq. (2.25) tends to the Born approximation, and Eq. (2.27) tends to the Ochkur result [8]. Let us note the general properties of Eq. (2.27) and of the Ochkur result which can be obtained by replacing $f(\nu, x)$ and $f(\nu, 1/4)$ by unity in (2.27). In the immediate vicinity of the threshold it follows that for $k_1 \to 0$ and $q^2 \to k_0^2 = \Delta\varepsilon$ we have $x \to 1/4$; i.e., the exchange excitation is equal to the direct excitation. In this case (2.27) coincides with (2.25). In the converse limiting case for $k_1 \gg 1$, the main contribution to the integral is made by the range of small $q \sim \Delta\varepsilon/k_0$, and under these conditions $f(\nu, x) \gg f\left(\nu, \frac{1}{4}\right)\frac{q^2}{k_0^2}$, i.e., the exchange excitation is negligibly small, and (2.27) similarly tends to (2.25). Under these conditions it is essential that consideration of exchange always leads to a decrease of the cross section; i.e., Eq. (2.27) tends to (2.25) from below in the limiting cases.

Since $x < 1$, the real function $f(\nu, x)$ can be calculated by means of conventional series for hypergeometric functions. Therefore, the presence of correction factors in the integrand expressions (2.25) and (2.27) does not lead to any significant complication of the numerical calculations.

Based on the equations obtained, numerical calculations were carried out for the cross sections of the $1s - 2p$, $1s - 2s$, and $4s - 5p$ transitions.

The results of the calculations for $\nu = 1/(k_0 + \sqrt{\varepsilon_0})$ with and without consideration of exchange, and for $\nu = 1/k_0$ without consideration of exchange are compared in Figs. 1a and 1b. Under these conditions, as is evident from these figures, consideration of exchange has practically no effect on the results. The introduction of the effective charge, as might be expected, has a considerably stronger influence, especially near the excitation threshold.

In Figs. 2a and 2b the results of calculations of the $1s - 2p$ and $1s - 2s$ transitions according to Eq. (2.27) with $\nu = 1/(k_0 + \sqrt{\varepsilon_0})$ are compared with the experimental data cited in [10-12]

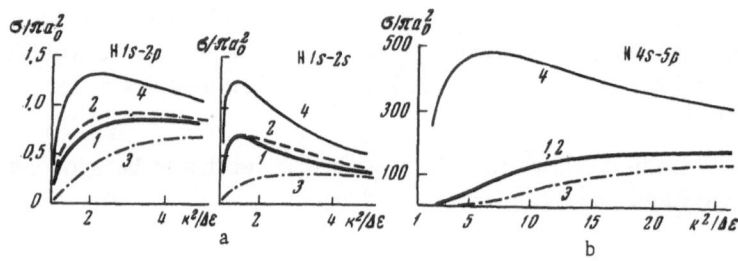

Fig. 1. Effective cross sections of the 1s − 2p, 1s − 2s, and
4s − 5p transitions in the hydrogen atom. 1) With allowance
for exchange and the effective charge; 2) neglecting exchange,
but with allowance for the effective charge; 3) neglecting both
exchange and the effective charge; 4) the Born approximation.

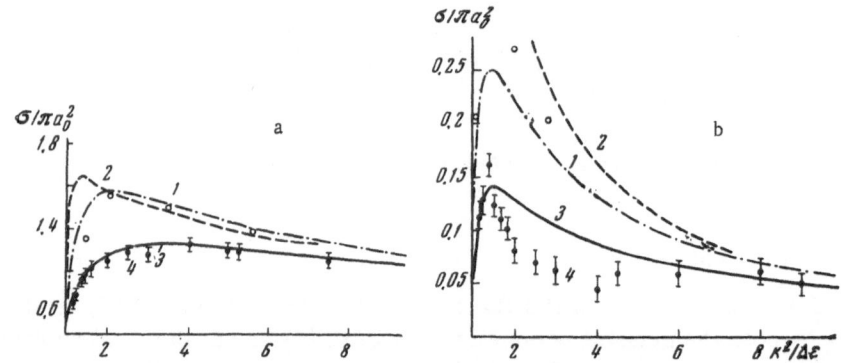

Fig. 2. Comparison of the calculations of the effective cross sections
of the 1s − 2p (a) and 1s − 2s (b) transitions in the hydrogen atom with
experimental data. 1) The Born approximation; 2) the method of distorted waves; 3) calculation according to Eq. (2.27); 4) experimental
data [10-12]; ∘) represents the strong-coupling method [14, 15].

and with the results of calculation by other methods [14-15]. As is evident, the method examined here yields very good agreement with experiment in these cases. As far as transitions between excited states are concerned, it follows that using the example of the 4s − 5p transition
(see Fig. 2b) it is obvious that the method given leads to a considerably stronger reduction of
the cross section than that obtained using the Born method. Moreover, the similarity of the
cross section in threshold energy units, which is typical of the Born method, is violated. Regrettably, almost no experimental data are available for transitions of this type. Therefore,
the extent to which this result corresponds to the true state of affairs is unclear. From an
analysis of the equations given here we can, however, conclude that the rule used to select the
effective charge is satisfactory only for transitions from the ground state. As far as transitions between excited states are concerned, it follows that in connection with the appearance of
the first order poles in the factor $(\nabla_1 \ln \varphi(\mathbf{r}_1))$ in the right side of (2.10) this problem requires
additional investigation.

§3.    Excitation of an Arbitrary Atom

The method developed can be transferred directly to an arbitrary atom, provided that the
one-electron approximation is used to describe the atom (i.e., if it is assumed that the optical

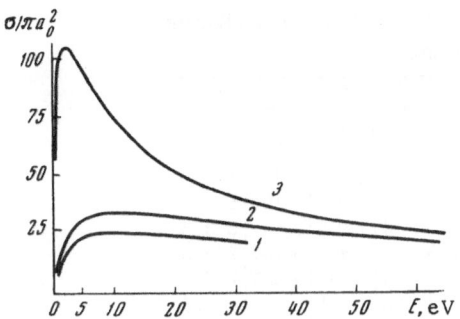

Fig. 3. Comparison of the calculation of the effective cross section of the 3s − 3p transition in the sodium atom with experimental data. 1) Experimental data [9]; 2) calculations according to Eq. (3.2); 3) the Born approximation.

electron moves in an atomic-core field which does not change when the optical electron makes a transition). The function g $(\mathbf{R}, \rho)$, as previously, is described by Eq. (2.10), but we should consider the incomplete screening of the nucleus by the atomic core in the term $\xi/R$. Then $\xi = \xi(R)$ tends to the form (2.14) only when $R \to \infty$. However, in order to preserve the simple analytical form for g we henceforth make use of the following relationship for all R:

$$\zeta = \text{const} = k_0 (k_0 + \sqrt{\varepsilon_0})^{-1}. \tag{3.1}$$

With allowance for what has been said above, all of the results of §2 can be carried over directly to the case of atoms of the alkali elements. In the case of more complex atoms a certain complication develops due to the necessity of considering the addition of angular and spin momenta. Under these conditions it turns out that the method described in §2 for calculating the exchange part of the transition integral requires separate processing of the spatial and spin parts of the matrix elements. Therefore, the general equation for the cross section of the $\gamma (L_p S_p) n_0 l_0 L_0 S_0 — \gamma (L_p S_p) n_1 l_1 L_1 S_1$ transition can be written merely by isolating the radial, angular, and spin factors in explicit form. Without dwelling on the rather cumbersome mathematical operations, we give the results:

$$\sigma_{01} = \frac{8\pi a_0^2}{k_0^2} \sum_\lambda c_\lambda \int_{k_0-k_1}^{k_0+k_1} \frac{dq}{q^3} \left| \int_0^\infty P_0(r) P_1(r) j_\lambda(qr)\, dr \right|^2 \Phi(q); \tag{3.2}$$

$$\Phi(q) = \delta_{S_0 S_1} \left[ f^2(\nu, x) - \frac{q^2}{k_0^2} f(\nu, x) f\left(\nu, \frac{1}{4}\right) \right] + \frac{2S_1+1}{2(2S_p+1)} \frac{q^4}{k_0^4} f^2\left(\nu, \frac{1}{4}\right); \tag{3.3}$$

$$f(\nu, z) = \frac{\pi\nu}{\operatorname{sh}\pi\nu} F(-i\nu, i\nu, 1, z), \quad x = \left[ \frac{\Delta\varepsilon + q^2}{\Delta\varepsilon + 3q^2} \right]^2, \quad \nu = \frac{1}{k_0 + \sqrt{\varepsilon_0}}; \tag{3.4}$$

$$c_\lambda = (2L_1+1) \left\{ \begin{matrix} \lambda & l_1 & l_0 \\ L_p & L_0 & L_1 \end{matrix} \right\}^2 (2\lambda+1)(2l_0+1)(2l_1+1) \begin{pmatrix} \lambda & l_0 & l_1 \\ 0 & 0 & 0 \end{pmatrix}^2. \tag{3.5}$$

Here $P_0$, $P_1$ are radial functions of the optical electron; $j_\lambda$ is a spherical Bessel function. In deriving (3.2) the same assumptions were made as in §2. Moreover, orthogonality of all one-electron atomic functions of the initial and final states was assumed. If we set $\Phi(q) = 1$, then (3.2) goes over into the Born approximation. For $f = 1$ Eqs. (3.2), (3.3) yield the generalization of the Ochkur results for a complex atom.

Using these formulas, the cross sections of a large number of transitions in the atoms of alkali elements were calculated [16]. Table 1 gives the results in the Born approximation and in the model adopted (in $\pi a_0^2$ units). The quantity $x_1$ is the momentum of the scattered electron in threshold units:

$$x_1 = \frac{k_1}{k_{\text{thr}}} = \sqrt{\frac{k_0^2 - \Delta\varepsilon}{\Delta\varepsilon}}. \tag{3.6}$$

Table 1 also indicates the oscillator forces $f_{ik}$ of the corresponding transitions.

TABLE 1. Cross Sections for the Excitation of Resonance Transitions
in Atoms of the Alkali Elements *

| $x_1$ | Li $2s-2p$ | | Na $3s-3p$ | | K $4s-4p$ | | Rb $5s-5p$ | | Cs $6s-6p$ | |
|---|---|---|---|---|---|---|---|---|---|---|
| | 1 | 2 | 1 | 2 | 1 | 2 | 1 | 2 | 1 | 2 |
| 0.141 | 2 414 | 1 514 | 2 307 | 1 392 | 2 547 | 1 485 | 2 585 | 1 479 | 2 750 | 1 520 |
| 0.2 | 2 571 | 1 723 | 2 426 | 1 551 | 2 757 | 1 683 | 2 811 | 1 675 | 3 104 | 1 733 |
| 0.283 | 2 771 | 2 101 | 2 578 | 1 769 | 3 103 | 1 960 | 3 110 | 1 948 | 3 141 | 2 103 |
| 0.4 | 2 998 | 2 140 | 2 756 | 2 106 | 3 136 | 2 134 | 3 146 | 2 133 | 3 186 | 2 146 |
| 0.566 | 3 120 | 2 188 | 2 932 | 2 145 | 3 166 | 2 187 | 3 180 | 2 188 | 3 230 | 2 208 |
| 0.8 | 3 131 | 2 243 | 3 104 | 2 197 | 3 187 | 2 265 | 3 203 | 2 271 | 3 260 | 2 305 |
| 1.131 | 3 124 | 2 302 | 2 103 | 2 263 | 3 186 | 2 374 | 3 202 | 2 392 | 3 260 | 2 452 |
| 1.6 | 3 103 | 2 352 | 2 884 | 2 324 | 3 159 | 2 490 | 3 176 | 2 524 | 3 226 | 2 624 |
| 2.262 | 2 753 | 2 333 | 2 666 | 2 342 | 3 120 | 2 547 | 3 133 | 2 593 | 3 172 | 2 719 |
| 3.2 | 2 501 | 2 317 | 2 454 | 2 302 | 2 821 | 2 503 | 2 913 | 2 551 | 3 117 | 2 682 |
| 4.52 | 2 312 | 2 239 | 2 287 | 2 228 | 2 520 | 2 392 | 2 590 | 2 433 | 2 746 | 2 544 |
| 6.4 | 2 186 | 2 161 | 2 173 | 2 153 | 2 314 | 2 257 | 2 351 | 2 289 | 2 452 | 2 379 |
| $f_{ik}$ | 0 748 | | 0 973 | | 1 104 | | 1 107 | | 1 112 | |

*Notation: 2 414 = 0.414 × $10^2$. 1) Results in the Born approximation; 2) results in the
model adopted.

Figure 3 shows the excitation cross section of a resonance transition in the sodium atom.
As is evident from the figure, the calculation is in satisfactory agreement with the experimental data cited by Zapesochnyi and Shimon [9], especially if the error of the experimental data is
taken into account.

## §4. Ionization of Atoms by Electron Impact

It is well known that for the problem of ionization within the framework of methods connected with expansion in partial waves an excursion beyond the limits of the Born approximation (more precisely, the distorted-wave approximation) is connected with substantial mathematical difficulties. Practically, heretofore no method was known which allowed the ionization cross section to be considered in any other but the first order of perturbation theory.

At the same time the method considered in the present paper does not require that the final state of the atom belong to a discrete spectrum. Therefore, the results of §§ 2 and 3 can be carried over directly to the case of ionization of atoms by electron impact. Thus, the ionization cross section of the hydrogen atom (or the atom of an alkali element described in the one-electron approximation) can be described in the form

$$\sigma_{0i} = \sigma_{0i}^+ + \sigma_{0i}^-, \tag{4.1}$$

$$\sigma_{0i}^\pm = c^\pm \frac{8\pi a_0^2}{k_0^2} \int d\mathbf{k}_i \int_{k_0-k_1}^{k_0+k_1} \frac{dq}{q^3} |\langle \mathbf{k}_i| e^{iqr} |0\rangle|^2 \left\{ f(\nu, x) \pm \left(\frac{q}{k_0}\right)^2 f\left(\nu, \frac{1}{4}\right) \right\}^2, \quad k_0^2 = k_1^2 + k_i^2 + I, \tag{4.2}$$

where $k_0$, $k_1$, $k_i$ are the wave numbers of the incident, scattered, and knock-on electrons; I is the ionization potential of the atom in the initial state; $\langle k_i|$ is the wave function of the atomic electron in the continuous spectrum, normalized to the $\delta$-function $\delta(\mathbf{k}_i - \mathbf{k}_i')$;

$$f(\nu, x) = \pi\nu [\text{sh } \pi\nu]^{-1} F(-i\nu, i\nu, 1, x), \quad \nu = (k_0 + \sqrt{I})^{-1}.$$

$$x = [(\Delta\varepsilon + q^2)/(\Delta\varepsilon + 3q^2)]^2, \quad \Delta\varepsilon = k_0^2 - k_1^2, \quad c^\pm = \tfrac{1}{2} \mp \tfrac{1}{4}. \tag{4.3}$$

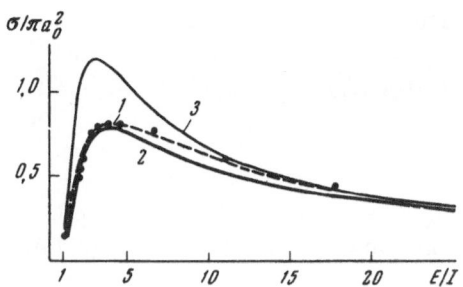

Fig. 4. Effective cross sections for the ionization of the hydrogen atom from the ground state. 1) Experimental data [13]; 2) calculations according to Eqs. (4.1)-(4.3); 3) the Born approximation.

The calculation of the cross section by means of Eqs. (4.1)-(4.3) is not much more complex than it is in the Born approximation. Figure 4 shows the results for the ionization of the hydrogen atom from the ground state.

## §5.  The Excitation of a Hydrogen-like Ion by Electron Impact

For collisions of electrons with a hydrogen-like ion the Schrödinger equation describing the process has the form

$$\left\{\frac{1}{2}\Delta_1 + \frac{1}{2}\Delta_2 + \frac{Z}{r_1} + \frac{Z}{r_2} - \frac{1}{|\mathbf{r}_2 - \mathbf{r}_1|} + E\right\}\Psi_0(\mathbf{r}_1, \mathbf{r}_2) = 0. \tag{5.1}$$

This equation must be solved with the boundary condition

$$\Psi_0(\mathbf{r}_1, \mathbf{r}_2) \underset{r_2 \to \infty}{\sim} \varphi_0(\mathbf{r}_1)\exp[ik_0 r_2 + iv_0 \ln(k_0 r_2)] + \sum_n \varphi_n(\mathbf{r}_1) f_n(\theta, \Phi) r_2^{-1}\exp[ik_n r_2 - iv_n \ln(k_n r_2)]. \tag{5.2}$$

Here $v_n = (Z - 1)/k_n$ is the Coulomb parameter; Z is the charge of the nucleus. In the first approximation of perturbation theory (the so-called Born – Coulomb approximation) the following function is used as $\Psi_0$:

$$\Phi_0(\mathbf{r}_1, \mathbf{r}_2) = N_0 \varphi_0(\mathbf{r}_1)\exp[ik_0 r_2] F(iv_0, 1i, ik_0 r_2 - i\mathbf{k}_0 \mathbf{r}_2),$$

$$N_0 = N_0(v_0) = e^{\frac{\pi v_0}{2}}\Gamma(1 - iv_0); \tag{5.3}$$

this function satisfies the boundary conditions (5.2) and is the solution of the equation

$$\left\{\frac{1}{2}\Delta_1 + \frac{1}{2}\Delta_2 + \frac{Z}{r_1} + \frac{Z-1}{r_2} + E\right\}\Phi_0(\mathbf{r}_1, \mathbf{r}_2) = 0. \tag{5.4}$$

We shall seek the solution of Eq. (5.1) in the form

$$\Psi_0(\mathbf{r}_1, \mathbf{r}_2) = \Phi_0(\mathbf{r}_1, \mathbf{r}_2)\chi(\mathbf{R}, \rho), \quad \mathbf{R} = \frac{1}{2}(\mathbf{r}_2 + \mathbf{r}_1), \rho = \frac{1}{2}(\mathbf{r}_2 - \mathbf{r}_1). \tag{5.5}$$

Under these conditions $\chi$ is the solution of the equation

$$\left\{\frac{1}{2}\Delta_R + \frac{1}{2}\Delta_\rho + \frac{\zeta}{R} - \frac{\zeta}{\rho} + i\mathbf{k}_0(\nabla_R + \nabla_\rho)\right\}\chi(\mathbf{R}, \rho) =$$

$$= \left\{\frac{\zeta}{R} - \frac{2}{|R + \rho|} + \frac{1 - \zeta}{\rho} - (\nabla_1 \ln \varphi_0)(\nabla_R \ln \chi - \nabla_\rho \ln \chi) + \right.$$

$$\left. + (\nabla_2 \ln F(iv_0, 1, ik_0 r_2 - i\mathbf{k}_0 \mathbf{r}_2))(\nabla_R \ln \chi + \nabla_\rho \ln \chi)\right\}\chi(\mathbf{R}, \rho). \tag{5.6}$$

The solution of this equation without a right side has the form

$$\chi(\mathbf{R}, \rho) = N(\nu) F(i\nu, 1, ik_0 R - i\mathbf{k}_0 \mathbf{R}) F(-i\nu, 1, ik_0 \rho - i\mathbf{k}_0 \rho),$$

$$\nu = \frac{\zeta}{k_0}, \qquad N(\nu) = \Gamma(1 - i\nu)\Gamma(1 + i\nu), \tag{5.7}$$

i.e., it differs from the previously derived function g (R, ρ) by the factor $\exp[-ik_0 (R + \rho)]$, which has already been considered in the Born − Coulomb function $\Phi_0(\mathbf{r}_1, \mathbf{r}_2)$ (see (2.9), (2.16)).

Equation (5.6) differs from the analogous equation for a neutral atom by the presence of an additional term $\tilde{V} = (\nabla_2 \ln F(i\nu_0, 1, ik_0 r_2 - i\mathbf{k}_0 \mathbf{r}_2))(\nabla_R \ln \chi + \nabla_0 \ln \chi)$ in its right side. However, it is not difficult to confirm the fact that

$$\tilde{V} \sim \begin{cases} O\left(\frac{1}{R^2}\right), & R \to \infty \\ O\left(\frac{1}{\rho^2}\right), & \rho \to \infty. \end{cases} \tag{5.8}$$

In calculating the effective charge we restricted the analysis above to the terms $O(1/R)$, $O(1/\rho)$. throughout. Repeating this procedure here, we may drop $\tilde{V}$ as being a term of higher order smallness, after which we obtain

$$\zeta = \frac{k_0}{k_0 + \sqrt{\varepsilon_0}}, \qquad \nu = \frac{1}{k_0 + \sqrt{\varepsilon_0}}, \tag{5.9}$$

where $\varepsilon_0$ is the ionization energy of the initial state.

Thus, the solution of Eq. (5.1) in the approximation given has the form

$$\Psi_0(\mathbf{r}_1, \mathbf{r}_2) = N_0(\nu_0) N(\nu) \varphi_0(\mathbf{r}_1) e^{i\mathbf{k}_0 \mathbf{r}_2} F(i\nu_0, 1, ik_0 r_2 - i\mathbf{k}_0 \mathbf{r}_2) F(i\nu, 1, ik_0 R - i\mathbf{k}_0 \mathbf{R}) F(-i\nu, 1, ik_0 \rho - i\mathbf{k}_0 \rho),$$

$$N_0(\nu_0) = e^{\frac{\pi\nu_0}{2}} \Gamma(1 - i\nu_0), \quad N(\nu) = \Gamma(1 - i\nu)\Gamma(1 + i\nu), \tag{5.10}$$

$$\nu_0 = \frac{Z - 1}{k_0}, \qquad \nu = \frac{1}{k_0 + \sqrt{\varepsilon_0}}.$$

It is easy to see that for Z = 1, we have $\nu_0 = 0$, and (5.10) goes over into the corresponding Eq. (2.16) for a neutral hydrogen atom.

The calculation of the matrix element in the case given is complicated by the presence of four degenerate hypergeometric functions under the integral sign:

$$T_{01} = N(\nu) N_0 N_1^* \iint d\mathbf{r}_1 d\mathbf{r}_2 \varphi_1^*(\mathbf{r}_1) \varphi_0(\mathbf{r}_1) \frac{e^{i\mathbf{q}\mathbf{r}_1}}{|\mathbf{r}_2 - \mathbf{r}_1|} F(i\nu_1, 1, ik_1 r_2 + i\mathbf{k}_1 \mathbf{r}_2) \times$$

$$\times F(i\nu_0, 1, ik_0 r_2 - i\mathbf{k}_0 \mathbf{r}_2) F(i\nu, 1, ik_0 R - i\mathbf{k}_0 \mathbf{R}) F(-i\nu, 1, ik_0 \rho - i\mathbf{k}_0 \rho). \tag{5.11}$$

The structure of (5.11) is most conveniently investigated by resorting to the Fourier transform

$$\tilde{\varphi}_{01}(\varkappa) = \int \varphi_1^*(\mathbf{r}_1) \varphi_0(\mathbf{r}_1) e^{-i\varkappa \mathbf{r}_1} d\mathbf{r}_1, \tag{5.12}$$

$$\widetilde{F}_{01}(\varkappa) = \int F(i\nu_1, 1, ik_1r_2 + i\mathbf{k}_1\mathbf{r}_2) F(i\nu_0, 1, ik_0r_2 - i\mathbf{k}_0\mathbf{r}_2) e^{i\varkappa\mathbf{r}_2} d\mathbf{r}_2. \tag{5.13}$$

Under these conditions the matrix element has the form

$$T_{01} = 4N(\nu, \nu_0, \nu_1) \iint ds\,ds_1 \widetilde{\varphi}_{01}(\mathbf{q} - \mathbf{s} - \mathbf{s}_1) \widetilde{F}_{01}(\mathbf{s}) \int d\mathbf{R}\, e^{i\mathbf{s}_1\mathbf{R}} F(i\nu, 1, ik_0R - i\mathbf{k}_0\mathbf{R}) \times$$

$$\times \int d\boldsymbol{\rho}\, \frac{\exp[i(2\mathbf{q} - 2\mathbf{s} - \mathbf{s}_1)\boldsymbol{\rho}]}{\rho} F(-i\nu, 1, ik_0\rho - i\mathbf{k}_0\boldsymbol{\rho}). \tag{5.14}$$

Just as in §2, we replace the slowly varying function $\widetilde{\varphi}_{01}(\mathbf{q} - \mathbf{s} - \mathbf{s}_1)$ by $\widetilde{\varphi}_{01}(\mathbf{q} - \mathbf{s})$ and carry out integration over the space $\mathbf{s}_1$, after which

$$T_{01} = N(\nu, \nu_0, \nu_1) \int ds\, \frac{\widetilde{\varphi}_{01}(\mathbf{q} - \mathbf{s})}{|\mathbf{q} - \mathbf{s}|^2} \widetilde{F}_{01}(\mathbf{s}) F\left(-i\nu,\ i\nu,\ 1\left[\frac{k_0(\mathbf{q} - \mathbf{s})}{(\mathbf{q} - \mathbf{s})^2 + k_0(\mathbf{q} - \mathbf{s})}\right]^2\right). \tag{5.15}$$

In view of the fact that $0 \leqslant \left[\frac{k_0(\mathbf{q} - \mathbf{s})}{(\mathbf{q} - \mathbf{s})^2 + k_0(\mathbf{q} - \mathbf{s})}\right]^2 < 1$, the hypergeometric function contained in (5.15) is real, positive, and bounded for all values of $\mathbf{s}$. It is not difficult to verify the fact that the point $\mathbf{s} = \mathbf{q}$ is not a pole of the integrand expression. Actually, as a consequence of the neutral orthogonality of the wave function of the discrete and continuous spectras we obtain

$$\widetilde{\varphi}_{01}(\mathbf{q} - \mathbf{s}) \underset{\mathbf{s}\to\mathbf{q}}{\longrightarrow} c_1(\mathbf{q} - \mathbf{s})^n,\ n \geqslant 1;\quad \widetilde{F}_{01}(\mathbf{s}) \underset{\mathbf{s}\to\mathbf{q}}{\longrightarrow} c_2(\mathbf{q} - \mathbf{s})$$

from (5.12) and (5.13); i.e.,

$$\frac{\widetilde{\varphi}_{01}(\mathbf{q} - \mathbf{s}) \widetilde{F}_{01}(\mathbf{s})}{|\mathbf{q} - \mathbf{s}|^2} \underset{\mathbf{s}\to\mathbf{q}}{\longrightarrow} c(\mathbf{q} - \mathbf{s})^{n-1},\ n \geqslant 1. \tag{5.16}$$

Taking into consideration the fact that the integrand expression decreases rapidly with an increase of $\mathbf{s}$ and that the function $\widetilde{F}_{01}(\mathbf{s})$ increases ad infinitum for $\mathbf{s} \to 0$, it is possible to take out the hypergeometric function at the point $\mathbf{s} = 0$ from under the integral sign. After this we obtain

$$T_{01} = N(\nu) F\left(-i\nu,\ i\nu,\ 1\left[\frac{k_0\mathbf{q}}{q^2 + k_0\mathbf{q}}\right]^2\right) N_0(\nu_0) N_1(\nu_1) \int ds\, \frac{\widetilde{\varphi}_{01}(\mathbf{q} - \mathbf{s}) \widetilde{F}_{01}(\mathbf{s})}{|\mathbf{q} - \mathbf{s}|^2}. \tag{5.17}$$

The remaining integral together with the normalizing factor $N_0(\nu_0) N_1(\nu_1)$ is exactly equal to the matrix element calculated in the Born—Coulomb approximation, as can easily be verified by applying inverse Fourier transformations. Thus, in this case also the results have the form

$$T_{01} = T_{01}^{BC} f(\nu, x). \tag{5.18}$$

In exactly the same way the following equation is valid for the exchange matrix element within the framework of the method given:

$$T_{01} = T_{01}^{BC} \frac{q^2}{k_0^2} f\left(\nu, \frac{1}{4}\right). \tag{5.19}$$

Here $T_{01}^{BC}$ is the matrix element of the transition in the Born – Coulomb approximation and has·
the form

$$T_{01}^{BC} = N_1^*(\nu_1) N_0(\nu_0) \iint d\mathbf{r}_1 d\mathbf{r}_2 \varphi_1^*(\mathbf{r}_1) \varphi_0(\mathbf{r}_1) \frac{e^{iq\mathbf{r}_2}}{|\mathbf{r}_2 - \mathbf{r}_1|} F(i\nu_1, 1, ik_1 r_2 + i\mathbf{k}_1\mathbf{r}_2) F(i\nu_0, 1, ik_0 r_2 - i\mathbf{k}_0\mathbf{r}_2). \quad (5.20)$$

For the subsequent analysis it is convenient to go over to "hydrogen" units for measuring
the momenta and the energy; i.e., it is convenient to carry out the substitution

$$k_0 \to Z k_0, \qquad k_1 \to Z k_1, \qquad q \to Z q. \quad (5.21)$$

Under these conditions

$$\nu_1 = \frac{1 - Z^{-1}}{k_1}, \qquad \nu_0 = \frac{1 - Z^{-1}}{k_0}, \qquad \nu = Z^{-1} \frac{1}{k_0 + \sqrt{\varepsilon_0}} = \frac{\nu'}{Z} \quad (5.22)$$

(Z is the charge of the ion nucleus), and the cross section can be written in the form

$$\sigma_{01} = \sigma_{01}^+ + \sigma_{01}^-,$$

$$\sigma_{01}^{\pm} = c^{\pm} \frac{|N(\nu_0, \nu_1)|^2}{Z^4} \cdot \frac{8\pi a_0^2}{k_0^2} \int\limits_{k_0-k_1}^{k_0+k_1} q dq \left| \int ds \frac{\widetilde{\varphi}_{01}(\mathbf{q} - \mathbf{s})}{|\mathbf{q} - \mathbf{s}|^2} \widetilde{F}_{01}(\mathbf{s}) \right|^2 \left\{ f\left(\frac{\nu'}{Z}, x\right) \pm \frac{q^2}{k_0^2} f\left(\frac{\nu'}{Z}, \frac{1}{4}\right) \right\}^2, \quad (5.23)$$

$$N(\nu_0, \nu_1) = N_1^*(\nu_1) N_0(\nu_0).$$

For Z = 1 we have $\nu_0 = \nu_1 = 0$ and $\widetilde{F}_{01}(\mathbf{s}) = \delta(\mathbf{s})$, and (5.23) goes over into Eq. (2.25) for the
excitation cross section of the hydrogen atom. It follows from the properties of the correction
function $(\nu'/Z, x)$ which were considered above that the maximum difference between the
cross section (5.23) and the Born – Coulomb approximation occurs for the range of minimal
values of collision energy. In §2 it was shown that in the immediate vicinity of threshold the
direct excitation is equal to exchange excitation. It is precisely this region, where the cross
section in the Born – Coulomb approximation reaches a maximum, which is of greatest interest.
The effect of the correction to the Born – Coulomb approximation, which has been introduced
within the framework of the method given, can be established without carrying out detailed
numerical calculations. With allowance for what has been said above, we may write the following inequality for the quantity R = $\sigma_{01}/\sigma_{01}^{BC}$:

$$\left[ f\left(\frac{\nu'}{Z}, \frac{1}{4}\right) \right]^2 \leqslant R \leqslant 1. \quad (5.24)$$

Since $f(0, x) = 1$, it follows that with increasing Z the role of the correction decreases more
and more. The maximum difference between the results of the method given and the Born – ·
Coulomb approximation should be expected for the helium ion. Let us give the estimates for
the helium ion and a hydrogen-like carbon ion (excitation of the 1s – 2p and 1s – 2s transtions):

$$0.85 \leqslant R_{\mathrm{He\,II}} \leqslant 1,$$

$$0.97 \leqslant R_{\mathrm{CVI}} \leqslant 1. \quad (5.25)$$

Thus, for a transition from the ground state of a hydrogen-like ion the correction introduced by the method given is of little practical consequence. This result is not difficult to understand from physical considerations. The Born — Coulomb approximation considers the long-range portion of the interaction of the incident electron with the ion. This Coulomb part increases as $(Z-1)$ with increasing Z. The remaining short-range part, which does not depend on Z at all, is considered within the framework of the model given. Therefore, obviously, the role of the correction must decrease with increasing Z. It is well known that short-range interactions against the background of long-range interactions leads to very weak complementary effects. It is precisely for this reason that even in the case of the helium ion the correction to the Born — Coulomb approximation does not exceed 15%.

## Appendix I

Let us show that the simplified version of the model allows the final result of the momentum approximation [1] to be obtained. Let us make two assumptions which lead to a substantial deterioration of the quality of the wave function $\Psi_0$ (2.16): first, we shall consider only interelectron repulsion; second we avoid screening of this repulsion by the effective charge. Neglecting the term $1/R$, in the left side of Eq. (2.7) (this corresponds to a description of the motion of the center of inertia of the electrons by means of a plane wave), we obtain the wave function of the system in the form

$$\Psi_0'^{(0)} = \varphi_0(\mathbf{r}_1) e^{-\frac{\pi}{2k_0}} \Gamma\left(1 + \frac{i}{k_0}\right) e^{i\mathbf{k}_1\mathbf{r}_2} F\left(-\frac{i}{k_0}, 1, ik_0\rho - i\mathbf{k}_0\boldsymbol{\rho}\right). \tag{A.1}$$

The matrix element with the function (A.1) is calculated exactly:

$$T_{01} = 4e^{-\frac{\pi}{2k_0}} \Gamma\left(1 + \frac{i}{k_0}\right) \int \varphi_1^*(\mathbf{r}_1) \varphi_0(\mathbf{r}_1) e^{i\mathbf{q}\mathbf{r}_1} d\mathbf{r}_1 \int d\rho \frac{e^{i2q\rho}}{\rho} \times$$

$$\times F\left(-\frac{i}{k_0}, 1, ik_0\rho - i\mathbf{k}_0\boldsymbol{\rho}\right) = \frac{4\pi}{q^2} \langle 1 | e^{i\mathbf{q}\mathbf{r}} | 0 \rangle e^{-\frac{\pi}{2k_0}} \Gamma\left(1 + \frac{i}{k_0}\right) \left[\frac{q^2}{qk_0 + q^2}\right]^{i/k}. \tag{A.2}$$

Equation (A.2) coincides with the result obtained by Akerib and Borowitz [1] in the momentum approximation by means of a number of approximations made in calculating the multiple integrals. Making use of (A.2), we obtain the final result of the Akerib-Borowitz paper for the cross section without consideration of exchange:

$$\sigma_{01} = e^{-\frac{\pi}{k_0}} \frac{\pi/k_0}{\text{sh}\,[\pi/k_0]} \sigma_{01}^B, \tag{A.3}$$

where $\sigma_{01}^B$ is the cross section in the Born approximation.

Thus, the function (A.1) having improper asymptotic behavior leads to a purely kinematic correction to the cross section, which depends solely on $k_0$. Equation (A.3), as is easily verified, leads to an extremely strong reduction of the cross sections for the $1s-2s$ and $1s-2p$ transitions in the hydrogen atom compared with experiments at energies below 40 eV.

## Appendix II

Let us set out with the aim of expressing the Born — Coulomb matrix element

$$T^{BC} = \int \frac{d\mathbf{s}\,\widetilde{\varphi}_{01}(\mathbf{q} - \mathbf{s})}{|\mathbf{q} - \mathbf{s}|^2} \widetilde{F}_{01}(\mathbf{s}) \tag{A.4}$$

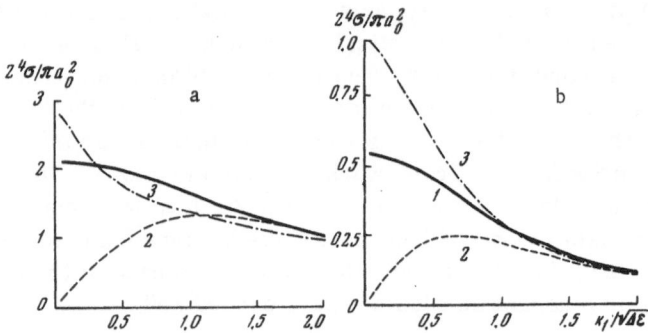

Fig. 5. The effective cross sections of 1s − 2p (a) and
1s − 2s (b) transitions in the C VI ion. 1) The Born −
Coulomb approximation; 2) the Born approximation;
3) Eqs. (A.6)

in the form of the product of the Born integral and a universal correction describing the order
of magnitude of the difference between the Born − Coulomb approximation and the Born approximation. The function $\widetilde{\varphi}_{01}(\mathbf{q} - \mathbf{s})$ and $\widetilde{F}_{01}(\mathbf{s})$ differ essentially in their dependence on their arguments if the states 0 and 1 of the discrete spectrum are not highly excited.

For the case in which these states are directly contiguous with the continuous spectrum
the differences between $\widetilde{\varphi}_{01}$ and $\widetilde{F}_{01}$ vanish. The entire subsequent discussion does not apply to
this case. For transitions between states having small values of the principal quantum number
the more slowly varying $\widetilde{\varphi}_{01}(\mathbf{q}-\mathbf{s})$ can be taken out from under the integral sign at the point
s = 0. The remaining integral can be calculated exactly:

$$T^{BC} \simeq \widetilde{\varphi}_{01}(\mathbf{q}) \int ds \frac{\widetilde{F}_{01}(\mathbf{s})}{|\mathbf{q} - \mathbf{s}|^2} = \widetilde{\varphi}_{01}(\mathbf{q}) \frac{4\pi}{q^2} e^{-\pi v_0} \left(\frac{q^2}{k_0 - k_1}\right)^{iv_0}\left(\frac{(k_0 + k_1)^2}{k_0^2 - k_1^2}\right)^{-iv_1} F\left(1 - iv_0, iv_1, 1, 1 - \frac{q_{min}^2}{q^2}\right),$$

$$q_{min} = |\mathbf{k}_0 - \mathbf{k}_1|. \tag{A.5}$$

Under these conditions the cross section has the form

$$\sigma^{\pm} = \frac{8\pi a_0^2}{k_0^2 Z^4} \int\limits_{k_0 - k_1}^{k_0 + k_1} \frac{dq}{q^3} \left| <1|e^{iqr}|0> \right|^2 \exp\left[\frac{\pi(v_1 - v_0)}{2}\right] \Gamma(1 - iv_0) \times$$

$$\times \Gamma(1 + iv_1)|^2 | F\left(1 - iv_0, iv_1, 1, 1 - \frac{(k_0 - k_1)^2}{q^2}\right)|^2 \left[f\left(\frac{v'}{Z}, x\right)\right]^2; \tag{A.6}$$

$$v_0 = \frac{1 - Z^{-1}}{k_0}, \qquad v_1 = \frac{1 - Z^{-1}}{k_1}, \qquad v' = \frac{1}{k_0 + \sqrt{\varepsilon_0}}. \tag{A.7}$$

Figure 5 shows the results of calculating the cross sections according to Eq. (A.6) and
by means of summing partial waves. As is evident, for the 1s − 2s and 1s − 2p transitions the
difference does not exceed a factor of two. Thus, Eq. (A.6) can be used for estimating the
order of magnitude of the excitation cross sections of ions.

References

1.    R. Akerib and S. Borowitz, Phys. Rev., 122:1177 (1961).
2.    L. Vainshtein, L. Presnyakov, and I. Sobel'man, Zh. Éksperim. i Teor. Fiz., 45:2015
      (1963).

3.  L. Presnyakov, Zh. Éksperim. i Teor. Fiz., 47:1134 (1964).
4.  L. Vainshtein, V. Opykhtin, and L. Presnyakov, Zh. Éksperim. i Teor. Fiz., 47:2306 (1964).
5.  D. Crothers and R. McCarroll, Proc. Phys. Soc., 86:753 (1965).
6.  L. Presnyakov, I. Sobel'man [Sobelman], and L. Vainshtein, Proc. Phys. Soc., 89:511(1966).
7.  A. Nordsieck, Phys. Rev., 93:785 (1964).
8.  V. I. Ochkur, Zh. Éksperim. i Teor. Fiz., 45:734 (1963).
9.  I. P. Zapesochnyi and L. L. Shimon, Opt. i Spektr., 19:480 (1965).
10.  W. L. Fite and R. T. Brackmann, Phys. Rev., 112:215 (1958).
11.  W. L. Fite, R. F. Stebbings, and R. T. Brackmann, Phys. Rev., 116:356 (1959).
12.  R. F. Stebbings, W. L. Fite, D. G. Hummer, and R. T. Brackmann, Phys. Rev., 119:1939 (1960); 124:2051 (1961).
13.  W. L. Fite and R. T. Brackmann, Phys. Rev., 122:1139 (1958).
14.  R. Damburg and R. Peterkop, Proc. Phys. Soc., 80:563 (1962).
15.  R. Peterkop and R. Damburg, Zh. Éksperim. i Teor. Phys., 43:1763 (1962).
16.  L. Vainshtein and L. Presnyakov, Preprint of the Physics Institute, Academy of Sciences of the USSR: Cross Sections for the Excitation of Atoms of the Alkali Elements by Electron Impact, (1966).

# THE ROLE OF INTERMEDIATE STATES IN
# THE EXCITATION OF ATOMS BY ELECTRON IMPACT

## L. A. Vainshtein and L. P. Presnyakov

**1.** The effective cross sections for the excitation of atoms by electrons are calculated in the Born approximation in the majority of cases (at least in applications). This approximation, as is well known, has been rigorously substantiated only for high energies of the imping- ing electrons. Nevertheless, available experimental data show that even at low energies $E \sim (2-3) \Delta E$ ($\Delta E$ is the excitation energy) the errors of the Born method are not too large: the cross section in the region of the maximum turns out to be overestimated by a factor of 1.5 to 2.0. It is convenient to call such an error the standard error of the Born method for purposes of the subsequent analysis. For many applications such an accuracy is fully acceptable.

Heretofore success has not been achieved in creating a substantially more exact method of calculating cross sections which would be applicable for a broad class of transitions and a broad range of energies. Therefore, it is important to indicate, starting from clear physical considerations, those situations in which it is possible to expect errors due to the Born method which differ from the standard error, and also to formulate the least complex method possible for estimating cross sections in such cases.

One of the well-known examples is the so-called violation of the normalization of cross sections, when the partial cross section $\sigma(l)$ in the Born approximation considerably exceeds the theoretical limit deriving from the condition of particle flux conservation. Such a situation usually develops in the case of transitions between close levels. A fully satisfactory device for eliminating this shortcoming is the "normalization" method based on the use of the R- matrix [1]. The "normalization" procedure allows the cross sections of transitions between close levels to be calculated with an error not exceeding the standard error of the Born method, while without normalization the cross section may be overestimated by an order of magnitude.

In the present paper we shall consider another important case of violation of the standard error indicated above — transitions for which the Born method yields a small cross section for some reason. Under these conditions transitions will occur with a higher probability via an intermediate state in the second order of perturbation theory. For example, the cross sec- tion of the quadrupole $1s - 3d$ transition in the first Born approximation will be smaller than the cross section of the $1s - 2p - 3d$ transition in the second order (a dipole transition at each state). It is important to emphasize the fact that complete calculation in the second order of perturbation theory includes summation over all intermediate states, which is a very complex problem. But here we are speaking of isolating one-two intermediate levels on the basis of physical considerations. For excitation from the ground state of the atom the resonance level, which is excited with a probability close to unity during the time of flight of the electron, is especially important.

Below we examine the excitation of atoms via an intermediate level in a simplified modification of the second Born approximation.

**2.** Let us consider the excitation of an atom by electron impact from the "zero" state to the "one" state. The effective cross section can be written in the form (we use atomic units with the Ry energy unit)

$$\sigma = \frac{1}{8\pi^2 g_0 k_0^2} \int_{k_0-k_1}^{k_0+k_1} q\,dq \sum_{M_0 M_1} \left| \sum_a \int d\mathbf{r}\, e^{i\mathbf{q}\mathbf{r}}\, W_{1a}(\mathbf{r})\, U_{a0}(\mathbf{r}) \right|^2 (\pi a_0^2), \tag{1}$$

where $\mathbf{q} = \mathbf{k}_0 - \mathbf{k}_1$, $\mathbf{k}$ is the wave vector of the outer electron, M is the projection of the angular momentum of the atom onto the z axis, $g_0$ is the statistical weight of the initial state. The matrix elements of the potentials U and W are taken in the variables $\mathbf{r}_j$ of the atomic electrons. $W_{1a}$ includes the nondiagonal polarization potential [2].

$$W = I + W^{II} + \ldots, \qquad W^{II} = \int d\mathbf{x}\, U(\mathbf{r} - \mathbf{x})\, G_a(x)\, e^{i\mathbf{k}_1 \mathbf{x}}, \tag{2}$$

$$G_a(x) = -\frac{e^{ik_a x}}{4\pi x}, \qquad k_a^2 + E_{ab} = k_1^2 + E_{10} = k_0^2, \tag{3}$$

where $G_a$ is the Green's function of the operator $\nabla^2$, $E_i$ is the energy of the i-th atomic state, $E_{ij} = E_i - E_j$. The first term in W yields the first Born approximation. In accordance with the problem stated in Sect. 1, we retain one term in the sum over the intermediate states $a$ in (1). Moreover, we shall assume that $E_0 < E_a < E_1$.

Let us begin by considering a one-electron atom, and let us drop the interaction with the nucleus.*

$$U = \frac{2}{|\mathbf{r} - \mathbf{r}_1|} = \frac{2}{\rho}, \qquad W^{II} = \int d\mathbf{x}\, \frac{2}{|\rho - \mathbf{x}|}\, G_a(x)\, e^{i\mathbf{k}_1 \mathbf{x}}. \tag{4}$$

The calculation of the second order term $W^{II}$ is connected with serious computational difficulties. For $k_1 \to 0$ the expression for $W^{II}$ can be simplified considerably. Using the spectral representation of the Green's function $G_a$, we obtain

$$\int \frac{d\mathbf{x}}{|\rho - \mathbf{x}|}\, G_a(x) = \frac{-1}{8\pi^3} \int \frac{d\mathbf{x}}{|\rho - \mathbf{x}|} \int d\mathbf{k}\, \frac{e^{i\mathbf{k}\mathbf{x}}}{k^2 - k_a^2 - i\varepsilon} = -\frac{e^{ik_a \rho} - 1}{k_a^2 \rho}.$$

Therefore, for $k_1 \to 0$ and $k_a^2 \to E_{1a}$ we have

$$W^{II} = -\frac{2}{E_{1a}\rho} [\exp(i\sqrt{E_{1a}})\rho - 1]. \tag{5}$$

This equation, by the way, indicates the range of applicability of the well-known adiabatic approximation in which it is assumed that [3]

---

*The factor 2 develops due to the use of the Ry unit for energy.

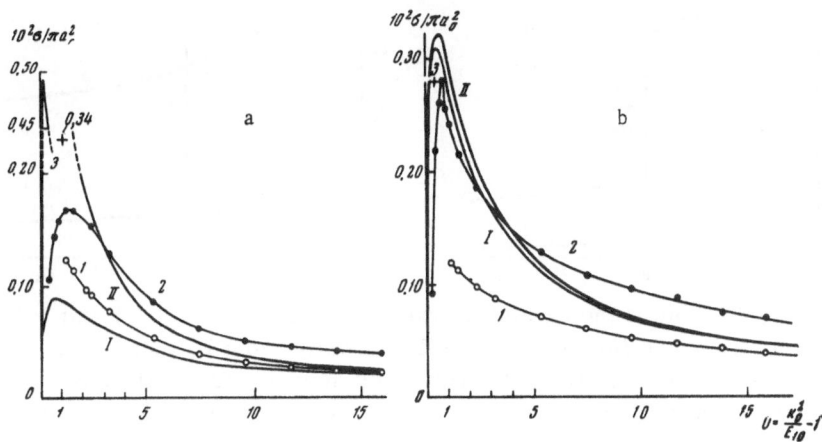

Fig. 1. Effective cross sections for the excitation of $1^1S - 4^1D$ (a) and $1^1S - 4^1S$ (b) transitions in He. Calculation: I) the first Born approximation; II) with allowance for transitions via the $2^1P$ level in the second order. Experiment: 1) [5]; 2) [6]; 3) [7].

$$G_a(x) = \frac{1}{E_{1a}} \delta(x), \qquad W^{II} = \frac{1}{E_{1a}} U. \tag{6}$$

If $E_{1a}$ is sufficiently large, the rapidly oscillating exponential in (5) can be neglected, and (5) goes over* into (6). The adiabatic approximation is widely used in the theory of elastic scattering, but it has essentially not been used heretofore in problems of inelastic scattering.

For large k we cannot obtain an expression as simple as (5). We can, however, show (for example, using the quasi-classical approximation) that the operator $W^{II}$ becomes purely imaginary and decreases as $k^{-1}$ [4]. The simplest extrapolation formula, which satisfies this condition and goes over into (3) for $k_1 \to 0$, has the form

$$W^{II} = -2 \left[ \frac{\cos \sqrt{E_{1a}}\,\rho - 1}{k_a^2 \rho} + i\, \frac{\sin \sqrt{E_{1a}}\,\rho}{k_a \sqrt{E_{1a}}\,\rho} \right]. \tag{7}$$

A comparison with the results of numerical calculations using a number of examples shows that Eq. (7) yields good results throughout the entire energy range [4].

Generalization for a multielectron atom and the inclusion of interaction with the nucleus (the latter is necessary if the atomic wave functions are not orthogonal) are accomplished in elementary fashion:

$$U \to \sum_j \left[ \frac{1}{\rho_j} - \frac{1}{r} \right], \qquad W^{II} \to \sum_j [W^{II}(\rho_j) - W^{II}(r)]. \tag{8}$$

In order to perform specific calculations it is necessary to separate the angular and radial parts in $U_{a0}$ and $W_{1a}^{II}$. When the approximation (7) is used this problem does not present

_____

*In the case of elastic scattering, $E_1 = E_0 < E_a$. Under these conditions it is necessary to replace $E_{1a}$ by $E_{a0}$.

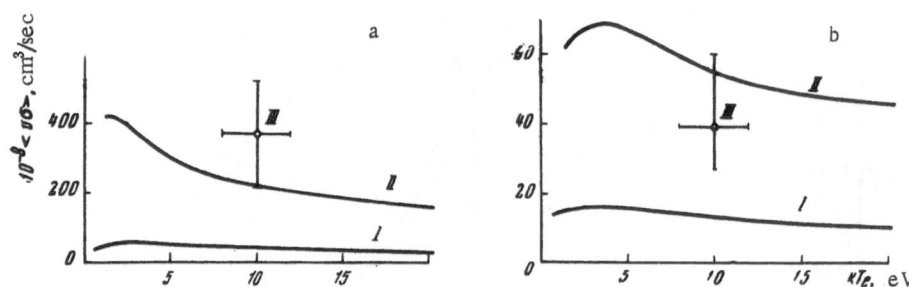

Fig. 2. Coefficient of the excitation rate $<\nu\sigma>$ for 5s' − 5d' (a) and 5s' − 6d' (b) transitions in Ne. I) first Born approximation; II) consideration of the transition via the 5p' level in the second order; III) experiment [8] with indication of the error in the measurement of both $<\nu\sigma>$ and T.

substantial difficulties, since the expansion of expressions of the type (7) in spherical functions is well known. The final computation formulas for $\sigma$ are given in the Appendix.

**3.** In Fig. 1 the cross sections for the excitation of the $4^1D$ and $4^1S$ levels from the ground state, as calculated in the first Born approximation ($\sigma^I$) and in the second order via the intermediate $2^1P$ ($\sigma^{II}$), are compared with experimental data ($\sigma_{exp}$). Regrettably, the results of experiments carried out by different authors differ significantly from each other. Nevertheless, the situation is qualitatively clear. In the case of excitation of the $4^1S$ level $\sigma_{exp}$ is less than $\sigma^I$ by approximately a factor of 1.5 (in the region of the maximum), which corresponds to the standard error of the Born method. As might have been expected under these conditions, consideration of the second order does not affect the results $\sigma^{II} \approx \sigma^I$. This is connected with the fact that the probability of the $2^1P − 4^1S$ transition is low (the general law for transitions with $\Delta L = -1$).

In the case of excitation of the $4^1D$ level even the smallest of $\sigma_{exp}$ noticeably exceeds $\sigma^I$ in the region of the maximum. $\sigma^{II}$, on the contrary, is larger than $\sigma_{exp}$, which at least qualitatively corresponds to the standard error of the Born method. The magnitude of the error is rather large: at the maximum $\sigma^{II}$ exceeds $\sigma_{exp}$ by almost a factor of four in comparison with the results of [5], and by a factor of 2-3 in comparison with [6, 7]. Since the cross sections in [6, 7] do not become the Born cross sections at high energies, we should evidently give preference to the results of [5].

The fact that the error in the second order turns out to be larger than the magnitude of the standard error of the conventional Born method is not surprising. Here a sort of accumulation of errors occurs at two stages of the transition. However, the constant sign of the error provides the possibility of considering it qualitatively in applications.

In Fig. 2 a comparison is made between the results of the calculation and experimental data [8] for transitions between excited states of Ne. Since the experiment was carried out in plasma, values are given for the excitation rates $<\nu\sigma>$ averaged over the Maxwellian distribution. In this case $\sigma_{exp}$ exceeds $\sigma^I$ by considerably more than it does for He. The consideration of transitions via the intermediate 5p' level basically eliminates this divergence (within the limits of the experimental accuracy).

Other intermediate levels possibly yield a noticeable although smaller contribution here.

In [8] the cross sections of transitions between excited and at the same time nonneighboring levels ($\Delta n^* > 1$ where $n^* = E_n^{-1/2}$ is the effective principal quantum number) were measured for the first time, as far as we know. The comparison made above shows that for such transitions the first Born approximation may be unsuitable even for u = E/ΔE ~ 20. At the same time, the cross sections of transitions between neighboring excited levels (Δn*< 1) can evidently already be described well by the Born method with correction for normalization already being made at u ~ 5-10 [9]. Consideration of the selective intermediate level in the required manner makes the error of the calculated cross section become of the order of the standard error for Δn* > 1. However, the problem of transitions with Δn* ≫ 1 remains open.

**4.** In calculating $\sigma^{II}$ the errors from each stage of the transition are multiplied. Moreover, one of the stages as a rule includes a transition between close levels, which leads to a substantial violation of the normalization of the cross section. Therefore, it is natural to expect that the theoretical cross section will considerably exceed the experimental one; this, for example, is just what comes about for the $1^1S - 2^1P - 4^1D$ transition in the He atom. Below we present a method (true, a rather coarse one) which allows qualitative elimination of the shortcoming indicated.

Let us write the wave function of the system describing the atom (for simplicity a one-electron atom) and the outer electron in the form

$$\Psi(\mathbf{r}_1, \mathbf{r}) = \Psi^M(\mathbf{r}_1, \mathbf{r}) + \psi_1(\mathbf{r}_1, \mathbf{r}). \tag{9}$$

Here $\Psi^M$ is the model wave function [10], $\psi_1$ describes the transition via the intermediate state, $\mathbf{r}_1$ and $\mathbf{r}$ are the coordinates (position vectors) of the atomic and outer electrons.

Let us expand the function (9) into a series in the total ensemble of the atomic wave functions

$$\Psi(\mathbf{r}_1, \mathbf{r}) = \sum_n \varphi_n(\mathbf{r}_1)\{\tau_n(\mathbf{r}) + F_n(\mathbf{r})\}. \tag{10}$$

Under these conditions the expansion coefficients $\tau_n$ of the model wave function have the form

$$\tau_n(\mathbf{r}) = \int \varphi_n^*(\mathbf{r}_1)\,\Psi^M(\mathbf{r}_1, \mathbf{r})\,d\mathbf{r}_1 = e^{i\mathbf{k}_0\mathbf{r}}\int d\mathbf{r}_1\varphi_n^*(\mathbf{r}_1)\,\chi(\mathbf{r}_1, \mathbf{r})\,\varphi_0(\mathbf{r}_1) = e^{i\mathbf{k}_0\mathbf{r}}\langle n|\chi|0\rangle, \tag{11}$$

$$\chi = \frac{\pi\nu}{\operatorname{sh}\pi\nu}F(i\nu, 1, ik_0R - i\mathbf{k}_0\mathbf{R})\,F(-i\nu, 1, ik_0\rho - i\mathbf{k}_0\boldsymbol{\rho}); \tag{12}$$

$$\nu = (k_0 + \sqrt{\varepsilon_0})^{-1},\; \mathbf{R} = \tfrac{1}{2}(\mathbf{r} + \mathbf{r}_1),\; \boldsymbol{\rho} = \tfrac{1}{2}(\mathbf{r} - \mathbf{r}_1),$$
$$\tau_n(\mathbf{r}) \underset{r_1\to\infty}{\sim} \delta_{0n}\left[e^{i\mathbf{k}_0\mathbf{r}} + O\left(\frac{1}{r}\right)\right] + O\left(\frac{1}{r^2}\right). \tag{13}$$

For the functions $F_n$ we can obtain the system of integral equations

$$F_n(\mathbf{r}) = \int d\mathbf{r}' G_n(|\mathbf{r} - \mathbf{r}'|)\left\{\langle n\left|\frac{\chi(\mathbf{r}_1, \mathbf{r})}{|\mathbf{r} - \mathbf{r}'|}\right|0\rangle e^{i\mathbf{k}_0\mathbf{r}'} + \sum_m \langle n\left|\frac{1}{|\mathbf{r}' - \mathbf{r}_1|}\right|m\rangle F_m(\mathbf{r}')\right\} + [\delta_{0n}e^{i\mathbf{k}_0\mathbf{r}} - \tau_n(\mathbf{r})] \tag{14}$$

by the usual method. Below we shall use the matrix element of the transition $T_{01}(\mathbf{q})$ corresponding to the integral under the modulus sign in (1) along with the potentials U and W. The matrix element $T_{0n}(\mathbf{q})$ of the transition derives from the asymptotic behavior of $F_n$ for $r \to \infty$; in particular, limiting consideration to the first term in the braces in (14), we obtain the model matrix element of the transition [10]. In solving the system (14) by the method of iterations it is expedient to neglect the terms in the square brackets. As is evident from (13), they are non-vanishing only for finite values of r, and with allowance for approximation which will be made later their retention would be a considerable exaggeration of the accuracy of the results. In the second order of perturbation theory, retaining only one term in the sum over m in the right side of (14), we can obtain the following results for the $0 \rightarrow 1$ transition:

$$T_{01}(\mathbf{q}) = T_{01}^{I} + T_{01}^{II} = f(\mathbf{q}) \, [T_{01}^{B}(\mathbf{q}) + T_{0 \rightarrow a \rightarrow 1}^{B}(\mathbf{q})] \equiv \qquad (15)$$

$$\equiv f(\mathbf{q}) \int d\mathbf{r} e^{i\mathbf{q}\mathbf{r}} W_{1a}(\mathbf{r}) U_{a0}(\mathbf{r}). \qquad (16)$$

Here $f(\mathbf{q})$ is a factor which differentiates the model results from the Born approximation,

$$f(\mathbf{q}) = \frac{\pi v}{\operatorname{sh}\pi v} F\left(-iv, iv, 1, \left[\frac{\Delta E_{10} + q^2}{\Delta E_{10} + 3q^2}\right]^2\right); \qquad (17)$$

$T_{0 \rightarrow 1}^{B}$ and $T_{0 \rightarrow a \rightarrow 1}^{B}$ are Born matrix elements of the first and second orders; the potential U and W are defined by Eqs. (2)-(4).

In deriving (15) the integration is carried out in the same approximation as previously [10]. Note that the second term in (15) corresponding to a transition via an intermediate level includes only one model factor $f(\mathbf{q})$. The interaction between the levels 1 and $a$ is thus not considered in any way in (15). In order to evaluate the role of this interaction we consider all possible methods of transition of an electron form level $a$ to level 1. The subsequent derivation is strictly in the nature of an estimate, and for this reason it is given without details.

Using (14), we write the formal solution for $F_1$ in all orders of perturbation theory. Then in the solution obtained we retain only the terms $0 \rightarrow a \rightarrow 1$ in the second order, $0 \rightarrow a \rightarrow n \rightarrow 1$ in the third order, and $0 \rightarrow a \rightarrow n \rightarrow m \rightarrow 1$ in the fourth order, where n, m take all possible values, etc. Then the matrix element $T_{01}^{II}$ of the transition, which considers the coupling between levels $a$ and 1 in all orders, has the form

$$T_{01}^{II} = \int d\mathbf{r} e^{i\mathbf{k_0}\mathbf{r}} V_{0a} \hat{G}_a \{ U_{a1} e^{-i\mathbf{k_1}\mathbf{r}'} + \sum_n U_{an} \hat{G}_n U_{n1} e^{-i\mathbf{k_1}\mathbf{r}''} + \sum_n U_{an} \hat{G}_n \sum_m U_{nm} \hat{G}_m U_{m1} e^{-i\mathbf{k_1}\mathbf{r}'''} + \dots \}; \quad (18)$$

$$V_{0a} = \langle 0 | \frac{\varkappa(\mathbf{r}, \mathbf{r_1})}{|\mathbf{r} - \mathbf{r_1}|} | a \rangle, \qquad U_{mn} = \langle m | \frac{1}{|\mathbf{r} - \mathbf{r_1}|} | n \rangle;$$
$$\hat{G}_l B = \int d\mathbf{r}' G_l(|\mathbf{r} - \mathbf{r}'|) B(\mathbf{r}'). \qquad (19)$$

Let us adopt the local approximation for all Green's functions in the braces; i.e., we set

$$G_{\mathbf{k}}(|\mathbf{r} - \mathbf{r}'|) = -\lambda \sigma(\mathbf{r} - \mathbf{r}'), \quad \lambda = \frac{1}{k_a}. \qquad (20)$$

In this case, using the completeness of the atomic wave functions, it is possible to carry out summation in each term of the series in the braces, the n-th term of the series acquiring the form

$$\langle a | \frac{(-1)^{n-1} \lambda^{n-1}}{|\mathbf{r} - \mathbf{r_1}|^n} | 1 \rangle e^{-i\mathbf{k_1}\mathbf{r}}.$$

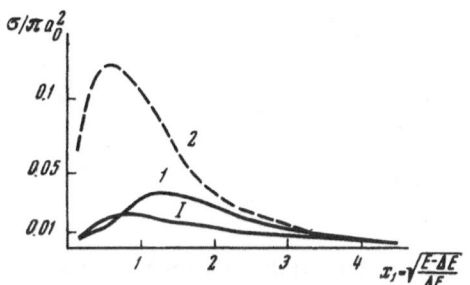

Fig. 3. Effective cross sections for the excitation of the 1s −3d transition in the H atom. I) The first Born approximation; 1) Eq. (22); 2) Eq. (15).

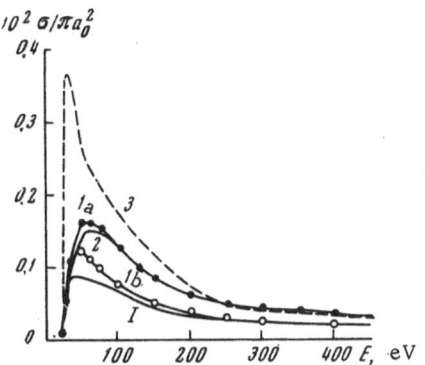

Fig. 4. Effective cross sections for the excitation of the $1^1S - 4^1D$ transition in the He atom. 1a) Experiment [6]; 1b) experiment [5]; 2) Eq. (22); 3) Eq. (15); I) first Born approximation.

Carrying out the summation by means of the geometric progression obtained, we have

$$T_{01}^{II} = \iint dr dr' \exp\left(i\mathbf{k}_0 \mathbf{r} - i\mathbf{k}_1 \mathbf{r}'\right) \langle 0 \left| \frac{\varkappa(\mathbf{r}, \mathbf{r}_1)}{|\mathbf{r} - \mathbf{r}_1|} \right| a\rangle \times$$

$$\times G_a(|\mathbf{r} - \mathbf{r}'|) \langle a \left| \frac{1}{|\mathbf{r} - \mathbf{r}'| + \lambda} \right| 1\rangle. \quad (21)$$

From a comparison of the potentials included in this equation it is evident that the $0 \to a$ transition is "normalized" by the model correction, while the $a \to 1$ transition is normalized due to the approximate summation of the series in all orders. Finally, carrying out the integration in the approximation which we adopted earlier in [10] and have already used here in deriving (15), we finally obtain

$$T_{01}(\mathbf{q}) = f(\mathbf{q})[T_{0 \to 1}^B(\mathbf{q}) + T_{0 \to a \to 1}^B(\mathbf{q})\Phi], \quad (22)$$

where

$$\Phi = 1 - x[\mathrm{ci}\,x \sin x - \mathrm{si}\,x \cos x], \qquad x = \frac{2q}{k_a}. \quad (23)$$

Equation (22) differs from (15) in the presence of the additional factor $\Phi$, which is of the order of unity in the case of weak coupling between levels $a$ and $1$. In the range of applicability of the Born approximation $f(\mathbf{q})$ and $\Phi$ tend to unity, and (22) goes over into the second Born approximation.

From Eqs. (15) and (22) we calculated the effective cross sections of a number of transitions in atoms of hydrogen, helium, and the alkali elements. Under these conditions the approximation (7) was used for the Green's function $G_a$. Figures 3 and 4 display the cross sections for the excitation of s − d transitions in H and He atoms in the first Born approximation, and via the intermediate level $2^1P$ as an illustration. The second Born approximation (not shown in these figures) without consideration of the corrections (15) and (22) exceeds the first Born approximation by an order of magnitude. In the case of the He atom the corrections (15), (21) lead to qualitative agreement with the experimental data of [5, 6]. Regrettably, experimental data are not available for the H atom.

The methods examined in the present section for normalizing the cross sections in the second order are rather coarse, but they undoubtedly indicate the possibility of such normalization without expansion into partial waves.

Appendix

In order to separate the radial and angular variables in U and W let us use the well-known equations

$$\frac{1}{\rho} \equiv \frac{1}{|\mathbf{r} - \mathbf{r_1}|} = \sum_{\lambda\mu} \frac{4\pi}{2\lambda + 1} \frac{r_<^\lambda}{r_>^{\lambda+1}} Y_{\lambda\mu}(\hat{r}) Y_{\lambda\mu}^*(\hat{r_1}),$$

$$\frac{\cos \alpha\rho}{\rho} = - 4\pi\alpha \sum_{\lambda\mu} j_\lambda(\alpha r_<) n_\lambda(\alpha r_>) Y_{\lambda\mu}(\hat{r}) Y_{\lambda\mu}(\hat{r_1}),$$

$$\frac{\sin \alpha\rho}{\rho} = 4\pi\alpha \sum_{\lambda\mu} j_\lambda(\alpha r) j_\lambda(\alpha r_1) Y_{\lambda\mu}(\hat{r}) Y_{\lambda\mu}^*(r_1).$$

For the wave functions of the atom we use the one-electron approximation. Below we present the equations for the cross section summed over all terms for the transition between two electron configurations: $l_0^{\bar{N}} - l_0^{N-1} l_1$ with allowance for the first and second orders with one intermediate level $l_0^{N-1} l_a$:

$$\sigma = \sum_\varkappa \frac{2N}{k_0^2} \int_{k_0-k_1}^{k_0+k_1} q\,dq \left| \int_0^\infty r^2 dr j_\varkappa(qr) [C_\varkappa y_{10}^\varkappa + \sum_{\lambda\lambda'} \frac{C_{\varkappa\lambda\lambda'}}{k_a} \left( \frac{1}{k_a} y_{1a}^\lambda + \frac{\sqrt{E_{1a}}}{k_a} Z_{1a}^\lambda - i V_{1a}^\lambda \right) y_{a0}^{\lambda'}] \right|^2,$$

$$y_{1a}^\lambda = 2 \int_0^\infty dr_1 P_1(r_1) P_a(r_1) \left[ \frac{r_<^\lambda}{r_>^{\lambda+1}} - \delta_{\lambda 0} \frac{1}{r} \right]',$$

$$Z_{1a}^\lambda(r) = 2(2\lambda+1) \int_0^\infty P_1(r_1) P_a(r_1) [j_\lambda(x_<) n_\lambda(x_>) - \delta_{\lambda 0} n_0(x)] dr_1,$$

$$V_{1a}^\lambda(r) = 2(2\lambda+1) j_\lambda(x) \int_0^\infty P_1(r_1) P_a(r_1) [j_\lambda(x_1) - \delta_{\lambda 0}] dr_1,$$

$$x_i = \sqrt{E_{1a}} r_i,$$

$$C_\varkappa = \left( \frac{2l_1+1}{2\varkappa+1} \right)^{1/2} \begin{pmatrix} \varkappa l_0 l_1 \\ 0 \, 0 \, 0 \end{pmatrix},$$

$$C_{\varkappa\lambda\lambda'} = (-1)^{l_a} (2l_a+1) \begin{pmatrix} \varkappa\lambda\lambda' \\ 0\,0\,0 \end{pmatrix} \begin{pmatrix} \lambda l_1 l_a \\ 0\,0\,0 \end{pmatrix} \begin{pmatrix} \lambda' l_a l_0 \\ 0\,0\,0 \end{pmatrix} \begin{Bmatrix} \varkappa\lambda\lambda' \\ l_a l_0 l_1 \end{Bmatrix}.$$

## References

1. M. J. Seaton, Proc. Phys. Soc., 77:184 (1961).
2. L. A. Vainshtein, Izv. Akad. Nauk SSSR, Seria Fiz., 27:1021 (1963); Atomic Collisions, Riga (1963).
3. I. I. Sobel'man, Introduction to the Theory of Atomic Spectra, Fizmatgiz (1963).
4. I. L. Beigman, L. A. Vainshtein, and A. V. Vinogradov, Preprint of the Physics Institute, Academy of Sciences of the USSR, No. 5 (1967).
5. H. B. Moustafa Mousa, J. de Heer, and J. Schutten, V th International Conference on the Physics of Electrons and Atomic Collisions, Leningrad (1967), p. 489.
6. R. M. St. John, E. I. Miller, and S. S. Lin, Phys. Rev., 134:A888 (1964).
7. I. P. Zapesochnyi, Astronom. Zh., 43:954 (1966).
8. A. S. Khaikin, Zh. Éksperim. i Teor. Fiz., 54:52 (1968); see also this volume, p. 93.
9. L. A. Vainshtein, M. A. Mazing, and P. D. Serapinas, Preprint of the Physics Institute, Academy of Sciences of the USSR, No. 64 (1967).
10. L. Vainshtein, L. Presnyakov, and I. Sobel'man, Zh. Éksperim. i Teor. Fiz., 45:2015 (1963); see also L. P. Presnyakov, this volume, p. 19.

# THE QUASI-CLASSICAL METHOD IN
# THE THEORY OF COLLISIONS OF ATOMS
# WITH HEAVY PARTICLES AND ELECTRONS

## A. V. Vinogradov

### §1. Introduction

The quasi-classical (or parametric) method in collision theory is the method in which the internal motion of the colliding systems is described quantum-mechanically, while their relative motion is described classically. Along with the quantum-mechanical method, this method is frequently used in studying collisions of atoms with heavy particles (ions, other atoms). For practical calculations one method may turn out to be more convenient than the other. Thus, the quantum-mechanical Born calculation is considerably simpler computationally than the quasi-classical method. Conversely, the quasi-classical method in the approximation of a finite number of states leads to a system of ordinary differential equations of the first order whose investigation is much simpler than an investigation of second order quantum-mechanical equations. A large quantity of numerical results has been obtained within the framework of the quasi-classical approximation, and many of these results are in good agreement with available experimental data. However, the problem of the relation between the results obtained within the framework of the quasi-classical method and using the simple quantum-mechanical statement of the problem has long remained open.*

The clarification of the connection between the quasi-classical and quantum-mechanical approaches allows substantiation of the application of the quasi-classical approach not only to collisions between heavy particles but also to some extent to electron – atom collisions. The latter fact is especially important, since these collisions are essential in various physical processes in plasma.

For many types of collisions there exists a range of energies (usually it is bounded from below) in which the quasi-classical and quantum-mechanical descriptions lead to identical results. However, not infrequently the quasi-classical method turns out to be more convenient from both the methodological and the computational points of view (the problem of the excitation of two-electron transitions can serve as a characteristic example). The quasi-classical method allows the influence of the normalization effect, which is especially important in calculating the

---

*In [1] (see also [2]) the equality of the total cross sections calculated quantum-mechanically and quasi-classically has been proved for only one particular case: the excitation of the $s-s$ transition in the first Born approximation.

cross sections of optically allowed transitions between close levels, to be considered fairly simply and without going beyond the framework of the first order of perturbation theory. Finally, the quasi-classical method frequently facilitates the investigation of the analytical dependence of the cross section on energy and other parameters [3-5].

In §2 a detailed examination is made of the transition to the limit $M = \infty$ ($M$ is the mass of the imping ing particle) in the exact quantum-mechanical expression for the excitation cross section of the atom of a heavy particle. The basic result of this section is to establish a relationship between the quantum-mechanical scattering amplitude and the quasi-classical probability amplitude, which is satisfied in the limit $M = \infty$ in all orders of perturbation theory. In §§ 3 and 4 the modification of the quasi-classical approximation for the description of electron–atom collisions is discussed, and two cases are examined in which the application of the quasi-classical method considerably simplifies the solution of the problem: calculation of the cross sections of two-electron transitions and cross sections of optically allowed transitions between close levels.

## §2. Collisions of Atoms with Heavy Particles

a. The Quasi-classical Method. Let us consider excitation of the atomic $0 \to 1$ transition during collisions of an atom with a structureless particle having a velocity $v_0$. Under these conditions it is assumed within the framework of the quasi-classical method that the particle moves along the straight line $\mathbf{R}(t) = \rho + \mathbf{v}_0$ ($\rho$ is the impact parameter), while the wave function of the atom satisfies the equation*

$$i \frac{\partial \Psi}{\partial t} = [H_a(\mathbf{r}) + V(\mathbf{r}, \mathbf{R}(t))] \Psi(\mathbf{r}, t) \tag{1}$$

with the initial condition

$$\Psi(\mathbf{r}, t) \xrightarrow[t \to -\infty]{} e^{-iE_0 t} \Phi_0(\mathbf{r}).$$

Here $\mathbf{r}$ is the ensemble of coordinates (position vectors) of the atomic electrons; $H_a(\mathbf{r})$ is the Hamiltonian of the atom; $E_a$, $\Phi_a(\mathbf{r})$ are the atomic wave functions and their energies; $V(\mathbf{r}, \mathbf{R}(t))$ is the interaction of the atom with the impinging particles.

The excitation cross section is determined in the following manner:

$$\sigma_{\mathrm{cl}}(v_0) = \int |c(\rho, v_0)|^2 \, d^2\rho = \int w_{0 \to 1}(\rho) \, d^2\rho, \tag{2}$$

$$c(\rho) = \lim_{t \to \infty} \langle e^{-iE_1 t} \Phi_1(\mathbf{r}) | \Psi(\mathbf{r}, t) \rangle. \tag{3}$$

In order to find the probability amplitudes $c(\rho)$ it is frequently the practice to expand $\Psi(\mathbf{r}, t)$ in unperturbed atomic functions:

$$\Psi(\mathbf{r}, t) = \sum_a c_a(t) \Phi_a(\mathbf{r}) e^{-iE_a t}. \tag{4}$$

---

*The atomic system of units is used in this paper.

Then the coefficients $c_a(t)$ satisfy the equations

$$ic_a(t) = \sum_b V_{ab}(\mathbf{R}(t)) e^{i\omega_{ab}t} c_b(t), \tag{5}$$

where

$$V_{ab}(\mathbf{R}) = <\Phi_a(\mathbf{r}) | V(\mathbf{r}, \mathbf{R}(t)) | \Phi_b(\mathbf{r})>,$$

$$\omega_{ab} = E_a - E_b.$$

Based on the system (5), various approximate approaches [4–8] have been developed. Later on we shall need merely the so-called normalized Born approximation which is analogous to the normalization methods used by Seaton [9]:

$$\omega_{01}(\rho) = \frac{w_{01}^B(\rho)}{1 + \omega_{01}^B(\rho)}, \tag{6}$$

$$\omega_{01}^B(\rho) = \left| \int_{-\infty}^{\infty} V_{01}(t) e^{i\omega_{01}t} dt \right|^2. \tag{7}$$

The transition probability calculated in the Born approximation (7) can exceed unity, which frequently leads to greatly overestimated values of the total cross section (see Sec. 4).

The use of Eq. (6) eliminates this shortcoming naturally and improves the results considerably.*

b. The Connection between the Quantum-Mechanical and Quasi-classical Methods.†
From the above it is obvious that the mass of the impinging particle does not appear at all in the quasi-classical method. Therefore, it is natural to hope to obtain the quasi-classical equations from the quantum-mechanical ones by the transition in the limit $M \to \infty$.

As is well known, the effective excitation cross section can be written in the form

$$\sigma = \frac{1}{4\pi^2 v_0^2} \int_0^{2\pi} d\varphi \int_{k_0 - k_1}^{k_0 + k_1} q\, dq\, |f(\mathbf{q}, v_0)|^2, \tag{8}$$

$$f(\mathbf{q}, v_0) = <e^{i\mathbf{k}_1\mathbf{R}}\Phi_1(\mathbf{r}) | V - VGV + VGVGV - \ldots | e^{i\mathbf{k}_0\mathbf{R}}\Phi_0(\mathbf{r})>, \tag{9}$$

where $\mathbf{k}_0 = Mv_0$, $\mathbf{k}_1 = Mv_1$ are the initial and final momenta of the particle; $\mathbf{q} = \mathbf{k}_1 - \mathbf{k}_0$, $\varphi$ is the polar angle of the vector $\mathbf{k}_1$ (or $\mathbf{q}$); $E = k_0^2/2M + E_0$ is the total energy of the system; the function $\Psi(\mathbf{r}, \mathbf{R})$ satisfies the equation

$$\left[ -\frac{1}{2M}\Delta_\mathbf{R} + H_a(\mathbf{r}) + V(\mathbf{r}, \mathbf{R}) - E \right] \Psi(\mathbf{r}, \mathbf{R}) = 0,$$

---

*A discussion of the various approximate methods of collision theory is provided in the paper by L. A. Vainshtein (see above, p. 1).
†In this subsection we shall follow [10] (see also [11]).

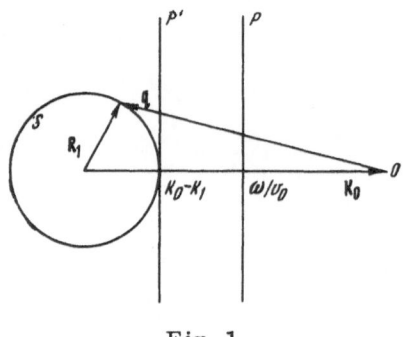

Fig. 1

and the Green's function G has the form

$$G = \left[-\frac{1}{2M}\Delta R + H_a(\mathbf{r}) - E - i\varepsilon\right]^{-1} = \frac{1}{(2\pi)^3}\sum_a \Phi_a(\mathbf{r})\Phi_a^*(\mathbf{r}') \times$$

$$\times \int \frac{d\mathbf{k}\exp[i\mathbf{k}(\mathbf{R}-\mathbf{R}')]}{\frac{k^2}{2M} - (E - E_a + i\varepsilon)}.$$

We shall show that for any transition the relationship

$$\frac{1}{v_0}f(\mathbf{q},v_0^-) = \int d^2\rho\, e^{-i\mathbf{q}\rho}c(\rho,v_0) \tag{10}$$

is fulfilled for $M \to \infty$. In Eq. (8) the region of integration with respect to the vector $\mathbf{q}$ depends on M besides the amplitude $f(\mathbf{q}, v_0)$. As long as M is finite, the tip of the vector $\mathbf{q}$ occupies all possible positions on the sphere S (Fig. 1). For $M \to \infty$ ($\mathbf{q}$ is specified) the center of the sphere S moves to the left, while the point $k_0 - k_1$ moves to the right and tends to its limiting value $\frac{\omega}{v_0} = \lim\limits_{M\to\infty}(k_0 - k_1)$; under these conditions as M increases, the sphere S adjoins the plane $P$ ($\mathbf{q}v_0 + \omega = 0$). more closely. In the limit $M = \infty$ the cross section (8) is the integral over this plane:

$$\sigma = \int_P d^2\mathbf{q}\,\frac{1}{4\pi^2 v_0^2}|f(\mathbf{q},v_0)|^2. \tag{11}$$

Let us now consider the term having the following intermediate states in the (n + 1)-th term of the expansion (9):

$$\Phi_{a_1}(\mathbf{r}), \ldots, \Phi_{a_n}(\mathbf{r}):$$

$$f_{a_1\ldots a_n}^{(n+1)}(\mathbf{q},v_0^-) = (-1)^n \int d\mathbf{R}\,d\mathbf{R}_1\ldots d\mathbf{R}_n V_{1a_1}(\mathbf{R})\ldots V_{a_n 0}(\mathbf{R}_n)\,I(\mathbf{R},\mathbf{R}_1,\ldots,\mathbf{R}_n), \tag{12}$$

where

$$I(\mathbf{R},\mathbf{R}_1,\ldots,\mathbf{R}_n) = e^{-i\mathbf{k}_1\mathbf{R}}\int \frac{d\mathbf{p}_1\ldots d\mathbf{p}_n}{(2\pi)^{3n}}\,\frac{\exp[i\mathbf{p}_1(\mathbf{R}-\mathbf{R}_1)]}{\frac{p_1^2}{2M}-(E-E_{a_1}+i\varepsilon)} \times \ldots \times \frac{\exp[i\mathbf{p}_n(\mathbf{R}_{n-1}-\mathbf{R}_n)]}{\frac{p_n^2}{2M}-(E-E_{a_n}+i\varepsilon)}\,e^{i\mathbf{k}_0\mathbf{R}_n}. \tag{13}$$

Making the substitution $\mathbf{p}_i = \mathbf{p}_i' + \mathbf{k}_0$ and noting that

$$\lim_{M\to\infty}\left[\frac{(\mathbf{p}_i+\mathbf{k}_0)^2}{2M}-E-E_{a_i}+i\varepsilon\right]^{-1}[\mathbf{p}_i v_0 + \omega_{a_i 0} - i\varepsilon]^{-1} = i\int_{-\infty}^{\infty}\exp\{i(\mathbf{p}_i v_0 + \omega_{a_i 0} - i\varepsilon)t_i\}\,dt_i, \tag{14}$$

we obtain

$$I(\mathbf{R},\mathbf{R}_1,\ldots,\mathbf{R}_n) = (-i)^n e^{-i\mathbf{q}\mathbf{R}}\int_{-\infty}^{0}\ldots\int_{-\infty}^{0}dt_1\ldots dt_n \times$$

$$\times \delta(\mathbf{R}-\mathbf{R}_1+v_0 t)\ldots\delta(\mathbf{R}_{n-1}-\mathbf{R}_n+v_0 t_n)\exp\{i\omega_{a_10}t_1+i\omega_{a_10}t_2+\ldots\}. \tag{15}$$

Making use of (15), we carry out the integration with respect to $R_1 \ldots R_n$ in (12) and make the substitution $R = \rho + v_0 t$ ($\rho v_0 = 0$), $t_1 = t_1' - t$, $t_2 = t_2' - t_1'$, ..., $t_n = t_n' - t_{n-1}'$ ), then,

$$f^{n+1}_{a_1 \ldots a_n}(q) = v_0(-i)^n \int d^2\rho \, e^{-iq\rho} \int\limits_{-\infty}^{\infty} dt V_{1a_1}(R) \, e^{i\omega_{1a_1}t} \int\limits_{-\infty}^{t} dt_1' \, e^{i\omega_{a_1a_2}t_1'} V_{a_1a_2}(R_1') \int\limits_{-\infty}^{t_{n-1}'} dt_n' \, e^{i\omega_{a_n0}t_n'} V_{a_n0}(R_n), \qquad (16)$$

where $R_i' = \rho + v_0 t_i'$.

Taking (5) into consideration, it is easy to see that the (n + 1)-th approximation in perturbation theory for the quasi-classical amplitude $c(\rho)$ appears in the right side of (16); i.e.,

$$f^{n+1}_{a_1 \ldots a_n}(q) = v_0 \int d^2\rho \, e^{-iq\rho} c^{(n+1)}_{a_1 \ldots a_n}(\rho). \qquad (17)$$

Summing (17) over the intermediate states $a_1 \ldots a_n$ and then over all n, we obtain Eq. (10). Thus, in the limit $M = \infty$ the quantum-mechanical scattering amplitude $f(q)$ is the Fourier transform of the quasi-classical probability amplitude $c(\rho)$. From this the equations for the total quantum-mechanical and quasi-classical cross sections are easily derived according to the Plancherel equation:

$$\frac{1}{4\pi^2 v_0^2} \int d^2q \, |f(q,v_0)|^2 = \int d^2\rho \, |c(\rho,v_0)|^2. \qquad (18)$$

## §3. Collisions of Atoms with Electrons

Let us now consider the problem of applying the quasi-classical method to electron–atom collisions. For this purpose let us turn again to the derivation of Eqs. (10), (18). In §2 it was shown that the possibility of a quasi-classical description presumes, strictly speaking, the validity of Eq. (18). The latter in turn can be treated as the corollary of two assumptions: a) replacement of the sphere S by the plane P; b) calculation of the amplitude $f(q, k_0)$ on the assumption that $M = \infty$. Let us clarify the degree to which it is possible to use these assumptions for calculating cross sections for finite masses.

Let us assume initially that the scattering amplitude $f(q, k_0)$ depends solely on $q$. Then assumption b) is automatically satisfied. As far as assumption a) is concerned, it is legitimate only at high velocities. In fact, for $v_0 \to \infty$

$$k_0 - k_1 = M v_0 \left[ 1 - \sqrt{1 - \frac{2\omega}{M v_0^2}} \right] \to \frac{\omega}{v_0}.$$

Moreover, for $v_0 \to \infty$ the radius of the sphere S (see Fig. 1) increases without limit. Therefore, at high velocities, just as for $M \to \infty$, the plane P adjoins the sphere S.

Assume now that the velocity $v_0$ is also the threshold velocity. Then the radius of the sphere S tends to vanish, and its center is situated at a distance $\sqrt{2\omega M}$ from the point O. The plane P recedes farther and farther from the sphere S as the velocity decreases. Precisely at threshold it is situated halfway between the center of the sphere and the point O.

From the reasoning given it is clear that we cannot use assumption a) for the approximate calculation of the integral (8) at low velocities. However, in this case, as is evident from Fig. 1, it is considerably better to replace the sphere S by the plane P': $\omega + qv = 0$ (where

$v = \frac{\omega}{k_0 - k_1} \frac{v_0}{v_0}$), which is tangent to the sphere S, rather than by the plane P. Then in accordance with (18),

$$\int_{P'} d^2\mathbf{q} \, |f(\mathbf{q})|^2 = 8\pi^2 v^2 \int |c(\rho,v)|^2 \, d^2\rho. \tag{19}$$

Substituting (19) into (8) and comparing with (2), we find

$$\sigma = 2\left(\frac{v}{v_0}\right)^2 \int_0^\infty |c(\rho,v)|^2 \rho \, d\rho \pi \, a_0^2 = \left(\frac{v}{v_0}\right)^2 \sigma_{\text{cl}}(v),$$
$$v = \frac{\omega}{k_0 - k_1} = \frac{v_0 + v_1}{2}, \tag{20}$$

where $\sigma_{\text{cl}}(v)$ is the conventional quasi-classical cross section defined by Eq. (2).

Note that in the first Born approximation the differential cross section averaged over the magnetic quantum numbers depends solely on $|\mathbf{q}|$. In this case it is sufficient to replace the upper limit of the integration with respect to q in Eqs. (8) by $\infty$ in order to derive (20).

It is Eq. (20) which represents the extrapolation of the quasi-classical method to the case of the excitation of the atom of a particle having finite mass at low collision velocities. In the derivation of (20) we started from the exact quantum-mechanical statement of the problem and did not use any classical concepts. In quantum language the use of Eq. (20) instead of (2) implies refinement of the range of possible values of momentum $\mathbf{q}$ transfer; i.e., it implies a more accurate examination of the kinematic part of the problem. This is especially essential for optically allowed transitions where the principal contribution to the total cross section is made by small q. The extension of the region of integration with respect to q to infinity as a rule does not introduce substantial errors, since $f(\mathbf{q})$ decreases rapidly for large q.

In order to solve the dynamic part of the problem — calculation of the amplitude $c(\rho,v)$ — it is possible to use approximate methods of quasi-classical theory in each specific case or, if it is convenient, to begin by finding the scattering amplitude $f(\mathbf{q})$ and then to use Eq. (10).

In deriving (20) we touched on only the first of the two assumptions indicated above. In the first Born approximation assumption b) in calculating $f(\mathbf{q}, k_0)$ in the limit $M = \infty$ is unessential, since $f$ depends solely on $\mathbf{q}$. No additional concepts, besides the indicated kinematic ones, exist for carrying over Eqs. (20) to the more general case in which the amplitude $f$ depends not only on $\mathbf{q}$ but also on $v_0$. However, it turns out that an indispensible condition for a transition to quasi-classical expressions of the type (20) is the quality of the velocity v, which determines the region of integration with respect to $\mathbf{q}$, and the velocity included in the argument of the scattering amplitude $f$. Since the equation $v = (v_0 + v_1)/2$ is justified by kinematic concepts, while $f(\mathbf{q}, v)$ is a smooth function of the velocity, we shall hold to this definition of v hereafter, thus replacing $f(\mathbf{q}, k_0)$ by $f\left(\mathbf{q}, \frac{v_0 + v_1}{2}\right)$. The scattering amplitude as a function of the transfer of momentum q, conversely, is usually either singular or reaches a maximum near $q = 0$; however, it is precisely the range of small q which is correctly considered by Eq. (20).

In the case of excitation by electrons, Eq. (20) describes the behavior of the cross section near threshold considerably better than Eq. (2) does. As an illustration, Figs. 2 and 3 compare the excitation cross sections of the H(1s – 2p), H(1s – 2s), and Na(3s – 3p) transitions in the quantum-mechanical and quasi-classical Born approximations. As is evident from the figures, the two methods lead to results which are in agreement over a wide range of energies,

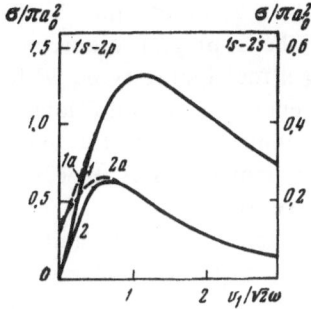

Fig. 2. Excitation cross sections of the $1s - 2p$ (curves 1 and 1a), $1s - 2s$ (curves 2 and 2a) transitions of the hydrogen atom in the Born approximation. 1, 2) Quantum-mechanical calculation; 1a, 2a) quasi-classical method (Eq. (20)).

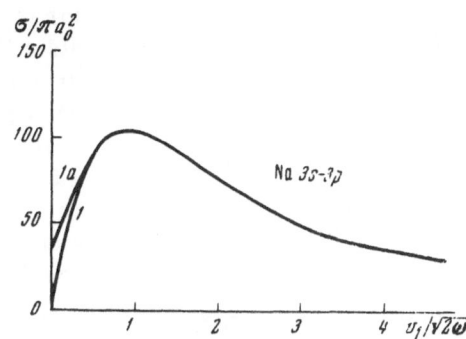

Fig. 3. Excitation cross section of the $3s - 3p$ transition of the Na atom in the Born approximation. 1) Quantum-mechanical calculation; 1a) quasi-classical method (Eq. (20)).

including the region of the cross section maximum. In the case of the $1s - 2s$ transitions the agreement is somewhat worse in view of the large role played by large q for optically forbidden transitions.

Thus, the description of the motion of the electrons by means of a classical trajectory does not lead to substantial errors in the cross section (at least in the first order of perturbation theory) even near the excitation threshold. This is a conclusion favoring the use of the modification (20) of the quasi-classical method for various refinements of the Born approximation. Under these conditions the consideration of some effects within the framework of the quasi-classical method often turns out to be simpler than it is in the quantum-mechanical statement of the problem. As an example, let us consider the excitation of the $4p - 4d$ transition of the Ne atom by electron impact. This is a "strong" optically allowed transition: a characteristic feature is the smallness of the excitation energy $\omega = E(4p) - E(4d)$ compared with the energy $E(4p)$ of the initial state. In calculating the excitation cross section of such transitions in the Born approximation the transition probabilities and partial cross sections may frequently exceed the upper theoretical limit dictated by the law of conservation of the number of particles. Thus, consideration of the normalization effects is necessary. In the quantum-mechanical approach this requires going over to expansion in partial waves with a large number (20-30) of partial waves contributing to the total cross section; this increases the volume of computational work considerably.*

The difficulty indicated does not arise when the quasi-classical method is used. Consideration of normalization reduces simply to redefining the transition probability. The generalization of Eq. (6) for the case of degeneration of the initial state has the form

$$w_{n_0 \to 1} = \frac{w_{n_0 \to 1}^B(\rho)}{1 + \frac{1}{g_0} \sum_{n_0} w_{n_0 \to 1}^B(\rho)} , \qquad (21)$$

where $g_0$ is the statistical weight of the initial level, while $n_0$ is one of its sublevels. Figure 4 shows the result of a numerical calculation of the cross section for excitation of the $4p - 4d$

*The results of the quantum-mechanical calculation will be published in another paper.

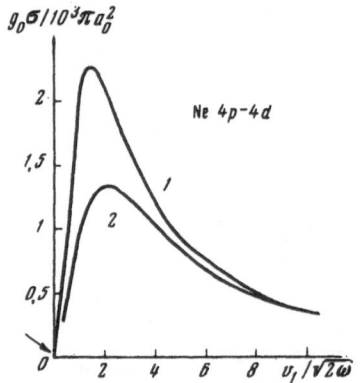

Fig. 4. Cross section for the excitation of the 4p − 4d transition of the Ne atom by electrons. 1) Born approximation (for $v_1 = 0$ the difference between the quantum-mechanical and quasi-classical cross sections is maximal; its value is indicated on the figure by the arrow); 2) quasi-classical Born approximation with allowance for the normalization (21).

transition of the Ne atom by electrons. In the expression for the interaction of the impinging electron with the atom the term having a multipolarity equal to three was discarded; the error under these conditions does not exceed 5%. The experimental value of the excitation rate of of this transition, as measured by A. S. Khaikin [13] at a temperature of T = 10 eV, is equal to $g_0 \langle \sigma v \rangle = 1.23 \cdot 10^{11}$ sec$^{-1}$, whereas calculations with and without consideration of normalization respectively yield $1.43 \cdot 10^{11}$ sec$^{-1}$ and $1.74 \cdot 10^{11}$ sec$^{-1}$. A consideration of normalization (21), as might be expected, improves the agreement between theory and experiment.

## § 4. Double Excitation

Recently, interest has increased in collisions as a result of which the states of two atomic electrons change. These include double ionization, ionization with simultaneous excitation of an ion, and double excitation. Here we shall consider only double excitation from the ground state of atoms having two optical electrons (transitions of the type $s^2 - l_a l_b$). Formulation of this problem within the framework of the quasi-classical method turns out to be simple and obvious.

We shall assume that the optical electrons move in a stipulated field of the core and that their motion can be described by a Hamiltonian with separated variables. Then the wave function of the system can be represented in the form of a product of identical (since the initial state is $s^2$) one-electron functions which depend on time:

$$\Psi(\mathbf{r}_1, \mathbf{r}_2, t) = \varphi(\mathbf{r}_1, t)\, \varphi(\mathbf{r}_2, t). \tag{22}$$

The function $\varphi(\mathbf{r}, t)$ satisfies the equation

$$i\, \frac{\partial \varphi(\mathbf{r}, t)}{\partial t} = \left[ H_a(\mathbf{r}) - \frac{1}{|\mathbf{r} - \mathbf{R}(t)|} \right] \varphi(\mathbf{r}, t), \tag{23}$$

where $H_a(\mathbf{r})$ is the effective one-particle Hamiltonian; $\mathbf{R}(t) = \rho + \mathbf{v}t$. The excitation probability of the level $l_a l_b$ is equal to

$$w = \sum_{m_a m_b} \lim_{t \to \infty} |\langle \Phi_{m_a m_b}(\mathbf{r}_1, \mathbf{r}_2) \,|\, \Psi(\mathbf{r}_1, \mathbf{r}_2, t) \rangle|^2 =$$

$$= \sum_{m_a m_b} \lim_{t \to \infty} \left| \frac{1}{\sqrt{2}} \left[ \varphi_{m_a}(\mathbf{r}_1) \varphi_{m_b}(\mathbf{r}_2) + \varphi_{m_a}(\mathbf{r}_2) \varphi_{m_b}(\mathbf{r}_1) \right] \,\Big|\, \varphi(\mathbf{r}_1, t)\, \varphi(\mathbf{r}_2, t) \rangle \right|^2 = 2 w_a w_b, \tag{24}$$

where $\Phi m_a m_b(\mathbf{r}_1, \mathbf{r}_2)$ is the unperturbed wave function of the final state with the projections of the angular momenta $m_a$ and $m_b$; $w_a$, $w_b$ are the total probabilities of the one-electron transitions $s - l_a$ and $s - l_b$, which are determined by Eq. (23). Analogously, considering the identity of the electrons, it is easy to show that the probability of the $s^2 - l^2$ transition is equal to

$$w = w_1^2, \tag{25}$$

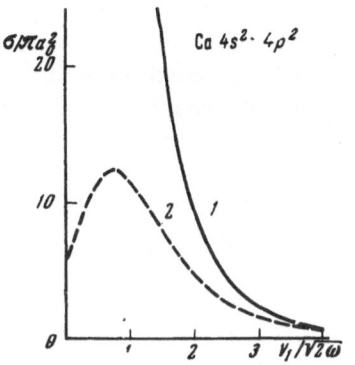

Fig. 5. Cross sections for excitation of the $4s^2 - 4p^2$ transition of the Ca atom. 1) Quasiclassical Born approximation; 2) quasi-classical Born approximation with allowance for normalization (21).

where $w_l$ is the total probability of the one-electron $s - l$ transition, which is determined by Eq. (23). Note that in those cases when the probability $w_l$ calculated in the Born approximation exceeds unity, the consideration of the normalization (21) for two-electron transitions is more important than for one-electron transitions.

In order to apply the quasi-classical method to excitation by electron impact it is necessary to use Eq. (20). Under these conditions, as was indicated in §3, only the shortcoming of the quasi-classical approximation involving integration with respect to $\mathbf{q}$ is correct. In the case of two-electron transitions the scattering amplitude is calculated in the second order of perturbation theory, where, unlike the Born approximation, a dependence of the amplitude on $k_0$ (or M) appears. This factor may be the cause of an additional divergence between the quantum-mechanical and quasi-classical calculations.

Figure 5 shows the results of calculating the cross section for the excitation of the $4s^2 - 4p^2$ transition of Ca according to Eqs. (20) and (25). The probability $w_l$ was calculated in the Born approximation with and without allowance for the normalization (21). In accordance with the assumption that the motion of atomic electrons is independent, the energy of the one-electron transition $4s - 4p$ was assumed equal to half the energy of the two-electron transition $4s^2 - 4p^2$.

The double excitation mechanism which we have considered is a particular case of a transition through an intermediate level $(sl)$. Such a mechanism is possible only in the second order of perturbation theory in $\lambda /v$ ($\lambda$ is a parameter which characterizes the interaction of an atom with an impinging particle; v is the velocity of relative motion). The expression for the probability of transition to an intermediate level in the form of the simple product of the probabilities of one-electron transitions is a distinguishing property of the double process specifically. In order to clarify this, we write the equations of motion in the interaction representation

$$i \frac{\partial \Phi (\mathbf{r}_1, \mathbf{r}_2, t)}{\partial t} = \widetilde{V}(t) \Phi (\mathbf{r}_1, \mathbf{r}_2, t), \qquad (26)$$

where $\Phi (\mathbf{r}_1, \mathbf{r}_2, t)$ is the wave function in the interaction representation and

$$\widetilde{V}(t) = \sum_i \widetilde{V}_i = \sum_i \exp [iH_a(\mathbf{r}_i)t] V(\mathbf{r}_i t) \exp [-iH_a(\mathbf{r}_i)t]. \qquad (27)$$

In the second order the transition probability is equal to

$$w = |<\Phi_1 | S^{II} | \Phi_0>|^2 = |<\Phi_1 | \iint_{-\infty}^{\infty} dt dt' P[\widetilde{V}(t)\widetilde{V}(t')] | \Phi_0>|^2, \qquad (28)$$

where $S^{II}$ is the scattering matrix in the second order, and

$$P[\widetilde{V}(t)\widetilde{V}(t')] = ] \begin{cases} \widetilde{V}(t) \widetilde{V}(t'), & t' < t, \\ \widetilde{V}(t') \widetilde{V}(t), & t' > t. \end{cases}$$

It is easy to see that the commutator $[\widetilde{V}(t)\,\widetilde{V}(t')]$ is a one-particle operator. Since in the case of double excitation the functions $\Phi_1$ and $\Phi_0$ differ in the states of both electrons, the operators $\widetilde{V}(t)$ and $\widetilde{V}(t')$ in Eq. (28) can be assumed to be commutated. Under these conditions the time-ordered product reduces to a conventional product, and integration with respect to t and t' is carried out independently. Thus, we obtain

$$w = |<\Phi_1|\,S^{II}\,|\Phi_0>|^2 = <\Phi_1|\frac{1}{2}\int_{-\infty}^{\infty}\widetilde{V}(t)\,dt \int_{-\infty}^{\infty}\widetilde{V}(t')\,dt'\,|\Phi_0>|^2 = |<\Phi_1|\frac{1}{2}S^I S^I|\Phi_0>|^2, \quad (29)$$

where $S^I$ is the scattering matrix in the first order of perturbation theory. The expression obtained can easily be reduced to a product of probabilities of one-electron transitions. However, in the case of conventional excitation the functions $\Phi_1$ and $\Phi_0$ differ in the states of only one electron; therefore, the operators $\widetilde{V}(t)$ and $\widetilde{V}(t')$ cannot be considered commutative, and Eq. (29) is not valid at all.

From the theories proved in Sec. 3 it follows that all equations obtained in Sec. 4 can be derived from general quantum-mechanical expressions.* However, in this case the quasiclassical method leads to the goal faster.

In conclusion I express my thanks to L. A. Vainshtein, as well as to I. L. Beigman and L. P. Presnyakov, for their valuable discussions.

References

1.  J. W. Frame, Proc. Cambridge Philos. Soc., 27:511 (1931).
2.  A. M. Arthurs, Proc. Cambridge Philos. Soc., 57:904 (1961).
3.  M. Rosen and C. Zener, Phys. Rev., 40:502 (1932).
4.  L. A. Vainshtein, L. P. Presnyakov, and I. I. Sobel'man, Zh. Éksperim. i Teor. Fiz., 43:518 (1962).
5.  Yu. N. Demkov, Zh. Éksperim. i Teor. Fiz., 45:195 (1963).
6.  D. R. Bates, Proc. Phys. Soc., 77:59 (1961).
7.  J. Callaway and E. Baur, Phys. Rev., 140A:1072 (1965).
8.  L. P. Presnyakov, Tr. Fiz. Inst. Akad. Nauk SSSR, 30:236 (1964).
9.  M. J. Seaton, Proc. Phys. Soc., 77:174 (1961).
10. A. V. Vinogradov, Preprint of the Physics Institute, Academy of Sciences of the USSR: On the Excitation of Atoms by Heavy Particles (1966).
11. A. V. Vinogradov, Opt. i Spektr., 22:663 (1967).
12. B. L. Moiseiwitsch, Proc. Phys. Soc., 87:885 (1966).
13. A. S. Khaikin, Zh. Éksperim. i Teor. Fiz. 54:52 (1968).
14. I. L. Beigman, L. A. Vainshtein, and A. V. Vinogradov, Preprint of the Physics Institute, Academy of Sciences of the USSR, No. 5 (1967).

---

*This problem was discussed in greater detail in [14].

# CALCULATION OF EFFECTIVE SCATTERING CROSS SECTIONS USING ATOMIC FUNCTIONS WHICH CONSIDER THE INTERACTION OF CONFIGURATIONS

## I. L. Beigman

In certain cases the description of atoms and ions in the one-electron approximation is not sufficient, and it is necessary to consider the next terms in the expansion of the total atomic function in one-configuration functions. Here we shall consider atoms and ions with two optical electrons in the so-called two-configuration approximation. The effective cross sections of ordinary one-electron transitions change only slightly under these conditions (20-30%).

However, the effective cross sections of two-electron transitions are of special interest; in this approximation these cross sections already turn out to be nonvanishing in the first order of perturbation theory in the interaction of the incident electron with the atom.

## §1. Semiempirical Atomic Functions in the Two-Configuration Approximation

Let us consider an atom with two optical electrons in the ground state. The atomic core is replaced by a special effective field, and the wave function of the two equivalent optical electrons will be sought in the form

$$\Psi = \varphi_0\,(r_1, r_2)\,\Omega_{\,l_0^2 LM} + \lambda \varphi_1\,(r_1, r_2)\,\Omega_{\,l_1^2 LM}, \tag{1}$$

where $\Omega_{l^2 LM}$ is the eigenfunction of the total angular momentum operator and is composed of spherical functions having the angular momentum $l$, while $\varphi(r_1, r_2)$ depends solely on the length of the position vectors $r_1$ and $r_2$. For convenience we shall assume that $\varphi_0$ and $\varphi_1$ are normalized to unity. Substituting (1) into the Schrödinger equation, multiplying by $\Omega_{l_i^2 LM}$, and integrating with respect to the angles, we find that $\varphi_0$ and $\varphi_1$ satisfy the system of equations

$$\{T_0 + \tilde{U}\,(r_1) + \tilde{U}\,(r_2) + V_{00}\,(r_1, r_2) - E\}\,\varphi_0 = -\lambda\,V_{01}\varphi_1, \tag{2a}$$

$$\{T_1 + \tilde{U}\,(r_1) + \tilde{U}\,(r_2) + V_{11}\,(r_1, r_2) - E\}\,\varphi_1 = -1/\lambda\,V_{10}\varphi_0, \tag{2b}$$

$$T_i = -\tfrac{1}{2}\Big[1/r_1\,\frac{\partial^2}{\partial r_1^2}\,r_1 + \frac{1}{r_2}\,\frac{\partial^2}{\partial r_2^2}\,r_2 - \frac{l_i\,(l_i+1)}{r_1^2} - \frac{l_i\,(l_i+1)}{r_2^2}\Big], \tag{2c}$$

$$V_{ik} = \langle l_i^2 \, LM \, \left| \frac{1}{r_{12}} \right| \, l_k^2 \, LM \rangle .$$

(2d)

Multiplying (2b) by $\varphi_0$, considering (2a) and integrating with respect to $r_1$, $r_2$, we obtain a quadratic equation for $\lambda$:

$$\lambda \langle \varphi_1 | V_{01} | \varphi_1 \rangle + \langle \varphi_0 | T_0 - T_1 | \varphi_1 \rangle + \langle \varphi_0 | V_{00} - V_{11} | \varphi_1 \rangle = 1/\lambda \langle \varphi_0 | V_{01} | \varphi_0 \rangle .$$

(3)

The smaller root of the equation corresponds to the $l_0^2$ configuration with a $l_1^2$ "impurity"; the larger root corresponds to the $l_1^2$ configuration with a $l_0^2$ "impurity."

In accordance with the semiempirical method we shall now assume that the value of energy E is stipulated, while we introduce the parameter $\omega$, which must be chosen in such a way that the eigenvalue of the system of Eqs. (2) coincides with E, into the potential $\tilde{U}(r)$. This potential is denoted by $\tilde{U}_\omega(r)$.

The system of Eqs. (2), of course, is too complex and requires further simplification. For this purpose it is useful to rewrite (2) in the form

$$\{T_0 + \tilde{U}_\omega(r_1) + \tilde{U}_\omega(r_2) + V_{00} - G_1 - E\} \varphi_0 = 0,$$

(4a)

$$\{T_1 + \tilde{U}_\omega(r_1) + \tilde{U}_\omega(r_2) + V_{11} - \hat{G}_0 - E\} \varphi_1 = 0,$$

(4b)

$$G_i = V_{01} [T_i + \tilde{U}_\omega(r_1) + \tilde{U}_\omega(r_2) + V_{ii} - E]^{-1} V_{10}.$$

(4c)

It can be expected that $\varphi_i$ can be approximated by functions with separated variables with sufficient accuracy, so that

$$\Psi = f_0(r_1) f_0(r_2) \Omega_{l_0^2 LM} + \lambda f_1(r_1) f_1(r_2) \Omega_{l_1^2 LM} .$$

(5)

Under these conditions $f_i$ will satisfy the equations

$$\{t_0 + \tilde{U}_\omega + \langle f_0 | V_{00} | f_0 \rangle - \langle f_0 | \hat{G}_1 | f_0 \rangle - \varepsilon_0\} f_0 = 0,$$

(6a)

$$\{t_1 + \tilde{U}_\omega + \langle f_1 | V_{11} | f_1 \rangle - \langle f_1 | \hat{G}_0 | f_1 \rangle - \varepsilon_1\} f_1 = 0,$$

(6b)

$$t_i = -\frac{1}{2} \left\langle \frac{1}{r} \frac{d}{dr^2} r - \frac{l_i(l_i+1)}{r^2} \right\rangle,$$

(6c)

$$\varepsilon_i = E - \langle f_i | t_i + \tilde{U}_\omega | f_i \rangle.$$

(6d)

Multiplying Eqs. (6a) and (6b) by $f_0$ and $f_1$ and integrating, we obtain*

$$2\varepsilon_0 = E + \langle f_0^2 | V_{00} | f_0^2 \rangle - \langle f_0^2 | \hat{G}_1 | f_0^2 \rangle,$$

(7a)

* Here and further on the expression $\langle f^2 | V | f \rangle^2$ should be understood to mean $\langle f(r_1) f(r_2) | V(r_1, r_2) | f(r_1) f(r_2) \rangle$.

$$2\varepsilon_1 = E + \langle f_1^2 | V_{11} | f_1^2 \rangle - \langle f_1^2 | \hat{G}_0 | f_1^2 \rangle. \tag{7b}$$

From (2) it follows that for large r the functions $\varphi_0$ and $\varphi_1$ behave identically, and therefore $f_0$ and $f_1$ similarly must have an identical asymptotic behavior (to the extent to which the separated-variables approximation is valid for the function $\varphi_i$). Therefore the difference $|\langle f_0^2 | V_{00} | f_0^2 \rangle - \langle f_1^2 | V_{11} | f_1^2 \rangle|$ is considerably less than $|E|$. The matrix element $\langle f_0^2 | G_1 | f_0^2 \rangle$ represents a correction to the total energy of the two electrons due to the interaction of the configurations and is similarly much smaller than $|E|$. The quantity $\langle f_1^2 | \hat{G}_0 | f_1^2 \rangle$ requires less detailed examination. The point is that for $V_{01} = V_{10} \to 0$ this matrix element by no means tends to vanish. Nevertheless, it can be shown that it remains substantially smaller than $|E|$.

Actually, in order to estimate this matrix element we have

$$\langle f_1^2 | \hat{G}_0 | f_1^2 \rangle \approx \langle \varphi_1 | \hat{G}_0 | \varphi_1 \rangle = \langle \varphi_1 | V_{10} (H_0 - E_0 + \Delta E)^{-1} V_{01} | \varphi_1 \rangle. \tag{8}$$

Here $\Delta E = E_0 - E_1$, where $E_0$ is the eigenvalue of the operator

$$H_0 = T_0 + \tilde{U}_\omega(r_1) + \tilde{U}_\omega(r_2) + V_{00}(r_1, r_2).$$

If $V_{01}$ tends to vanish, then $\Delta E$ similarly tends to vanish, and it is possible to retain one term in the expansion of $V_{01} \varphi_1$ in eigenfunctions $H_0$:

$$(H_0 - E_0 + \Delta E)^{-1} | V_{01} \varphi_1 \rangle - \frac{\langle \varphi_0 | V_{01} | \varphi_1 \rangle}{\Delta E} | \varphi_0 \rangle. \tag{9}$$

From Eqs. (2) and (3) it follows that

$$\Delta E \approx - \lambda \langle \varphi_0 | V_{01} | \varphi_1 \rangle \approx - \frac{\langle \varphi_0 | V_{01} | \varphi_1 \rangle \langle \varphi_0 | V_{10} | \varphi_0 \rangle}{\langle \varphi_0 | T_0 - T_1 | \varphi_1 \rangle}. \tag{10}$$

And, finally,

$$\langle f_1^2 | \ddot{G}_0 | f_1^2 \rangle \approx \langle \varphi_1 | \hat{G}_0 | \varphi_1 \rangle \approx \frac{|\langle \varphi_0 | V_{01} | \varphi_1 \rangle|^2}{\Delta E} = - \langle \varphi_0 | T_0 - T_1 | \varphi_1 \rangle \frac{\langle \varphi_1 | V_{10} | \varphi_0 \rangle}{\langle \varphi_0 | V_{10} | \varphi_0 \rangle}. \tag{11}$$

The operators $T_0$ and $T_1$ differ solely in their centrifugal potentials. Naturally, the nondiagonal matrix elements of the difference between the centrifugal potentials is considerably smaller than $|E|$.

Thus, from what has been said above it follows that $|\varepsilon_0 - \varepsilon_1| \ll |\varepsilon_0|, |\varepsilon_1|$ and it is possible to set

$$\varepsilon_0 = \varepsilon_1 = \varepsilon. \tag{12}$$

In accordance with the semiempirical method we shall assume that $\varepsilon = \varepsilon_{exp}$ is the experimental energy of the levels.

Let us now replace the nonlocal potentials $\tilde{U}_\omega + \langle f_i | V_{ii} | f_i \rangle - \langle f_i | \hat{G}_k | f_i \rangle$ by certain effective local potentials $U_{\omega_i}(r) \equiv U(r / \omega_i)$, where $U(r)$ is the conventional one-electron potential of the atomic core and one-valence electron.

Thus, the functions $f_i$ are defined by independent linear differential equations:

$$\left\{-\frac{1}{2}\frac{1}{r}\frac{d^2}{dr^2}r + \frac{l_0(l_0+1)}{2r^2} + U_{\omega_0}(r) - \varepsilon\right\}f_0 = 0, \tag{12a}$$

$$\left\{-\frac{1}{2}\frac{1}{r}\frac{d^2}{dr^2}r + \frac{l_1(l_1+1)}{2r^2} + U_{\omega_1}(r) - \varepsilon\right\}f_1 = 0. \tag{12b}$$

The parameters $\omega_0$ and $\omega_1$ are chosen in such a way that the eigenfunctions of Eqs. (12) coincide with the experimental values of $\varepsilon$.

Equation (12a), and therefore the function $f_0$ is in no way different from the radial equation and the corresponding function of the one-electron semiempirical method [1]. Under these conditions the factor $\omega_0$ compensates the inaccuracy of the wave functions used in calculating U (r), as well as the approximate character of the equations proper (i.e., the comparatively small terms of the type $\langle f_0 | G_1 | f_0 \rangle$).

In Eq. (12b) $\omega_1$ also compensates the difference in the centrifugal potentials (12a) and (12b) besides the factors indicated, leading to equality of the eigenvalues of these equations. This is a reflection of the fact that the quantity $\langle f_1 | G_0 | f_1 \rangle$ in Eqs. (6b) is not small.

The parameter $\lambda$ is not included in Eq. (12).* In order to determine it we can use the variational principle which yields the following equations for $\lambda$:

$$\lambda^2 - 2A\lambda - 1 = 0, \tag{13a}$$

$$A = \frac{H_{00} - H_{11}}{\langle V_{10} \rangle}, \tag{13b}$$

$$H_{ii} = \langle f_i^2 | H | f_i^2 \rangle = \frac{\frac{\langle V_{00} \rangle - \langle V_{11} \rangle}{2} + \langle f_0 | \widetilde{U} - U_{\omega_0} | f_0 \rangle - \langle f_1 | \widetilde{U} - U_{\omega_1} | f_1 \rangle}{\langle V_{01} \rangle}. \tag{13c}$$

With an accuracy of up to terms of order $V_{00} - V_{11}$ it possible to assume that A is equal to $\langle f_1 | U_{\omega_0} - U_{\omega_1} | f_1 \rangle$.

Thus, the problem reduces to solving two independent equations (12a) and (12b) and determining $\lambda$ from (13). The computational difficulties under these conditions are of the same order as they are in the conventional one-electron semiempirical method [1].

## § 2. The Oscillator Forces

One of the criteria of the quality of the atomic functions is the coincidence of the oscillator forces calculated by means of them with the experimental oscillator forces. The oscillator forces of resonance transitions of elements of the second group of the periodic table, as calculated by means of the wave functions (5) for the ground state and by means of the one-configuration semiempirical functions [1] for the excited state, are given in Table 1. The results obtained are in better agreement with the experimental data than the calculations in the one-con-

---

*It would seem that in order to determine $\lambda$ we may use the identity (3) deriving from Eqs. (2). However, whereas Eqs. (13) already consider the approximate form of the functions $\varphi_0$, $\varphi_1$, the identity (3) is valid only for exact functions $\varphi_i$.

Specific calculations show that $\lambda$, as determined from Eqs. (3), is very sensitive to the quality of the functions by contrast with (13).

TABLE 1. Oscillator Forces of Resonance Transitions of Elements in the Second Group of the Periodic Table and Certain Ions Similar to Them

| Element | Experiment | Semiempirical method | | According to the Bates – Damgaard tables | The parameter $\lambda$ |
|---|---|---|---|---|---|
| | | two-configuration method | one-configuration method | | |
| Be | | 1.26 | 1.9 | 2.02 | 0.34 |
| Mg | 1.2[4], 1.85[11], 1.11[13] | 1.57 | 2.06 | 1.63 | 0.26 |
| Ca | 1.49[4], 1.8[9], 1.7[9] | 1.73 | 2.22 | 1.86 | 0.27 |
| Sr | 1.54[4], 1.92[9], 2.09[12] | 1.62 | 2.24 | 1.9 | 0.29 |
| Ba | 1.4[4], 1.58—1.48[12], 2.1[14] | 1.64 | 2.34 | 1.9 | 0.33 |
| Zn | 0.91[5], 1.46[9] | 1.52 | 1.82 | 1.87 | 0.2 |
| Cd | 1.19[6], 1.66[8], 1.66[10] | 1.6 | 2.04 | 1.61 | 0.24 |
| Hg | 1.19[7] | 1.54 | 2.24 | 1.58 | 0.33 |
| AlII | | 1.68 | 2.18 | 1.87 | 0.24 |
| O V | | 0.46 | 0.78 | 0.946 | 0.35 |
| Mn XIV | | 0.46 | 0.59 | 0.562 | 0.19 |

figuration approximation. The discrepancy between the calculated oscillator forces and the experimental ones are in general of the same order as in the one-configuration method for monovalent atoms, but it should be noted that the experiment for divalent atoms was evidently less reliable.

The last column in Table 1 represents the results obtained by calculations of the oscillator forces using the Bates – Damgaard tables. Since this method is a priori inferior to semiempirical methods of the type [1], the fact that these data agree better than those from [1] should be considered accidental.

Table 1 also gives the oscillator forces of the resonance lines of certain ions. The magnitude of the oscillator force for the OV resonance transition is in agreement with calculations of [2], while for the Be resonance transition it is in agreement with the calculations in [3].

It should be noted that the calculations in the two-configuration approximation based on semiempirical functions is much simpler than the method used in [3].

## §3. The Effective Cross Sections of One-Electron Transitions

It is of interest to estimate the error in the effective excitation cross sections which is connected with the use of one-electron atomic functions. This can be done by calculating cross sections, for example, using the two-configuration atomic functions (5). In the Born approximation the cross section of the $nl_0^2 \to nl_0 n'l'$ transition when the functions (5) are used has the form*

$$\sigma = \frac{8\pi}{gk} 2 \sum_{M'M} \int_{k-k'}^{k+k'} \frac{dq}{q^3} \frac{1}{1+\lambda^2} | \langle f_{l_0}(r_1) f_{l_0}(r_2) \Omega_{l_0^2 LM} + \lambda f_{l_1}(r_1) f_{l_1}(r_2) \Omega_{l_1^2 LM} | e^{iqr_1} | f_{l_0}(r_2) f_{l'}(r_1) \Omega_{l_0 l' L'M'} \rangle |^2 =$$

$$= \frac{16\pi}{gk^2} \int_{k-k'}^{k+k'} \frac{dq}{q^3} \frac{1}{1+\lambda^2} \left| \sum_{\varkappa} \langle f_{l_0} | j_{\varkappa}(qr) | f_{l'} \rangle a_{\varkappa}^{l_0 l'} (2L+1)(2L'+1) \left\{ \begin{matrix} l_0 L l_0 \\ L' l' \varkappa \end{matrix} \right\} (-1)^{l_0 + \varkappa - L'} + \lambda \langle f_{l_1} | f_{l'} \rangle \delta_{ll'} \times \right.$$

$$\left. \times \sum_{\varkappa} \langle f_{l_1} | j_{\varkappa}(qr) | f_{l_0} \rangle a_{\varkappa}^{l_1 l_0} (2L+1)(2L'+1) \left\{ \begin{matrix} l_1 L l_1 \\ L' l_0 \varkappa \end{matrix} \right\} (-1)^{l_1 + \varkappa - L'} \right|^2, \tag{14a}$$

___
*In §§ 3 and 4 we shall use a prime to denote quantities referring to the final state.

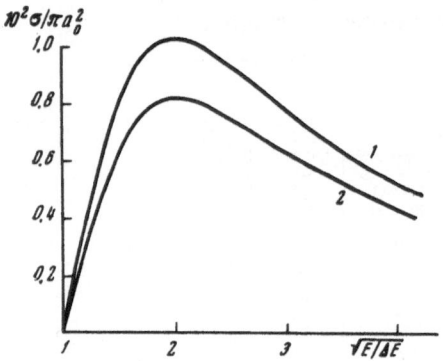

Fig. 1. Effective cross sections of the
$4s^2 - 4s4p$ transitions in the Ca atom
in the Born approximation. 1) Using
one-electron functions; 2) in the two-
configuration approximation.

$$a_\varkappa^{l_1 l_0} = (2l_0 + 1)(2l_1 + 1) \begin{pmatrix} l_1 & \varkappa & l_0 \\ 0 & 0 & 0 \end{pmatrix}, \qquad (14b)$$

where $j_\varkappa (qr)$ is a spherical Bessel function; $k, k'$ are
the momenta of the outer electron; $g$ is the statistical
weight of the initial state.

The effective cross section of the $4s^2 \rightarrow 4s4p$
transition in the Ca atom, calculated according to
Eq. (14a), is displayed in Fig. 1. It is evident that
the correction connected with consideration of the
interaction between configurations is in this case
the same as it is for the oscillator forces of a resonance transition* and amounts to 20%.

## §4.   The Effective Cross Sections
## of Two-Electron Transitions

The calculation of the effective cross sections
of two-electron transitions using atomic functions
of the type (5) is of special interest, since under these conditions the amplitudes of the two-
electron transitions may already turn out to be nonvanishing in the first order of perturbation
theory.† Here we shall restrict our analysis to the case of alkali-earth atoms and ions similar
to them.

According to (5) the two-electron wave function in the field has the form

$$\Psi = \sum_{i,k} a_{ik} f_i (r_1) f_k (r_2) \Omega_{l_i l_k LM},$$

$$\int f^2 (r) dr = 1, \qquad \sum_{i,k} a_{ik}^2 = 1. \qquad (15)$$

For convenience we shall assume that the first term in the expansion (15) (having the maximum
coefficient) corresponds to the function $\Psi$ in the one-electron approximation. Then the effective
cross section of the transition in the Born approximation will be (the prime is used to label the
final state)

$$\sigma = \frac{8\pi}{k^2} \frac{1}{g} \int\limits_{k-k'}^{k+k'} \sum_{M, M'} \frac{dq}{q^3} \; |\langle \Psi (r_1, r_2)| e^{iqr_1} + e^{iqr_2} | \Psi' (r_1, r_2)\rangle|^2 =$$

$$= \frac{8\pi}{gk^2} \int\limits_{k-k'}^{k+k'} N^2 \frac{dq}{q^3} \sum_\varkappa \left| \sum_{\substack{i,k \\ i',k'}} a_{ik} a_{i'k'} \left[ b_{l_{i'} l_{k'} l_i l_k}^{\varkappa LL'} + b_{l_{k'} l_{i'} l_k l_i}^{\varkappa LL'} \pm (b_{l_{i'} l_{k'} l_k l_i}^{\varkappa LL'} (-1)^{L-l_k-l_i} + b_{l_{k'} l_{i'} l_i l_k}^{\varkappa L, L'} (-1)^{L'-l_{k'}+l_{i'}}) \right] \right|,$$

$$(16)$$

---

*At high energies this derives from the Bethe approximation.

†Physically, this corresponds to a process in which the impinging electron excites one of the
optical electrons, and this electron in turn interacts strongly with the second optical electron
and excites it. The fairly large "impurity" coefficient of the $(p^2)$ configuration in the ground
state configuration $(s^2)$ serves as an indication of a rather strong interaction between optical
electrons in ions with two valence electrons. The impurity of other possible configurations
is much smaller.

TABLE 2. Effective Cross Section of the $4s^2 - 4p^2$ Transition in the Ca Atom and the $2s^2 - 2p^2$ Transition in the OV Ion

| $\sqrt{\dfrac{E-\Delta E}{\Delta E}}$ | Ca, $4s^2-4p^2$ | | OV, $2s^2-2p^2$ | | $\sqrt{\dfrac{E-\Delta E}{\Delta E}}$ | Ca, $4s^2-4p^2$ | | OV, $2s^2-2p^2$ | |
|---|---|---|---|---|---|---|---|---|---|
| | $^1S-^1S$ | $^1S-^1D$ | $^1S-^1S$ | $^1S-^1D$ | | $^1S-^1S$ | $^1S-^1D$ | $^1S-^1S$ | $^1S-^1D$ |
| 0.2 | $211^{+0}$ | $110^{+1}$ | $600^{+0}$ | $156^{+1}$ | 4 | $129^{+0}$ | $628^{+0}$ | $370^{+0}$ | $114^{+1}$ |
| 4 | $349^{+0}$ | $185^{+1}$ | $102^{+1}$ | $270^{+1}$ | 8 | $112^{+0}$ | $488^{+0}$ | $234^{+0}$ | $874^{+0}$ |
| 6 | $400^{+0}$ | $204^{+1}$ | $128^{+1}$ | $328^{+1}$ | 3.2 | $864^{-1}$ | $338^{+0}$ | $224^{+0}$ | $688^{+0}$ |
| 8 | $396^{+0}$ | $196^{+1}$ | $126^{+1}$ | $336^{+1}$ | 6 | $700^{-1}$ | $314^{+0}$ | $182^{+0}$ | $556^{+0}$ |
| 1.2 | $323^{+0}$ | $154^{+1}$ | $954^{+0}$ | $278^{+1}$ | 4.0 | $580^{-1}$ | $259^{+0}$ | $150^{+0}$ | $456^{+0}$ |
| 6 | $272^{+0}$ | $113^{+1}$ | $686^{+0}$ | $206^{+1}$ | 4 | $484^{-1}$ | $217^{+0}$ | $126^{+0}$ | $382^{+0}$ |
| 2.0 | $182^{+0}$ | $824^{+0}$ | $498^{+0}$ | $152^{+1}$ | 5.2 | $357^{-1}$ | $158^{+0}$ | $912^{-1}$ | $278^{+0}$ |

Note. $211^{+0}$ denotes $0.211 \times 10^0$.

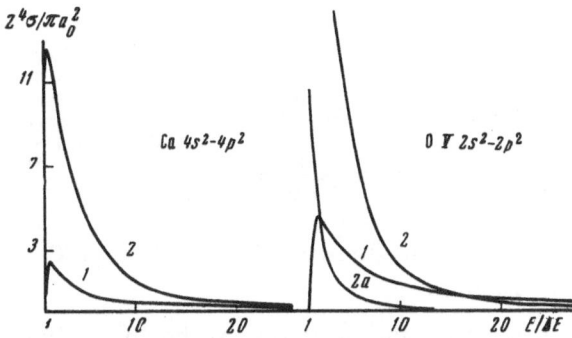

Fig. 2. Effective cross sections of two-electron transitions in the Ca atom and the OV ion. 1) First order of perturbation theory (15), (18); 2) second order of perturbation theory (quasi-classical approximation [15]) (2a is curve 2 reduced by a factor of 10).

where

$$b^{\varkappa LL'}_{l'_1 l'_2 l_1 l_2} = \langle f_{l'_1} | f_{l_1} \rangle \, \delta_{l'_1 l_1} \langle f_{l'_2} | j_\varkappa(qr) | f_{l_2} \rangle (-1)^{l_1 + \varkappa - L'} \times$$

$$\times [(2l'_2 + 1)(2l_2 + 1)(2\varkappa + 1)(2L + 1)(2L' + 1)]^{\frac{1}{2}} \begin{pmatrix} l'_2 \varkappa l_2 \\ 0\ 0\ 0 \end{pmatrix} \begin{Bmatrix} l_2\ L\ l_1 \\ L'\ l'_2 \varkappa \end{Bmatrix};$$

$N = \frac{1}{2}$ if both states contain equivalent electrons; $N = 1/\sqrt{2}$ if one state contains equivalent electrons; $N = 1$ if the electrons are nonequivalent in both states. The plus sign corresponds to a transition between singlet states; the minus sign corresponds to a transition between triplet states.

Only two-electron transitions from the ground state are of practical interest. Under these conditions two cases are possible:

1. For the excited two-electron level the effects of interaction between the configurations are small (for example the $4p^2\ ^1D$ level in the Ca atom). Then it is possible to restrict the

the analysis to functions of the type (5) for the ground state only, while the excited state can be described by conventional semiempirical function [1]. Under these conditions

$$a_{00} = 1/\sqrt{1+\lambda^2}, \qquad a_{11} = \lambda/\sqrt{1+\lambda^2}, \qquad a'_{00} = 1.$$

Examples of such transitions ($4s^2\ ^1S - 4p^2\ ^1D$ in the Ca atom; $2s^2\ ^1S - 2p^2\ ^1D$ in the OV ion) are given in Table 2.

2. The excited level contains a ground-state "impurity"* (for example, the $4p^2\ ^1S$ level in the Ca atom). In this case the excited state should similarly be described by functions of the type (5). Under these conditions the "impurity" coefficient may be determined from the orthogonality condition

$$\langle \Psi | \Psi' \rangle = \lambda |\langle f_0 | f_1' \rangle|^2 + \lambda' |\langle f_1 | f_0' \rangle|^2 = 0. \tag{17}$$

For the coefficients $a$ we find

$$
\begin{aligned}
a_{00} &= \frac{1}{\sqrt{1+\lambda^2}}, & a'_{00} &= \frac{1}{\sqrt{1+(\lambda')^2}}; \\
a_{11} &= \frac{\lambda}{\sqrt{1+\lambda^2}}, & a'_{11} &= \frac{\lambda'}{\sqrt{1+(\lambda')^2}};
\end{aligned}
\qquad
\lambda' = -\lambda \left| \frac{\langle f_1 | f_0' \rangle}{\langle f_0 | f_1' \rangle} \right|^2. \tag{18}
$$

Examples of transitions of the type indicated ($4s^2\ ^1S - 4p^2\ ^1S$ in the Ca atom; $2s^2\ ^1S - 2p^2\ ^1S$ in the OV ion) are also given in Table 2.

The resultant cross sections for the excitation of the $4p^2$ configuration in the Ca atom and the $2p^2$ configuration in the OV ion are displayed in Fig. 2. For comparison purposes these same figures display the effective excitation cross sections in the second order of perturbation theory in the quasi-classical approximation calculated by A. V. Vinogradov [15]. It is evident that the excitation cross section in the second order of perturbation theory decreases considerably faster than in the first order as Z and the energy of the incident electron increase.

## References

1. L. A. Vainshtein, Opt. i Spektr., 11:301 (1961).
2. G. R. La Paglia, Okt. Sinanoglu. J. Chem. Phys., 44:1888 (1966).
3. A. B. Bolotin and A. P. Yutsis, Zh. Éksperim. i Teor. Fiz., 24:537 (1963); V. V. Kibartas and A. P. Yutsis, Zh. Éksperim. i Teor. Fiz., 25:264 (1953); A. P. Yutsis, V. V. Kibartas, and I. I. Glembotskii, Zh. Éksperim. i Teor. Fiz., 27:425 (1953).
4. N. P. Penkin, Abstract of Doctoral Dissertation, Leningrad (1962).
5. J. Auslander, Helv. Phys. Acta, 11:562 (1932).
6. M. W. Zemansky, Z. Phys., 72:587 (1931).
7. A. Filippov, Sow. Fiz., 1:289 (1936).
8. C. G. Matland, Phys. Rev., 91:436 (1953).
9. A. Lurio, Phys. Rev., 140A:1505 (1964).
10. A. Lurio, Phys. Rev., 134A:1198 (1964).
11. A. Lurio, Phys. Rev., 136A:376 (1964).
12. E. Hulpke, E. Paul, and W. Paul, Z. Phys., 177:257 (1964).
13. W. Demtroder, Z. Phys., 166:42 (1961).
14. G. Wessel, Z. Phys., 126:440 (1949).
15. A. V. Vinogradov, this volume, p.45.

*It is easy to see that for the amplitude of the transition an "impurity" from another state leads to a correction of the second order in the "impurity" coefficient $\lambda$.

# CHARGE EXCHANGE AT
# INTERMEDIATE COLLISION ENERGIES

## I. A. Poluéktov and L. P. Presnyakov

### §1.  Introduction

Collisions of heavy particles (in particular, collisions which are accompanied by charge exchange) can be described over a wide energy range in the quasi-classical (parametric) approximation. This investigation consists in assuming that the coordinates of the nuclei are stipulated time functions rather than dynamic variables. Under these conditions the problem of charge exchange reduces to solving a system of equations in nonstationary perturbation theory for the probability amplitudes of the states.

In order to obtain these equations as they apply to the process $A + B \rightarrow A^+ + B$ the total wave function of the system consisting of an $A^+$ ion, a $B^+$ ion, and an electron is usually written in the form of an expansion in wave functions of the atoms* formed by the attachment of an electron to the $A^+$ ion (the eigenfunctions $\psi_i$ of the A atom) and to the $B^+$ ion (the eigenfunctions $\varphi_m$ of the B atom), respectively, [1, 2]. Since $\varphi$ and $\psi$ are eigenfunctions of different Hamiltonians and are nonorthogonal with respect to each other, the system of equations obtained is non-Hermite, while the transition probabilities are nonnormalized. This is evidently one of the causes of the fact that for integration of the system in general form it is usually necessary to restrict the analysis to the first approximation of perturbation theory. The special case of symmetrical resonance, and the charge exchange in the presence of a small resonance defect considered by Demkov [3] are exceptions. Note that in [3] the matrix elements of the transition were calculated in an approximation such that the system was artifically made Hermitian.

In the present paper we use a different formulation of the charge exchange problem, which was suggested by the authors and Sobel'man [4]; its formulation leads to a system with a Hermitian matrix. Along with charge exchange in the ground state, to which the two-level strong-coupling approximation [4] is applicable, the problem of charge exchange in the excited state is considered; this problem requires departure beyond the framework of the two-level approximation.

---

*In the present paper we restrict our analysis to the energy range above 1 keV. At lower energies (especially at energies of the order of thermal energies) it is necessary to carry out expansion in eigenfunctions of the quasimolecule which is formed in the collision process.

## §2. Hermite Formulation of the Problem.

## Charge Exchange in the Ground State

For simplicity let us begin by considering the charge exchange process which accompanies collision of a hydrogen atom with a proton. In this case the system consists of two protons, which we shall denote by the subscripts A and B, and an electron. The Hamiltonian of the system has the form

$$H = -\frac{1}{2}\Delta + V(\mathbf{r}_A) + V(\mathbf{r}_B) + W(\mathbf{R}), \tag{2.1}$$

where $\mathbf{r}_A$ and $\mathbf{r}_B$ are the radius vectors (position vectors) of the electron relative to the protons A and B; $\mathbf{R}$ is the distance between protons, which is a stipulated time function within the framework of the quasi-classical method:

$$V(\mathbf{r}_A) = -\frac{1}{r_A}, \qquad V(\mathbf{r}_B) = -\frac{1}{r_B}, \qquad W(\mathbf{R}) = \frac{1}{R}. \tag{2.2}$$

Assume that before collision the hydrogen atom consists of an electron and a proton A. We are interested in the charge exchange process during which the electron goes over into the proton B. Henceforth we shall consider the trajectory to be linear: $\mathbf{R} = \boldsymbol{\rho} + \mathbf{v}t$ where $\boldsymbol{\rho}$ is the impact parameter; $\mathbf{v}$ is the relative velocity ($\boldsymbol{\rho}\mathbf{v} = 0$). Under these conditions the frames of reference connected with the centers of inertia of atoms A and B are equally legitimate; it is convenient to choose the center of the B atom as the origin. Let us represent the total wave function $\Psi(t)$ of the system in the form of an expansion in wave functions $\varphi_m$ of the B atom (i.e., in eigenfunctions of the Hamiltonian $H_B$):

$$H_B = -\frac{1}{2}\Delta + V(\mathbf{r}_B), \qquad H_B\varphi_m = \varepsilon_m^B\varphi_m; \tag{2.3}$$

$$\Psi(t) = \left(\sum_m + \int\right) a_m(t)\varphi_m(\mathbf{r}_B)\exp(-i\varepsilon_m^B t), \tag{2.4}$$

where $\left(\sum_m + \int\right)$ denotes summation over the states of the discrete spectrum and integration over the states of the continuum. The expansion (2.4) is carried out in the complete system of eigenfunctions of the Hamiltonian (2.3) and is therefore exact. For $t \to -\infty$ the wave function (2.4) must describe a hydrogen atom consisting of an electron and a proton A which moves with the velocity $\mathbf{v}$:

$$\Psi(t) \underset{t\to-\infty}{\to} \psi_{0v}(t) = \psi_0(\mathbf{r}_B - \mathbf{R}(t))\exp\left\{i\mathbf{v}\mathbf{r}_B - i\left(\varepsilon_0^A + \frac{v^2}{2}\right)t\right\}. \tag{2.5}$$

Simultaneously,

$$a_m(t) \to a_m^0(t) = S_{m0}(t)\exp(i\omega_{m0}t), \quad \omega_{m0} = \varepsilon_m^B - \varepsilon_0^A, \tag{2.6}$$

$$S_{m0}(t) = \int \psi_{0v}\left(\mathbf{r} - \frac{\mathbf{R}(t)}{2}\right)e^{i\mathbf{v}\mathbf{r}}\varphi_m^*\left(\mathbf{r} + \frac{\mathbf{R}(t)}{2}\right)d\mathbf{r}. \tag{2.7}$$

Substituting (2.4) into the nonstationary Schrödinger equation, we obtain the conventional system of equations for $a_m(t)$ with the Hermitian matrix

$$i\,a_m = \left(\sum_{m'} + \int\right) H_{mm'}(t) \exp\{i\omega_{mm'}t\}\,a_m, \tag{2.8}$$

$$H_{mm'}(t) = \int \varphi_m^*(\mathbf{r}_B)\{V(\mathbf{r}_B - \mathbf{R}) + W(\mathbf{R})\}\varphi_m(\mathbf{r}_B)\,d\mathbf{r}_B, \qquad \omega_{mm'} = \varepsilon_m^B - \varepsilon_{m'}^B. \tag{2.9}$$

It is necessary to solve the system (2.8) with the initial conditions (2.6). As is evident from (2.8)-(2.9), the term W(R) merely introduces a nonessential phase factor into the solution. Therefore, henceforth it can be omitted.

The system (2.8) can easily be generalized for the case of charge exchange of arbitrary atoms on singly charged ions. Under these conditions (2.8) will include potentials which are averaged over the states of all the inner electrons.

Having substituted $a_{m'}(t) = a_{m'}^0(t)$ into the right side of (2.8) and using the completeness condition for the functions $\varphi_m(\mathbf{r}_B)$, it is not difficult to obtain the conventional Born equation for the charge exchange probability $|a_n(\infty)|^2$ :

$$|a_n(\infty)|^2 = |\int_{-\infty}^{\infty} dt\, \langle \varphi_n^* | V(\mathbf{r}_B - \mathbf{R}) | \Psi_{0v}\rangle\, e^{i\omega_{n0}t}|^2. \tag{2.10}$$

In a definite approximation it is also possible to derive a system of two equations for the probability amplitudes of the states (this system is analogous to the conventionally examined one in [1]) while preserving the Hermitian character of the matrix of the equations.

For m ≠ n we set

$$a_m(t) = b(t)\,a_m^0(t)[1 - |S_{n0}|^2]^{-\frac{1}{2}}, \qquad a_n(t) = a(t).$$

Substituting these expressions into (2.8), multiplying each of the equations having m ≠ n by $a_m^{0*}(t)\,[1 - |S_{n0}|^2]^{-1/2}$ and summing all of the equations having m ≠ n, it is possible to obtain the following system of two equations:

$$i\,a = \widetilde{V}_{nn}a + \widetilde{V}_{n0}e^{i\omega_{n0}t}b, \qquad b = \widetilde{V}_{0n}e^{i\omega_{0n}t} + \widetilde{V}_{00}b \tag{2.11}$$

with the initial conditions $b(-\infty) = 1$, $a(-\infty) = a_n^0(-\infty) = 0$ $[a_n^0(t) \to 0$ for $t \to -\infty$, since the functions of the discrete spectrum $\varphi_n(\mathbf{r}_B)$ and $\psi_0(\mathbf{r}_B - \mathbf{R}(t))\exp iv\mathbf{r}_B$ do not overlap for R (t) → ∞]. Here

$$\widetilde{V}_{nn} = \langle \varphi_n^* \big| \frac{1}{|\mathbf{r}_B - \mathbf{R}|} \big| \varphi_n \rangle; \tag{2.12}$$

$$\widetilde{V}_{n0} = [\langle \varphi_n^* | r_B^{-1} | \psi_{0v}\rangle - S_{n0}\widetilde{V}_{nn}]/(1 - |S_{n0}|^2)^{1/2}, \qquad \widetilde{V}_{0n} = \widetilde{V}_{n0}^*; \tag{2.13}$$

$$\widetilde{V}_{00} = \{\langle \psi_{0v} | r_B^{-1} | \psi_{0v} \rangle - \text{Re} [S_{n0} \langle \varphi_n^* | r_B^{-1} | \psi_{0v} \rangle - S_{0n} \langle \psi_{0v} | r_B^{-1} | \varphi_n \rangle] - |S_{n0}|^2 \widetilde{V}_{nn}\} [1 - |S_{n0}|^2]^{-1}. \quad (2.14)$$

It is not difficult to show that b (t) has the meaning of a probability amplitude of a state:

$$\widetilde{\psi}(t) = \{\psi_{0v} - S_{n0}\varphi_n\}(1 - |S_{n0}|^2)^{-\frac{1}{2}}.$$

For $|t| \to \infty$ we have $\widetilde{\psi}(t) \to \psi_{0v}$. The functions $\varphi_n$, $\widetilde{\psi}$ are orthogonal for any t, and therefore $|a(t)|^2 + |b(t)|^2 = 1$. Note that the system examined in the paper by Bates and McCarroll (Eqs. (71), (72) in [1]) can also be obtained from (2.8) if for m $\neq$ n we set $a_m(t) = b'(t) a_m^0(t)$ and $a_n(t) = a'(t) = a'(t) + b'(t) a_n^0(t)$. As has already been noted in Sec. 1, this system has a non-Hermitian matrix.

It is easy to show that the exact solution of the system (2.11) (a, b) and the exact solution of the system (71), (72) from [1] (a', b') are associated by the relationship

$$a(t) = a'(t) + b'(t) S_{n0}(t), \qquad b(t) = b'(t)(1 - |S_{n0}|^2). \quad (2.15)$$

Therefore, in those cases when it is possible to find an exact solution (for example, for symmetrical resonance) the charge exchange probabilities $|a(\infty)|^2$ and $|a'(\infty)|^2$ coincide. However, in those cases when we cannot find an exact solution, the system (2.11), which has a Hermitian matrix, has well-known advantages. In particular, for integration of the system (2.11) it is possible to use approximate methods developed for problems in the excitation of atoms. Below we shall make use of the approximate method found earlier [5] and examined in detail in [6].

In the general case the nondiagonal matrix elements in (2.11) are complex:

$$\widetilde{V}_{n0} = U e^{i\varphi(t)}, \qquad U = |\widetilde{V}_{n0}|. \quad (2.16)$$

This distinguishes the problem of charge exchange from the problem of excitation of atoms. In the latter problem the phase $\varphi(t)$ is zero or is independent of time. Therefore, in the case given, the approximate equation for the transition probability has the form (see also §3):

$$w(\rho, v) = |a(\infty)|^2 = |\int_{-\infty}^{\infty} U(t) \cos(\int_0^t [\omega + \varphi + \widetilde{V}_{nn} - \widetilde{V}_{00} + 4U^2]^{1/2} d\tau) dt|^2. \quad (2.17)$$

Equation (2.17) has been discussed in detail previously [5, 6]. It was shown that it yields the correct expressions in all of the known limiting cases (the Born approximation, the Landau – Zener case [7], the Zener – Rosen case [8], the Demkov case [3]).

It is well known [3, 4] that for large R the potential $\widetilde{V}_{n0}$ decreases as $e^{-\gamma R}$, where $\gamma = (2I_{min})^{1/2}$, and $I_m$ is the least of the ionization energies $\varepsilon_0^A$ and $\varepsilon_n^B$. In this case the integral in (2.17) can be written in the form

$$w(\rho, v) = \exp\{-2\sqrt{\left(\frac{\pi\omega_{n0}}{2\gamma v}\right)^2 + \left(\frac{\rho\omega_{n0}}{v}\right)^2}\} \sin^2 \int_{-\infty}^{\infty} U(\rho, v, t) dt. \quad (2.18)$$

For $\omega_{n0} = 0$ Eq. (2.18) takes the same form as the exact formula for the case of symmetrical resonance. The difference consists solely in the fact that instead of the expression

$$(V'_{n0} - S_{n0}H'_{nn})/(1 - |S_{n0}|^2) \qquad (2.19)$$

U (t) from (2.16) is included in (2.18).

As an example, the effective charge-exchange cross section was calculated by means of (2.18):

$$p + \mathrm{He}(1s^2) \rightarrow \mathrm{He}^+(1s) + \mathrm{H}(1s).$$

The Slater approximation $\psi_{\mathrm{He}} = \pi^{-1/2} \, a^{3/2}, \, e^{-ar} \, a = 1.7$ was used for the wave function of the 1s electron in the He atom. Under these conditions the integral appearing under the $\sin^2$ sign in (2.18) has the form [4]

$$\int_{-\infty}^{\infty} U(\rho, v, t)\,dt = I_1(\rho, v) + I_2(\rho, v), \qquad (2.20)$$

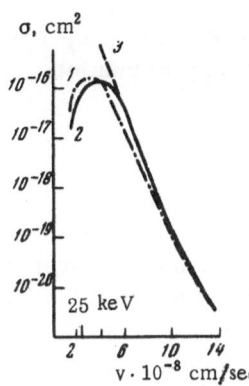

Fig. 1. Effective cross section for the charge exchange p + He (1s²) → H (1s) + He⁺(1s). 1) Experimental data [1]; 2) Eqs. (2.18)–(2.22); 3) Eq. (2.24).

$$I_1(\rho, v) = -\frac{8a^{3/2}}{v\,(a^2-1)}\left\{\frac{\rho}{\sqrt{1+\frac{v^2}{4}}} K_1\left(\rho\sqrt{1+\frac{v^2}{4}}\right) + \frac{2}{(a^2-1)}\left[K_0\left(\rho\sqrt{a^2+\frac{v^2}{4}}\right) - K_0\left(\rho\sqrt{1+\frac{v^2}{4}}\right)\right]\right\}, \qquad (2.21)$$

$$I_2(\rho, v) = \frac{32a^{5/2}}{\pi v}\int_{-\infty}^{\infty}\left[K_0(\rho|q|) - K_0\left(\rho\sqrt{4+q^2}\right) + \frac{2\rho}{\sqrt{4+q^2}}K_1\left(\rho\sqrt{4+q^2}\right)\right]\left\{\frac{2}{(q^2-1+2vq)^3}\times\right.$$
$$\times\left[K_0\left(\rho\sqrt{a^2+\left(q+\frac{v}{2}\right)^2}\right) - K_0\left(\rho\sqrt{1+\left(q-\frac{v}{2}\right)^2}\right)\right] + \frac{\rho}{2(a^2-1+2vq)^2}\times$$
$$\times\left[\frac{1}{-\sqrt{1+\left(q-\frac{v}{2}\right)^2}}K_1\left(\rho\sqrt{1+\left(q-\frac{v}{2}\right)^2}\right) + \frac{1}{\sqrt{a^2+\left(q+\frac{v}{2}\right)^2}}K_1\left(\rho\sqrt{a^2+\left(q+\frac{v}{2}\right)^2}\right)\right]\right\}dq. \quad (2.22)$$

Note that for $a = 1$ Eq. (2.21) goes over into the expressions derived by Brinkman and Kramers [1], while (2.22) goes over into the integral found by Murakhver [9]. Equation (2.18) with allowance for (2.20)–(2.22) requires numerical calculations in computing the cross section

$$\sigma(v) = 2\pi\int_0^{\infty}\rho\,d\rho\,w(v,\rho).$$

However, in the limiting cases of large and small velocities it turned out to be possible to use simple analytical estimates

$$\sigma(v) = \frac{2\pi v^2 \exp\left(-\frac{\pi\omega_{n_0}}{\gamma v}\right)}{\omega_{n_0}^2}\left[\frac{\pi\omega_{n_0}}{2\gamma v} + \frac{1}{2}\right], \qquad \omega_{n_0}\neq 0, \quad v\rightarrow 0; \qquad (2.23)$$

$$\sigma(v) = \frac{128a^{5/2}\pi}{5v^2\left(1+\frac{v^2}{4}\right)^5}, \qquad v\rightarrow\infty. \qquad (2.24)$$

The results of the calculations, as well as the experimental data, are displayed in Fig. 1. From the figure it is evident that both the numerical calculation and the analytical approximation (2.23)-(2.24) yields satisfactory agreement with experiment.

## §3.  The Role of the Intermediate Level.
### Charge Exchange in the Excited State

The approximation we used for the interaction of the initial and final states of the system is justified if we speak of charge exchange in the ground state of the atom or of a transition between any two neighboring levels of the system. In the case of charge exchange in an excited state such a two-level approximation is, generally speaking, inadequate. In calculating the charge-exchange cross section in an excited state it is necessary to consider the competing transitions via the intermediate level along with the direct charge exchange (the two-level approximation). For charge exchange in the first excited state the ground state is such an intermediate level. Recently, a number of experimental papers has appeared [10-12] in which charge exchange in the 2p- and 2s-states of hydrogen were investigated for collisions of protons with atoms of inert gases. The experimental cross sections have a number of qualitative features which are difficult to explain within the framework of the Born or two-level approximation. In particular, in the range of energies E which are not too high (E ≤ 20 keV) the cross sections can have two maxima, and furthermore at these energies the charge exchange in the 2p-state is considerably more effffective than in the 2s-state.

Here we shall consider a method of calculating the charge exchange in the excited state with allowance for a transition via an intermediate level; this method allows a qualitative explanation to be provided for the indicated features of the experimental cross sections. The collision is considered in the quasi-classical (parametric) approximation.

In order to take account of this transition via the intermediate level along with the direct transition it is necessary to solve a system of three equations

$$ia_m = \sum_{n \neq m} U_{mn} \exp\left[i \int^t \alpha_{mn}(\tau)\, d\tau\right] a_n, \quad m, n = 0, 1, 2; \tag{3.1}$$

$$|a_n(-\infty)| = \delta_{0n}, \quad \sum_n |a_n(t)|^2 = 1. \tag{3.2}$$

The subscripts 0, 1, 2 respectively denote the original, intermediate, and final states; $U_{mn}(t) = U_{nm}^*(t)$ are the nondiagonal matrix elements of the interaction, and their specific form depends, in particular, on the basis functions, atomic or molecular, in which the expansion of the total wave function of the system is carried out.* The phases $\alpha_{mn}(t)$ determine the difference between the energy levels of the system in the zero approximation.

Going over from the probability amplitudes $a_n(t)$ to the functions $R_n(t)$ [5, 6]

$$R_n(t) \equiv \mu_n(t) \exp\left[-i\,\Omega_{n0}(t) - i\,\frac{\pi}{2}\right] = \frac{a_n(t)}{a_0(t)}, \quad R_0(t) \equiv 1, \tag{3.3}$$

---

*The Hermite formulation of the problem of charge exchange in the two-level approximation (see Sec. 2) can easily be generalized for a larger number of levels [13].

where $\mu_n$, $\Omega_{n0}$ are real functions, we obtain a system of two nonlinear equations instead of (1):

$$i\dot{R}_n = U_{n0}\exp\left(i\int\limits^t \alpha_{n0}\,d\tau\right) - U_{0n}\exp\left(-i\int\limits^t \alpha_{n0}\,d\tau\right) + U_{nm}\exp\left(i\int\limits^t \alpha_{nm}\,d\tau\right)R_m -$$
$$- U_{0m}\exp\left(i\int\limits^t \alpha_{0m}\,d\tau\right)R_n R_m, \quad n \neq m \quad n, m = 1, 2; \tag{3.4}$$

here the normalization condition (3.2) is always satisfied:

$$|a_n(t)|^2 = \frac{\mu_n^2(t)}{1 + \mu_1^2(t) + \mu_2^2(t)}. \tag{3.5}$$

Having separated the real and imaginary parts in (3.4), it is easy to obtain the system of equations for $\mu_n(t)$:

$$\mu_n = U_{n0}(1 + \mu_n^2)\cos\Phi_{n0} + \mu_m U_{nm}\cos\Phi_{nm} + \mu_n\mu_m U_{0m}\cos\Phi_{0m}, \tag{3.6}$$
$$n, m = 1, 2; \quad n \neq m.$$

The phases $\Phi_{nm}$ are determined as follows:

$$\Phi_{nm}(t) = \int\limits^t \alpha_{mn}(t)\,d\tau + \Omega_{mn}(t), \quad \Omega_{mn} = \Omega_{m0} - \Omega_{n0}, \tag{3.7}$$

and under quasi-stationary conditions they determine the correction to the distance between the m and n terms caused by $U_{mn}$ interactions [5-6]. The corresponding equations for the phases $\Omega_{n0}$ were given in [6].

In the problem examined $\mu_2 < \mu_1 \leq 1$ (the probability of charge exchange in the ground state is much higher than in the excited state), and therefore in the right side of (3.6) we may neglect all terms containing $\mu_2$. Under these conditions $\mu_1$ has the form [6]

$$\mu_1(t) = \tan\int\limits_{-\infty}^{t} U_{10}(t)\cos\Phi_{10}dt \approx \int\limits_{-\infty}^{t} U_{10}\cos\Phi_{10}dt. \tag{3.8}$$

Substituting (3.8) into the equations for $\mu_2$, we obtain

$$\mu_2(\infty) = \int\limits_{-\infty}^{\infty} U_{20}\cos\Phi_{20}dt + \int\limits_{-\infty}^{\infty} U_{21}\cos\Phi_{21}dt \int\limits_{-\infty}^{t} U_{10}\cos\Phi_{10}d\tau. \tag{3.9}$$

Since under our conditions $\mu_2 < 1$, it follows that in (3.5) we may omit the normalizing denominator. The expression for the probability of a transition to state 2 takes the form

$$w_{02}(\rho, v) = |a_2(\infty)|^2 = |\mu_2(\infty)|^2 = \left|\int\limits_{-\infty}^{\infty} U_{20}\cos\Phi_{20}dt + \int\limits_{-\infty}^{\infty} U_{21}\cos\Phi_{21}dt \int\limits_{-\infty}^{t} U_{10}\cos\Phi_{10}d\tau\right|^2. \tag{3.10}$$

Here the first term describes the direct $0 - 2$ transition, while the second describes the transition $0 - 1 - 2$ via an intermediate level.

The transition probability (3.10) was obtained from the nonlinear system of equations (3.6) in the second order of perturbation theory. The consistent application of perturbation theory used here requires the substitution of the phases $\Phi_{mn}(t)$ in the right side of (3.10) in that form which they have on the assumption of pair interaction of the m and n terms*:

$$\Phi_{mn}(t) = \int_{t^0_{mn}}^{t} \sqrt{\alpha^2_{mn}(\tau) + 4U^2_{mn}(\tau)}\, d\tau. \tag{3.11}$$

The problem of choosing the lower limit of integration $t^0_{mn}$ was discussed earlier [6], it being shown that for the solution of the system (3.1) consisting of just two equations, $t^0 = 0$. As applied to the case examined here, this implies that for direct transitions $t^0_{10} = t^0_{20} = 0$. The quantity $t^0_{12} \neq 0$ depends on the characteristic parameters of the problem. Its calculation in general form is hardly possible. Therefore, we shall estimate the qualitative role of $t^0_{12}$. In squaring the right side of (3.10) a cross (interference) term arises. The presence of $t^0_{12} \neq 0$ leads to the appearance of a cofactor in this term, which oscillates rapidly with a change of the impact parameter and collision velocity. Its contribution to the total cross section is negligibly small, and in our approximation the interference term can be dropped. On the other hand, the specific value of $t^0_{12}$ has a weak influence on the magnitude of the square of the second term in the right side of (3.10). Therefore, in calculating this integral we assume $t^0_{12} = 0$. Under these conditions we have

$$\int_{\infty}^{\infty} U_{21} \cos \Phi_{21} dt \int_{-\infty}^{t} U_{10} \cos \Phi_{10} d\tau = \frac{1}{2} \int_{-\infty}^{\infty} U_{21} \cos \Phi_{21} dt \int_{-\infty}^{\infty} U_{10} \cos \Phi_{10} d\tau \tag{3.12}$$

as a consequence of the even character of the integrand expressions. The expression for the transition probability takes the following form under these conditions:

$$w_{02} = w_{0 \to 2} + \frac{1}{4} w_{0 \to 1} w_{1 \to 2}, \tag{3.13}$$

$$w_{m \to n} = \left| \int_{-\infty}^{\infty} U_{mn} \cos \Phi_{nm} dt \right|^2. \tag{3.14}$$

Thus, the total transition probability $w_{02}$ is the sum of the probabilities $w_{0 \to 2}$ and $w_{0 \to 1 \to 2} = {}^1/_4 w_{0 \to 1} w_{0 \to 2}$, of the direct transition and the term which approximately describes the transition via the intermediate level. There is no interference between the processes indicated, and the probability of the transition via the intermediate level is expressed in terms of the product of the probabilities of the transitions between neighboring states. At high collision velocities the expressions for $w_{m \to n}$ (3.14) go over into the Born probabilities of the first order [4, 6]. It is not difficult to verify the fact that in the Born region Eq. (3.13) coincides with the second Born approximation.

---

*It is well known that only two terms of a physical system can intersect at one point of a complex plane; a harmonic oscillator is an exception.

The effective charge exchange cross section in the excited state similarly consists of the sum of two terms which describe the processes indicated:

$$\sigma_{02} = \sigma_{0\to2} + \sigma_{0\to1\to2}, \tag{3.15}$$

$$\sigma_{0\to2} = 2\pi \int\limits_0^\infty \rho \, d\rho \, w_{0\to2}, \qquad \sigma_{0\to1\to2} = \frac{\pi}{2} \int\limits_0^\infty \rho \, d\rho \, w_{0\to1} w_{0\to2}, \tag{3.16}$$

where $\rho$ is the impact parameter.

The general equations (3.13)-(3.16) can also be used in calculating the cross sections of other inelastic processes which occur during collisions of heavy particles. The simplest example is the excitation of atomic levels situated above the resonance level from the ground state. Practically, the problem reduces to calculating the matrix elements $U_{mn}$ and the splittings $\Phi_{mn}$ between the terms.

Let us use the derived equations for calculating the cross sections of charge exchange of protons on inert gas atoms with the formation of hydrogen in the 2p-state. Let us limit the analysis to the energy range from 5 to 30 keV. The initial, intermediate, and final states are the ground state of the target atom B, and the 1s- and 2p-states of the hydrogen atom which is formed. The $B^+$ ion is assumed to be a structureless charged particle under these conditions. The "direct charge exchange" cross section $\sigma_{0\to2}$ has already been considered in §2. Therefore, only $\sigma_{0\to1\to2}$ shall be investigated in detail. The probabilities $w_{0\to1}$ and $w_{0\to2}$ are also known separately [3-6]; they have the form

$$w_{0\to1}(\rho, v) = \exp\left\{-2\sqrt{\left(\frac{\pi\omega_{10}}{2\gamma v}\right)^2 + \left(\frac{\rho\omega_{10}}{v}\right)^2}\right\} \sin^2 \frac{1}{v} \int\limits_{-\infty}^{\infty} U_{01} dx, \tag{3.17}$$

$$w_{1\to2}(\rho, v) = \exp\left\{-2\sqrt{\left(\frac{\lambda\omega_{21}}{2v^2}\right) + \left(\frac{\rho\omega_{21}}{2}\right)^2}\right\} \sin^2 \frac{1}{v} \int\limits_{-\infty}^{\infty} U_{12} dx, \tag{3.18}$$

$$x = vt.$$

Here v is the relative velocity; $\omega_{mn}$ is the resonance defect between the levels m and n; $U_{01}$ and $U_{12}$ are nondiagonal matrix elements which are connected respectively with charge exchange in the ground state and excitation from the ground state. At large distances, $U_{10} \sim \exp(-\gamma R)$. For the optically allowed transition, $U_{12} \sim 1/R^2$ [5, 6]. If the transition $1-2$ is a transition of the $s-s$ type (for example, $1s-2s$ in the hydrogen atom), then the excitation probability $w_{1-2}$ has the form (3.17), but $\gamma = \sqrt{2I_1} + \sqrt{2I_2}$ [3,6].

The exact calculation of $\sigma_{0\to1\to2}$ using (3.17) and (3.18) requires numerical integration. However, as in §2, it is possible to obtain a simple analytical approximation for $\sigma_{0\to1\to2}$, if consideration is given to the fact that the excitation probability $w_{1\to2}(\rho, v)$ is a smoother function than the probability of charge exchange in the ground state $w_{0\to1}(\rho, v)$. We take out the function $w_{1\to2}(\rho)$ from under the integral sign for $\rho = \rho_0$ in (3.16). The remaining integral yields the cross section of charge exchange in the ground state

$$\sigma_{0\to1\to2} = \frac{\pi}{2} \int\limits_0^\infty \rho \, d\rho \, w_{0\to1}(\rho, v) w_{1\to2}(\rho, v) \approx \frac{w_{1\to2}(\rho_0)}{4} \int\limits_0^\infty 2\pi\rho \, d\rho \, w_{0\to1}(\rho) = \frac{1}{4} w_{1\to2}(\rho_0) \sigma_{0\to1}. \tag{3.19}$$

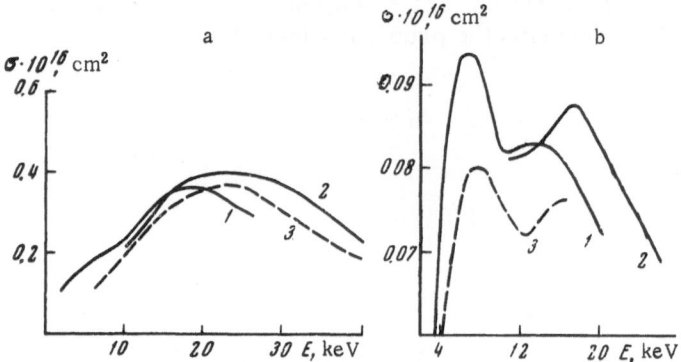

Fig. 2. Cross section of the processes $p + He\ (1s^2) \to H\ (2p)$ $+ He^+$ (a) and $p + Ne \to H\ (2p) + Ne^+$ (b). 1) Experimental data [10]; 2) experimental data [12]; 3) the results of the work described in the present paper.

Using the explicit form of $w_{0 \to 1}\ (\rho,\ v)$ (2.18) and preserving the correct dependence on velocity for low and high velocities in the approximate calculation, we can obtain the following value for the quantity $\rho_0 = \rho_0\ (v)$:

$$\rho_0(v) = \frac{\sqrt{1 + \frac{v^2}{4}}}{2I_B - 1}, \qquad 2I_B - 1 > 0, \tag{3.20}$$

where $I_B$ is the ionization potential of the target atom. Within the framework of the rather rough approximation made here, it makes no sense to use the exact form of the factor in front of the exponential in the probability $w_{1 \to 2}\ (\rho,\ v)$ (3.18). Since it is of the order of $1/2$ at low velocities and $\sim \left[\frac{\lambda}{v\rho_0}\right]^2$ at high velocities, it follows that for a qualitative description it may be replaced by $\dfrac{1}{\left[2 + \left(\frac{v\rho_0(v)}{\lambda}\right)^2\right]}$.

As a result, we obtain the very simple approximate expression

$$\sigma_{0 \to 1 \to 2} \approx \sigma_{01} \frac{\exp\left\{-2\sqrt{\frac{\lambda\omega_{21}}{2v} + \left(\frac{\rho_0\omega_{21}}{v}\right)^2}\right\}}{4\left[2 + \left(\frac{v\rho_0}{\lambda}\right)^2\right]} \tag{3.21}$$

for $\sigma_{0 \to 1 \to 2}$. Here $\sigma_{01}$ is the cross section of charge exchange in the ground state of the hydrogen atom; $\lambda = 0.7$, $\omega_{21} = 0.375$ for charge exchange in the 2p-state of the hydrogen atom.

The results of calculating the total cross section of charge exchange in the 2p-state of hydrogen for collision of a proton with He and Ne atoms are given in Fig. 2. It is evident that the method developed leads to satisfactory agreement with the experimental data cited in [10, 12].

In the case of charge exchange of protons on He the basic contribution to the cross section in the 6-40 keV range is made by charge exchange via the intermediate 1s level of the H atom, while direct charge exchange plays no role in the range indicated. In the case of the Ne

atom the maximum at the energy ~ 8 keV owes its origin to charge exchange via the intermediate 1s state, and the next maximum at the energy ~ 14-18 keV is caused by direct charge exchange. At energies below 10 keV direct charge exchange plays practically no role, and at
energies above 18 keV the contribution from charge exchange via the intermediate state is in
turn very small.

It is not possible to write a simple approximate equation of the type (3.21) for the cross
section $\sigma_{0 \to 1 \to 2}$ in the case of charge exchange in the 2s-state of hydrogen. We merely note the
basic qualitative features of such a cross section. The cross section for charge exchange via
an intermediate level in this case is similarly proportional to the cross section $\sigma_{01}$ of charge
exchange in the ground state, but the corresponding maximum of the cross section is reached
at higher energies. For collision energies $< 25\,\mathrm{KeV}, \sigma(2s) < \sigma(2p)$. The singularities indicated
occur due to the fact that the matrix element of the $1s - 2s$ transition has an exponential character rather than the dipole character associated with the $1s - 2p$ transition.

Thus, the competing transition via the intermediate level can lead to the appearance of an
additional maximum in the cross section for charge exchange in the excited state at intermediate energies. At higher collision energies (of the order of 100 keV and above) the effects
examined here, which consist of strong coupling between the initial, intermediate, and final
states, cease to play a role. At high collision energies we may use the first Born approximation to calculate the charge exchange cross section.

In conclusion let us note that a transition via an intermediate level can also play a substantial role in other problems in the theory of atomic collisions.

We express our sincere thanks to I. I. Sobel'man for his useful discussions.

## References

1. D. R. Bates and R. McCarroll, Adv. in Physics, 11:39 (1962).
2. Yu. N. Demkov, Zh. Éksperim. i Teor. Fiz., 38:1879 (1960).
3. Yu. N. Demkov, Zh. Éksperim. i Teor. Fiz., 45:195 (1963).
4. I. Poluéktov, L. Presnyakov, and I. Sobel'man, Zh. Éksperim. i Teor. Fiz., 47:181 (1964).
5. L. Vainshtein, L. Presnyakov, and I. Sobel'man, Zh. Éksperim. i Teor. Fiz., 43:518 (1962).
6. L. Presnyakov, Tr. Fiz. Inst. Akad. Nauk SSSR, 30:235 (1964).
7. L. D. Landau and E. M. Lifshits, Quantum Mechanics, Fizmatgiz (1960).
8. N. Rosen and C. Zener, Phys. Rev., 40:502 (1932).
9. Yu. E. Murakhver, Zh. Éksperim. i Teor. Fiz., 40:1080 (1961).
10. D. J. Jaecles, Van Zyl, and R. Geballe, Phys. Rev., 137:A340 (1965).
11. E. J. de Heer, J. Van Eck, and J. Kistemarker, Proceedings of the Sixth International Conference on Ionization Phenomena in Gases, Paris (1963).
12. E. P. Andreev, V. A. Ankudinov, and S. V. Babashov, Zh. Éksperim. i Teor. Fiz., 50:585 (1966).
13. I. A. Poluéktov and L. P. Presnyakov, Preprint of the Physics Institute, Academy of Sciences of the USSR, No. A-144 (1965).

# BROADENING OF SPECTRAL LINES
# OF NONHYDROGENIC IONS

## V. A. Alekseev and E. A. Yukov

**1.** In the theory of broadening of spectral lines of neutral atoms it is usually assumed that the perturbing particles move along straight classical trajectories [1-3]. For neutral atoms such an approximation is justified, since the calculation in this case leads to good agreement (±20%) with experiment. An analogous calculation of the width of the spectral lines of ions leads to highly underestimated values of the widths. It is obvious that in this case the effect of curvature of the trajectories of the perturbing electrons in the Coulomb field of the ions may play a substantial role. This is substantiated by the results of [4] in which the curvature of the trajectories of the perturbing electrons is considered for the case of hydrogenic ions.

The present paper calculates the quasi-classical scattering phases with allowance for the curvature of the trajectories of the perturbing electrons. The width and shift may easily be calculated for any spectral line of a nonhydrogenic ion using the compiled phase tables.

A specific calculation was carried out for the lines of ArII. Under these conditions we started from Eqs. (1) and (2) of the nonstationary theory of spectral line broadening, which have been given below.

**2.** Within the framework of the impact approximation the spectral line component $n \to k$ has a dispersion form, the width $\gamma$ and shift $\Delta$ being equal to [1]

$$\gamma = 2Nv2\pi \int_0^\infty \rho d\rho \, [1 - e^{\Gamma(\rho)} \cos \eta(\rho)], \tag{1}$$

$$\Delta = Nv2\pi \int_0^\infty \rho d\rho e^{-\Gamma(\rho)} \sin \eta(\rho), \tag{2}$$

where

$$\Gamma(\rho) = \Gamma_n(\rho) + \Gamma_k(\rho); \, \eta(\rho) = \eta_n(\rho) - \eta_k(\rho); \tag{3}$$

$$\Gamma_j = \frac{1}{2\hbar^2} \sum_s{}' \left| \int_{-\infty}^{+\infty} \langle j | V(t) | s \rangle e^{i\omega_{js}t} dt \right|^2, \qquad j = n, k; \tag{4}$$

$$\eta_j = \text{Im} \; \frac{1}{\hbar^2} \sum_s{}' \int\limits_{-\infty}^{+\infty} \langle j \,|\, V(t) \,|\, s \rangle \, e^{i\omega_{js}t} \, dt \int\limits_{-\infty}^{t} \langle s \,|\, V(t') \,|\, j \rangle \, e^{-i\omega_{js}t'} dt', \qquad j = n, k. \tag{5}$$

Here $\rho$ is the impact parameter; v is the velocity of the perturbing electron, and $\omega_{js}$ is the frequency of the transition. In (4) and (5) the summation is carried out over all states of the ions. Let us restrict our analysis to the dipole approximation; then, using a coordinate system which is stationary in space, we write the interaction V(t) in the form

$$V(t) = -\,e \, \frac{\mathbf{d}\,\mathbf{r}(t)}{|r(t)|^3}, \tag{6}$$

where e is the electron charge; $\mathbf{d}$ is the dipole moment of the ion; $\mathbf{r}(t)$ is the coordinate of the perturbing electron.

In order to obtain the resulting contour of the entire line, it is necessary to add the individual M → M' components of the lines which are broadened in accordance with (1) and (2). Within the limits of the accuracy which it is generally possible to count on within the framework of the approximations being used, it should be assumed that such a summation yields the dispersion contour, the width $\gamma$ and shift $\Delta$ of the resultant contour being obtained by averaging Eqs. (1) and (2) over M and M'.

The main contribution to the integrals (1) and (2) is made by collisions having large values of the impact parameter $\rho$. For such solutions it is possible to make use of an approximation which is linear in $\Gamma$ and $\eta$:

$$1 - e^{-\Gamma} \cos \eta \approx \Gamma, \qquad e^{-\Gamma} \sin \eta \approx \eta.$$

This means that the width $\gamma$ and the shift $\Delta$ of this resulting contour can be obtained with good accuracy by substituting the values of $\Gamma$ and $\eta$ averaged over M and M' into Eqs. (1) and (2). Let us average (4) and (5) over M and M', as well as over all possible alignments of the vectors $\rho$ and v and let us make use of the parametric dependence of the coordinates of the perturbing electrons on time in the Coulomb field of the ions:

$$
\begin{aligned}
x &= a\,(\varepsilon - \text{ch}\,u), & y &= a\,\sqrt{\varepsilon^2 - 1}\;\text{sh}\,u, \\
r &= a\,(\varepsilon\,\text{ch}\,u - 1), & t &= \sqrt{\frac{ma^3}{Ze^2}}\,(\varepsilon\,\text{sh}\,u - u), \\
a &= \frac{Ze^2}{mv^2}, & \varepsilon &= \sqrt{1 + \frac{\rho^2}{a^2}}
\end{aligned}
$$

(Z is the ion charge). To sum up, we obtain

$$\Gamma_j = \sum_s{}' 2 \, \frac{\beta_s^2}{b_s^2} \, A(z_s, \chi_s), \tag{7}$$

$$\eta_j = \sum_s{}' 2 \, \frac{\beta_s^2}{b_s^2} \, B(z_s, \chi_s) \, \frac{\Delta E_{js}}{|\Delta E_{js}|}, \tag{8}$$

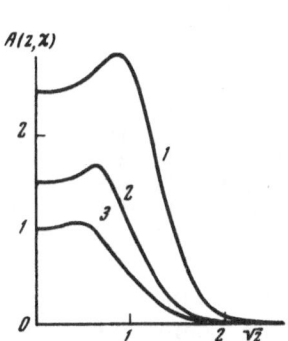

Fig. 1. Dependence of the function $A(z, \chi) = b^2 \dfrac{\Gamma(z, \chi)}{2\beta^2}$ on $\sqrt{z}$ for $\chi = 1.0$ (1); 0.25 (2), and 0 (3).

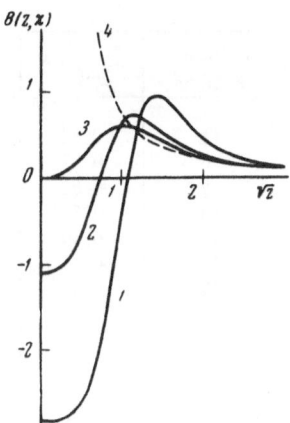

Fig. 2. Dependence of the function $B(z, \chi) = b^2 \dfrac{\eta(z, \chi)}{2\beta^2}$ on $\sqrt{z}$ for $\chi = 1.0$ (1); 0.25 (2), and 0 (3). Curve 4 corresponds to the asymptotic behavior $B(z, \chi) \approx \left(\dfrac{\pi}{4z}\right)$ for $z \to \infty$.

where

$$B(z, \chi) - iA(z, \chi) = -\frac{i}{2} b^2 \int\limits_{-\infty}^{+\infty} du \int\limits_{-\infty}^{u} du' \times$$

$$\times \frac{(\chi \operatorname{ch} u - b)(\chi \operatorname{ch} u' - b) + z^2 \operatorname{sh} u \operatorname{sh} u'}{(b \operatorname{ch} u - \chi)^2 (b \operatorname{ch} u' - \chi)^2} \exp\left[i(b \operatorname{sh} u - \chi u) - i(b \operatorname{sh} u' - \chi u')\right], \tag{9}$$

$$\beta_s^2 = \frac{f_{js}}{2} \frac{\omega_{js} e^4}{\hbar m v^4}, \qquad \chi_s = \frac{Z e^2 \omega_{js}}{m v^3}, \qquad z_s = \frac{\rho \omega_{js}}{v}, \qquad b_s^2 = z_s^2 + \chi_s^2, \tag{10}$$

m is the electron mass, and $f_{js}$ is the oscillator strength for the $j \to s$ transition.

In the case of neutral atoms $Z = 0$ (consequently, $\chi_s = 0$), and Eqs. (7)–(9) go over into the conventional formulas derived on the assumption of straight trajectories of the perturbing electrons. Analysis shows that for $\chi_s \neq 0$, $\Gamma_j$ and $\eta_j$ remain finite for $z \to 0$ whereas in the case of a straight trajectory ($\chi_s = 0$) the quantities $\Gamma_j$ and $\eta_j$ increase without limit for $z \to 0$. (The dependence of the functions $A = b^2 \dfrac{\Gamma(z, \chi)}{2\beta^2}$ and $B = b^2 \dfrac{\eta(z, \chi)}{2\beta^2}$ on $\sqrt{z}$ for various values of $\chi$ is displayed in Figs. 1 and 2.)

In order to find the quantities A and B it is necessary to carry out a numerical calculation. A comparison was made between the tables of the quantities $A(z, \chi)$ and $B(z, \chi)$ as functions of z for various values $0 \leq \chi \leq 10$, which exhausts practically all possible cases [5]. Using these tables for each specific case, the values of $\Gamma$ and $\eta$ are calculated according to Eqs. (7) and (8); then the width $\gamma$ and the shift $\Delta$ of the spectral lines considered are easily found from Eqs. (1) and (2).

**3.** A specific calculation was carried out for several lines of ArII. Averaging over the velocities of the perturbing electrons was not carried out. In Eqs. (10) the mean square values

TABLE 1

| Multiplet number | $\gamma_{exp}$, Å | $\gamma_1$, Å | $\gamma_2$, Å | $N_1 \cdot 10^{-17}$, cm$^{-3}$ | $N_2 \cdot 10^{-17}$, cm$^{-3}$ |
|---|---|---|---|---|---|
| 1 | 0.42 | 0.10 | 0.18 | 4.2 | 2.3 |
| 6 | 0.57 | 0.17 | 0.32 | 3.4 | 1.8 |
| 7 | 0.34 | 0.21 | 0.26 | 1.60 | 1.3 |
| 10 | 0.31 | 0.11 | 0.20 | 2.8 | 1.6 |
| 14 | 0.60 | — | 0.35 | — | 1.7 |
| 15 | 0.54 | 0.21 | 0.32 | 2.6 | 1.7 |
| 17 | 0.57 | 0.17 | 0.27 | 3.4 | 2.1 |
| 64 | 1.30 | 0.64 | 0.77 | 2.0 | 1.7 |
| 77 | 1.40 | 0.67 | 0.78 | 2.1 | 1.8 |

of the velocity of the perturbing electrons was used. In view of the weak dependence of the widths of the lines on temperature in the temperature range examined ($T_e \sim 3$ eV), such an approximation is fully justified.

The oscillator strengths of the transitions were calculated on an electronic computer by means of semiempirical wave functions [6]. The contribution of the ions to the width and the shift was calculated according to adiabatic theory. Broadening by electrons was calculated by two methods:

1) the trajectory of the perturbing electrons is assumed to be linear ($\chi_s = 0$);

2) the perturbing electron moves along a hyperbolic trajectory in the Coulomb field of the ion ($\chi_s \neq 0$). The results of calculations of the widths of ArII lines for an electron concentration $N_e = 10^{17}$ cm$^{-3}$ and a temperature $T_e = 3$ eV are given in Table 1. The average concentrations of the perturbing electrons, as obtained from a comparison of the calculated width $\gamma_1$ (straight trajectory) and $\gamma_2$ (hyperbolic trajectory) with experimental data, are respectively equal to $<N_1> = 2.8 \cdot 10^{17}$ cm$^{-3}$, $<N_2> = 1.8 \cdot 10^{17}$ cm$^{-3}$. As a consequence of compensation of the influence of various perturbing levels (the terms in (8) have different signs) the resultant shift of the ArII lines turns out to be much smaller than their widths. Under these conditions the magnitude of the calculated shift exceeds the limits of the accuracy to which calculations can generally be carried out within the framework of the approximation used. The 64-th and 77-th (according to the classification adopted in [7]) multiplets constitute an exception. In this case the upper level, which is common to both multiplets, is chiefly the one that broadens, and the perturbing levels, which yield the main contribution to the broadening, shift the line in the same direction. The shifts of the lines of the 64-th and 77-th multiplets practically coincide. Calculation by methods 1 and 2 respectively yields $\Delta_1 = 0.35$ Å, $\Delta_2 = 0.34$ Å ($N_e = 10^{17}$ cm$^{-3}$).

The corresponding values of the concentrations of the perturbing electrons are equal to $N_1 = 1.86 \cdot 10^{17}$ cm$^{-3}$, $N_2 = 1.9 \cdot 10^{17}$ cm$^{-3}$. Let us note that the slight difference between the shifts $\Delta_1$ and $\Delta_2$ obtained by methods 1 and 2 is not accidental. As a rule, the effect of curvature of the trajectory of the perturbing electron has considerably less effect on the shift of the line than on the line width. The values of concentration $N_1$ and $N_2$ obtained from the shift of the 64-th and 77-th multiplets are in good agreement with the average value of the concentrations $<N_2>$ (i.e., the concentration obtained by comparing the experimental values of the widths of the ArII lines with those calculated according to method 2).

Table 2 compares the widths of the ArII lines calculated according to method 2 with the calculations and experimental data cited in [4]. The widths calculated according to method 2 are in good agreement with both experiment and the calculation carried out in [4].

TABLE 2

| Multi-plet number | Tempera-ture, eV | 1 | 2 | 3 |
|---|---|---|---|---|
| | | $\gamma/2$ (exp.), Å | $\gamma/2$ (theor.) Å | $\gamma/2$ (theor.) Å |
| 6 | 1.1 | 0.15 | 0.13 | 0.17 |
| 6 | 1.55 | 0.19 | 0.13 | 0.17 |
| 6 | 2.7 | 0.21 | 0.13 | 0.16 |
| 7 | 1.55 | 0.17 | 0.15 | 0.13 |
| 7 | 2.7 | 0.17 | 0.12 | 0.13 |
| 10 | 1.55 | 0.11 | 0.12 | 0.10 |
| 14 | 1.7 | 0.17 | 0.16 | 0.18 |

Note. 1) Experimental widths of the ArII lines for an electron concentration $N_e = 10^{17}$ cm$^{-3}$ [4]; 2) the Griem calculation [4]; 3) widths of the lines obtained according to Eq. (1) with allowance of curvature of the trajectories of the perturbing electrons.

Thus, calculation shows that consideration of the effect of curvature of the trajectory of the perturbing electron leads to an increase of the width of ArII spectral lines by approximately a factor of 2, and yields good agreement with experiment. Similar results have recently been obtained in [8, 9].

We thank I. I. Sobel'man for his constant attention to our work.

References

1. I. I. Sobel'man, Introduction to the Theory of Atomic Spectra, Fizmatgiz (1963).
2. H. R. Griem, Plasma Spectroscopy, New York, McGraw-Hill Book Co. (1964).
3. Atomic and Molecular Processes, Izd. Mir (1964).
4. H. R. Griem, Phys. Rev. Letters, 17:509 (1966).
5. V. A. Alekseev and E. A. Yukov, Preprint of the Physics Institute, Academy of Sciences of the USSR, No. 87 (1968).
6. L. A. Vainshtein, Opt. i Spektr., 3:313 (1957).
7. M. A. Mazing, Tr. Fiz. Inst. Akad. Nauk SSSR, 15:55 (1961).
8. S. Brechot, Phys. Letters, 24A:476 (1967).
9. F. Feautrier, S. Brechot, and H. Van Regemorter, Fifth International Conference on Electron Physics and Atomic Collisions, Leningrad (1967), p. 525.

# ON THE PROBLEM OF THE INTERACTION OF
# EXCITED ATOMS OF ALKALI METALS WITH INERT GASES

## P. D. Serapinas

A large number of papers have been devoted to investigating the effect of spurious gases on the spectra of alkali metals. The basic results obtained before the 1960's were expounded in a review by Chen and Takeo [1]. The literature has described the investigation of the first doublets of the principal series (right up to spurious gas pressures of the order of hundreds of relative density units) and of the high terms of the series corresponding to transitions to levels having principal quantum numbers $n \sim 10$-20. In the intermediate range there are comparatively few measurements. Moreover, the results available in the literature were obtained at fairly high spurious gas pressures (relative density > 1), where the character of the interaction becomes complicated. Therefore it was decided to carry out the measurements of the broadening and shift of the lines of an alkali metal at a comparatively low spurious gas pressure for all of the terms in the series for the purpose of attempting a qualitative description of the character of the interaction for various n. It is well known that for the very first terms of the series the concept of van der Waals' interaction [1] is valid in the case of a series of perturbing gases; the behavior of the lines having very large quantum numbers is described by the mechanism advanced by Fermi [2, 3]. A transition from one type of interaction to another affects the width and the shift of the lines differently, and even a qualitative examination of the phenomenon is of interest.

Measurements were carried out of the width $\gamma$ and shift $\Delta$ of the absorption lines of the principal Cs series at an argon pressure of 190 mm Hg ($N = 6.7 \cdot 10^{18}$ cm$^{-3}$), beginning with the second doublet (n = 7) right up to $n \sim 40$, including n = 7-15 for both components of the fine structure. The literature describes investigations of the effect of argon on the width and shift of cesium lines only for the first terms of the series: in [4] the lines with n = 7-9 were investigated, and in [5] lines with n = 6.7 were investigated. Measurements of the shift were carried out for the high terms of the series in [3].

### Procedure and Results of the Measurements

Description of the Apparatus. The measurements were carried out in a sealed tube of molybdenum glass having thick-walled windows. The length of the tube was 300 mm. The tube had a branch whose temperature (the temperature of the tube itself was maintained 20 deg higher) determined the concentration of the cesium atoms in the absorptive volume. It was chosen in such a way that the absorption coefficient at the maximum of the line would not exceed unity — as a rule, it had a value near 0.5. The concentration of argon atoms was set before sealing the tube.

The light source consisted of a xenon GSVD-120 high-pressure lamp (the boundary of the principal absorption series of Cs corresponds to a wavelength $\lambda = 3183$ Å). A DFS-3 diffraction spectrograph was used for photographic recording of the spectrum. The spectrograph dispersion in the second order of the 1200 lines/mm grating amounted to 0.85-1.00 Å/mm in the spectral range studied. In order to monitor the absence of equipment shift of the spectrum (due to temperature change or other causes), the spectrum of a discharge in a hollow cathode was superimposed on the cesium spectrum before and after each photographing. The photographs were measured on an MF-4 microphotometer. The table of an IZA-2 comparator was built into it, and the accuracy of the table-displacement reading was 0.001 mm.

Measurement of the Width of the Line. In view of the fact that at low pressures of the perturbing gas the lines are narrow, special attention was devoted to considering all other effects leading to broadening. This consideration is substantially complicated by the presence of a hyperfine structure of the cesium lines. Fundamentally, the hyperfine structure is determined by the hyperfine splitting of the ground $6^2S_{1/2}$ level into two components with a distance 0.31 cm$^{-1}$ [6, 7]. The splitting of the $6^2P$ levels is approximately one order smaller, while that of the $7^2P$ levels is several times smaller yet [7].

The measured intensity distribution of an individual line of the fine structure can be represented as the convolution of two distributions – the apparatus function and the exponential which describes the contour of the absorption line investigated:

$$I(\nu) = \int A(\nu - x) e^{-\varkappa(x)_c} dx, \tag{1a}$$

where

$$\varkappa(x)_c = \int T_D(x_1) \varkappa_1(x - x_1) dx_1 + \int T_D(x_2) \varkappa_2(x - x_2) dx_2 \tag{1b}$$

represents the resultant absorption coefficient of the tube hyperfine structure components $\varkappa_1$ and $\varkappa_2$ which have been broadened due to the Doppler effect; A is the apparatus function of the instrument.

The apparatus function of the instrument is obtained in the second order of the spectrum from the absorption lines of pure cesium (t = 260°C) whose proper width was measured by means of a Fabry-Perot interferometer crossed with the spectrograph. It was found that $\gamma_{eq.} = (0.021 \pm 0.003)$ Å, which within the limits of the measurement error coincides with the results obtained from the emission lines of the discharge in the hollow cathode.

From the statistical weights of the hyperfine structure components (F = 4 and 3) it follows that the ratio between their intensities is equal to 9:7. This was verified by measurements with the separated component in pure cesium. In the subsequent analysis it was assumed that the contour of the absorption coefficients of an individual hyperfine structure component is the convolution of the dispersion and Gauss distributions, and has an identical form for both components (the analysis of asymmetrical contours will be discussed later). In such a case the contour of the individual hyperfine structure component is described by a Voigt integral. In adding two such contours having different original widths and an intensity ratio 9:7 at a distance 0.31 cm$^{-1}$ it was clarified that for widths $\gamma_c > 1$ cm$^{-1}$ the resultant contour is similarly of the Voigt type with fair accuracy. In accordance with Eqs. (1a) and (b) the distributions I($\nu$) were calculated for various values of the resultant absorption coefficient and various Voigt parameters of its distribution. The error introduced by deviation from the Voigt form of the contour was similarly considered. Making use of the calculated distributions, it should be possible

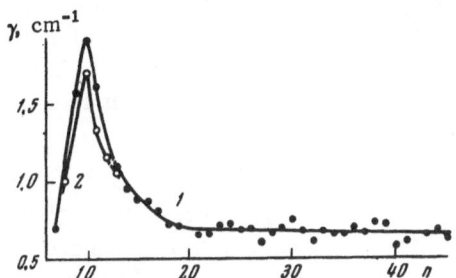

Fig. 1. Dependence of the width of Cs lines ($N_{Ar} = 6.7 \cdot 10^{18}$ cm$^{-3}$) on the principal quantum number.

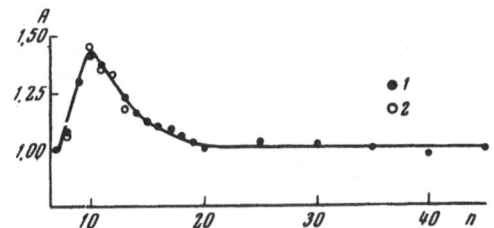

Fig. 2. Dependence of the asymmetry of Cs lines ($N_{Ar} = 6.7 \cdot 10^{18}$ cm$^{-3}$) on the principal quantum number. 1) The $6^2S_{1/2} - n^2P_{3/2}$ transition; 2) the $6^2S_{1/2} - n^2P_{1/2}$ transition.

to determine the widths of the individual hyperfine structure components from two parameters of the observed contour $I(\nu)$: its width and the magnitude of the absorption at the maximum.

In order to check the correctness of the procedure indicated we analyzed experimental contours $I(\nu)$ for one and the same line for different values of the absorption. The results were in good agreement. Moreover, when Eq. (1) was used for the inverse analysis of the final contours $\varkappa_1$ and $\varkappa_2$ obtained by this method, contours $I(\nu)$ resulted which coincided with the observed contours in both the width and the Voigt parameter, even though the latter was not used in the analysis.

The errors introduced when all of the factors indicated are considered do not exceed several percent, and the error of the result is caused by the scatter of individual measurements; the magnitude of this scatter reaches 10% of the average.

The method described above was used to analyze lines having $n > 12$. Lines corresponding to smaller n have a considerable asymmetry, and the true contours are not Voigt contours even above half intensity. But the widths of these lines on the energy scale, and the more so on the wavelength scale, increase (with decreasing n the wavelength of the transition increases noticeably), and the relative role played by the apparatus function drops: $\gamma_{eq.} < 0.1 \, \gamma_{\varkappa}$. Therefore, in analyzing these lines the apparatus function, in view of its smallness and Gaussian n nature, was neglected. Then, as can easily be seen from (1), $\log I(\nu)$ immediately gives the resultant contour of the absorption coefficient $\varkappa_i$. The contour was graphically divided into components under the conditions

$$\varkappa(x)_c = \varkappa_1(x) + \varkappa_2(x), \qquad \varkappa_1(x) = {}^7/_9 \varkappa_2(x - \delta), \tag{2}$$

where $\delta$ is the magnitude of the hyperfine splitting.

The results of the measurements are displayed in Fig. 1. Curve 1 corresponds to the $6^2S_{1/2} - n^2P_{3/2}$ transitions, while curve 2 corresponds to the $6^2S_{1/2} - n^2P_{1/2}$ transitions. In order to obtain the required degree of absorption for each line the spectra were taken at various cesium atom densities which corresponded to various temperatures of the branch tube (140-300°C). All of the results were reduced to the temperature 180°C from the dependence of the line width on the density of the intrinsic Cs vapors. The remaining contribution to the width, which is caused by the interaction of the Cs atoms themselves, was estimated not to exceed several percent.

No measurements were carried out for $n > 45$ since the superposition of the wings of neighboring lines can lead to a noticeable systematic error in the measured width.

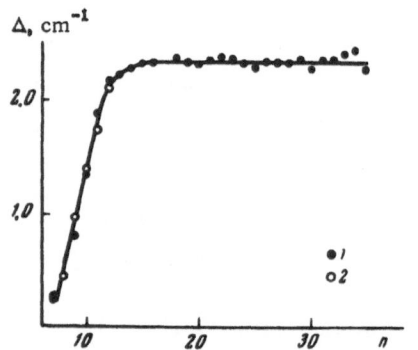

Fig. 3. Dependence of the shift of the Cs lines ($N_{Ar} = 6.7 \cdot 10^{18}$ cm$^{-3}$) on the principal quantum number. 1) The $6^2S_{1/2} - n^2P_{3/2}$ transition; 2) the $6^2S_{1/2} - n^2P_{1/2}$ transition.

The Asymmetry. Figure 2 shows the variation of the asymmetry of the Cs lines as a function of the principal quantum number n. The ratio between the red and blue parts of the width was taken as a measure of the asymmetry.

The Shift. In order to measure the shift the spectra were taken in the first order of the 1200 lines/mm grating, where the dispersion of the instrument is 2 Å/mm. The lines of the spectrum of the discharge in the hollow cathode served as references. The shift of the lines was measured relative to the positions of the lines at low cesium pressure in the absence of argon (see [3]). The absolute magnitudes of the wavelengths of cesium in argon were also measured; these were then compared with data obtained in pure cesium at low pressure [6]. In both cases identical (within the limits of the measurement error) values of shift were obtained. The results are displayed in Fig. 3. The shift caused by the intrinsic cesium pressure (based on the data given in [3]) was subtracted from the measured values of the shift.

For n > 10 the results obtained correspond to the results of [3], with the exception of the data for the intense component of the doublet for n = 10-15. The displacement of the two components of the doublet is identical within the limits of the measurement errors. In [3] the shift of the components of $^2P_{3/2}$ for n = 10-15 is overestimated in view of the fact that the measurements were carried out for an excessively intense absorption, and the position of the center of gravity was of the absorption lines [3a] registered rather than the position of the maximum.

Discussion of Results

**1.** All of the lines investigated in this paper can be split into three parts according to the character of the dependence of their width and shift on the principal quantum number n. For the first terms (n = 7-9) the width, shift, and asymmetry increases with an increase in the principal quantum number; for n = 10-17 the width and asymmetry decreases, while the shift continues to increase. Under these conditions all of the quantities indicated approach the values characteristic of large n, where they already change little or not at all as the principal quantum number changes. Below we shall consider each of these ranges and shall strive to provide a physical substantiation of such a subdivision.

Let us begin with the range of small n. In [2] a complete quantitative description is given of the effects of broadening and shift of the lines having large quantum numbers on the basis of the Fermi mechanism. The essence of the latter consists in the fact that when an atom is strongly excited the interactions of the perturbing atoms with the valence electron removed from the atomic core and with the core proper can be examined independently. The former reduces to elastic scattering of quasi-free electrons by the perturbing atom, while the latter reduces to polarization of particles by the field of the atomic core. The range of quantum numbers n in which the scattering and polarization effects are statistically independent shall be called the range of large n.

In broadening processes at low particle densities the effective interaction radius is the Weisskopf radius. Its magnitude $\rho_e$ for the interaction of the perturbing atoms with an electron is considerably less than the radius $\rho_i$ of the interaction of these atoms with the core. Therefore, statistical independence occurs if the largest of these radii is small compared with the orbital radius of the electron:

$$\rho_i = \left(\frac{\pi}{4}\frac{\alpha e^2}{\hbar V}\right) \ll a_0 n^2, \tag{3}$$

where $\alpha$ is the polarizability of the perturbing atom; V is the relative velocity. When the concentration of perturbing particles N is high $(\chi_i = \rho_i^3 N \gg 1)$, the impact approximation is inapplicable, while the broadening and shift are predominantly determined by atoms situated at a distance $\bar{R} = (^4/_3 \pi N)^{-1/3}$. Thus, in the case of high densities the condition of statistical independence is

$$\chi_i \gg 1, \qquad \bar{R} \ll a_0 n^2. \tag{4}$$

The experimental results given above were obtained at low pressure, where $\bar{R} > \rho_i$ and the statistical independence is determined by the inequality (3). Moreover, the conditions for the applicability of the impact approximation [2] for interaction with both an electron and the atomic core are satisfied:

$$\chi_e = \frac{\pi}{4}\alpha n N \ll 1, \qquad \chi_i = \frac{\pi}{4}\frac{\alpha e^2}{\hbar V} N \ll 1. \tag{5}$$

In this case the broadening $\gamma$ and the shift $\Delta$ are additive as a consequence of scattering (the subscript "s") and polarization (the subscript "p") and are determined by the equations

$$\gamma_s = N\frac{\hbar}{m}\int\left[\frac{4\pi}{q}\sum_l (2l+1)\sin^2\delta_l\right] W(q)\,dq = N\frac{\hbar}{m}\int q\sigma(q)\,W(q)\,dq. \tag{6a}$$

$$\gamma_p = 11.4\left(\frac{\alpha e^2}{2\hbar}\right)^{2/5} V^{1/5} N, \tag{6b}$$

$$\Delta_s = N\frac{\hbar}{m}\int\left[\frac{\pi}{q}\sum_l (2l+1)\sin 2\delta_l\right] W(q)\,dq, \tag{7a}$$

$$\Delta_p = -\frac{\sqrt{3}}{2}\gamma_p, \tag{7b}$$

where $W(q)$ is the electron velocity distribution function in the state n; $\delta_l$ is the phase; $\sigma(q)$ is the effective cross section for elastic scattering of an electron having the momentum $\hbar q$ by a perturbing atom. The assumption concerning the elastic character of the scattering is valid up to n of the order of 50. For larger n inelastic transitions due to the action of perturbing atoms become possible.

The discussion of the variation of the shift in the range of large n and its comparison with the theoretical calculation were carried out in [2, 3] where good agreement was established between theory and experiment. Here we shall dwell only on the width of the line. In Fig. 4 the straight line 4 corresponds to the polarization part of the width calculated for $\alpha_{Ar} = 1.65\,\text{Å}^3$, while lines 5 and 6 correspond to the total width. The width caused by scattering was calculated in [3] using Slater atomic wave functions for argon. The parameter $\beta$ of the function had the same value as it did for the calculation of the shift: curve 5 is for $\beta = 0.917$ and curve 6 is for $\beta = 0.900$.

As is evident from the figure, the main contribution to the width is made by the polarization effect. From the good agreement between the calculated and measured (curve 1) quantities

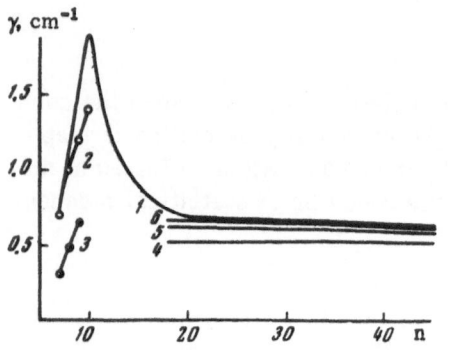

Fig. 4. Comparison of the experimental dependence of the width of the $6^2S_{1/2} - n^2P_{2/3}$ lines on n with the theoretical calculation. 1) Experimental curve; 2) doubled blue portion of the width; 3) calculated width in the van der Waals range; 4) polarization contribution to the width for large n; 5, 6) total width from [2]: $\beta = 0.917$ (5); $\beta = 0.900$ (6).

it follows that if the polarizability of the perturbing gas is known, the polarization contribution to the width and shift of the lines can be calculated reliably.

Thus, for large values of the principal quantum number both the width and the shift of the lines are described qualitatively and quantitatively by the Fermi mechanism.

Let us go over to an examination of the range of large n (i.e., the first terms of the series). From the empirical relationships established in experiments [1] it evidently follows that the interaction of a weakly excited cesium atom with an argon atom may be described by means of the van der Waals potential of the form $C_6/r^6$. Note that no criterion exists that allows one to determine with complete certainty what potential should be used to describe interaction with which of the excited alkali metal atoms. Some approximation can be chosen either in the presence of a complete quantum-mechanical calculation of the interaction energy at various distances, or on the basis of empirical facts.

Let us trace the extent to which the experimental results presented above correspond to the case of van der Waals interaction (i.e.. to a potential of the form $C_6/r^6$).

The experimental ratio of the doubled blue portion of the width to the shift for n = 7 is 3.0, and for n = 8 it is equal to 2.2. The theoretical value for this type of interaction in the impact approximation is equal to 2.8.

The width of the fine structure component $6^2S_{1/2} - n^2P_{3/2}$ is 1.05 -1.20 times as large as the width of $6^2S_{1/2} - n^2P_{1/2}$ which within the limits of the error is in agreement with the theoretical value of 1.12 obtained by Foley [1] while considering the splitting of the state $I = \frac{3}{2}$ into $|m| = \frac{3}{2}$ and $\frac{1}{2}$ for interaction of the type $r^{-6}$. For the same reasons there should be no difference in the shift of the components, which is in agreement with Fig. 3. (Since both components behave similarly, we shall hereafter speak solely of the more intense component $6^2S_{1/2} - n^2P_{3/2}$).

The absolute values of the width and shift of the lines in the case of van der Waals interaction in the impact approximation are equal to

$$\gamma = \frac{8.16}{2\pi c} C_6^{2/5} \, V^{3/5} \, N \ (\text{cm}^{-1}), \quad \Delta = -\frac{\gamma}{2.8} \ . \tag{8}$$

No data are available on the van der Waals interaction constants $C_6$ for the case in which one of the atoms is excited. However, investigations of the Stark effect exist according to which it is possible to estimate the polarizability of an atom in the excited state. Therefore, in order to calculate $C_6$ we made use of the approximate Slater – Kirkwood equations [8]

$$C_6 = \frac{3e\hbar}{2 \, (m_e)^{1/2}} \frac{\alpha_1 \alpha_2}{\left(\frac{\alpha_1}{N_1}\right)^{1/2} + \left(\frac{\alpha_2}{N_2}\right)^{1/2}} \ , \tag{9}$$

TABLE 1

| Excited state | $\alpha_1$, Å$^3$ | | $C_6 \cdot 10^{30}$, sec$^{-1}$ · cm$^{-6}$ | $\chi_{at}$ |
|---|---|---|---|---|
| | from [10, 11] | from [9] | | |
| $6p_{1/2}$ | $1.87 \cdot 10^2$ | $1.93 \cdot 10^2$ | | |
| $6p_{3/2}$ | $(1.96-2.73) \cdot 10^2$ | $2.32 \cdot 10^2$ | 0.58 | 0.08 |
| $7p_{1/2}$ | $4.15 \cdot 10^3$ | $4.30 \cdot 10^3$ | | |
| $7p_{3/2}$ | $5,34 \cdot 10^3$ | $5.37 \cdot 10^3$ | 2.88 | 0.21 |
| $8p_{3/2}$ | | $4.14 \cdot 10^4$ | 8.00 | 0.42 |
| $9p_{3/2}$ | | $1.91 \cdot 10^5$ | 17.2 | 0.67 |
| $10p_{3/2}$ | | $6.62 \cdot 10^5$ | 32.0 | 0.91 |
| $11p_{3/2}$ | | $1.85 \cdot 10^6$ | 53.5 | 1.25 |

where $\alpha_1$, $\alpha_2$ and $N_1$, $N_2$ are the polarizabilities and numbers of outer electrons of the atoms investigated, respectively. The polarizability of argon, as indicated above, is $\alpha_2 = 1.65$ Å$^3$. The polarizability of cesium in the excited state $\alpha_1$, however, was calculated from the theoretical values of the shift $\Delta\nu$ (in cm$^{-1}$) [9] of the corresponding level in a static electric field E:

$$\alpha_1 = \frac{2hc\Delta\nu}{E^2} \ .$$

The values of $\Delta\nu$ used are in good agreement with the results of experimental investigations of the Stark effect of the $6^2P_{1/2, 3/2}$ levels [10] and of the $7^2P_{1/2, 3/2} - 6^2S_{1/2}$ transitions [11]; this is evident from Table 1, where $\alpha_1$ is given as calculated from experimental and theoretical $\Delta\nu$.

The values obtained for $C_6$ are given in Table 1, and the values of width calculated from them are given in Fig. 4 (curve 3). The experimental curve 1 ($\gamma(n)$ for the $6^2S_{1/2} - n^2P_{3/2}$ component) has been transferred from Fig. 1. In the case of asymmetrical lines the blue part of the width was isolated for purposes of comparison with calculations (impact approximation) for n = 7-10. The variation of double its magnitude is displayed by curve 2. The calculated values vary as a function of n in accordance with the experimental curve, but their magnitude is approximately twice as small.

Table 1 gives the parameter $\chi_{at} = \frac{4}{3}\pi\rho_{at}^3 N$, which is the criterion of applicability of the impact theory ($\chi_{at} \ll 1$). In this expression $\rho_{at} = \left(\frac{3\pi}{8}\frac{C_6}{V}\right)^{1/5}$ is the Weisskopf radius for the interaction of atoms. Based on the fact that calculation yields the width ($\gamma \sim C_6^{1/5}$), which is underestimated by a factor of 2, the value of $C_6^{1/5}$ was increased by a factor of 1.4 in order to estimate $\chi_{at}$. As is evident from the table, for n = 7 the value of $\chi_{at}$ is substantially less than unity, while it increases rapidly with increasing n. This must lead to the appearance of a statistical wing of the line (i.e., to an increase of the asymmetry), which is what was observed in the experiment (Fig. 2).

For a further increase of the polarizability (and therefore of the effective interaction radius) we have $\chi_{at} \gtrsim 1$, with increasing n, and even the center of the line cannot be described by the impact approximation [12]. Several perturbing atoms participate simultaneously in the interaction. Since the condition $\Delta\omega \gg \Omega = V^{4/5}\overline{C}_6^{1/5}$ for a quasi-static process is satisfied only with a slight margin, the statistical theory also does not describe the contour of the line and is applicable only to the far wing. The general case of interaction of atoms was examined in the paper by Anderson and Tolman (see [1]) without using the impact or statistical approximations. Although the analytical expressions for the entire contour could not be obtained, the asymptotic expressions which were given show that for increasing $\chi_{at}$ the intensity at the maximum of the line must drop (in view of the presence of a term with a negative real part proportional to $\chi_{at}$ in the exponent of the exponential), while the line broadens. As is evident from Fig. 4, the width in this range of n actually does fall rapidly.

On the other hand, for n = 10 (where $\chi_{at} \approx 1$) the radius of the atom already becomes larger than the Weisskopf radius for the interaction of perturbing particles with an electron or with the atomic core, $a_0 n^{*2} > \rho_i \gg \rho_e$, and collisions of the Fermi type already take place. With increasing n their role increases rapidly. Then the width, asymmetry and shift must approach the limiting values characteristic of large n, where $\gamma \ll \Delta$; this corresponds to the experimental data.

Thus, already beginning with n for which $a_0 n^{*2} \gtrsim \rho_i$, interaction of the Fermi type turns out to be substantial and determines the qualitative variation of the contour as a function of n. It is natural that collisions of this kind occur even for smaller n, which leads to a reduction of the ratio $\gamma/\Delta$ in the van der Waals range as well (n = 7-9).

The following description of the interaction derives from the discussion carried out above. At low argon pressures and, probably, for heavier inert gases the effects of the broadening and shift of the spectral lines of weakly excited alkali atoms can be described satisfactorily by means of the van der Waals potential in the impact approximation. The presence of a limiting value of n for this type of interaction is connected with the beginning of the transition to the Fermi mechanism, which is determined by the condition

$$\rho_i \lesssim a_0 n^2. \tag{10}$$

The range of applicability of the Fermi mechanism (the range of large n) is determined by the condition (3).

At higher densities the Fermi range can begin at smaller n in accordance with the condition (4). In this case the end of the impact van der Waals range may be connected either with the beginning of the transition into the Fermi range for

$$\chi_i \gg 1, \quad \bar{R} \approx a_0 n^2, \tag{11}$$

or with the increase in the influence of many-particle collisions for

$$\chi_{at} \approx 1. \tag{12}$$

In the case of the interaction of excited atoms of alkali metals with heavy inert gases it is usually true that $\rho_{at} > \rho_i$, and the condition (12) turns out to be more rigorous. Then for a definite n many-particle collisions of the van der Waals type will predominate in a certain pressure range; for lower pressures we have the impact van der Waals range, and for higher pressures we have the transition to the Fermi range. Such a complex variation of the character of the interaction for weakly excited atoms as a function of the density of the perturbing particles can lead to a nonmonotonic variation of the characteristics of the line contours. The characteristics of the lines corresponding to those n for which condition (3) is satisfied must vary monotonically with a variation of the density of the perturbing particles.

Let us note that in conditions (3), (4), (10), and (11) $a_0 n^2$ denotes the orbital radius of the valence electrons. For small n it should be replaced by n* (the effective principal quantum

TABLE 2

| Excited state $(J = \frac{3}{2})$ | $C_6^{2/5}$ $10^{19}$, sec$^{-2/5}$· cm$^{-12/5}$ | $\Delta$, cm$^{-1}$/r.d. | | $N_I$ (r.d.) | | $N_{II}$ (r.d.) | |
|---|---|---|---|---|---|---|---|
| | | Theory (8) | Experiment [13] * | Theory | Experiment [13] | Theory | Experiment [13] |
| 6 $p$ | 1.84 | 0.4 | 0.21(32°) | 1.8 | 2.8 | — | — |
| 7 $p$ | 3.53 | 0.8 | 0.62(80°) | 0.8 | 1.0 | 37 | 5.5 |
| 8 $p$ | 5.3 | 1.3 | 1.30(117°) | 0.4 | 0.65 | 7.5 | 1.6 |
| 9 $p$ | 7.2 | 1.8 | 2.9(144°) | 0.3 | 0.3 | 2.5 | 0.5 |
| 10 $p$ | 9.2 | | 5.0(180°) | 0.2 | 0.15 | 1.0 | 0.2 |
| 11 $p$ | 11.4 | | | | | $\rho_i < a_0 n^{*2}$ | 0 |

*Temperature in °C.

number), although in this case $a_0 n^{*2}$ is also a very approximate characteristic of the most probable distance of the electrons from the nucleus.

2. Let us attempt to trace how the concepts presented above are in agreement with the results published in the literature for various pressures of the perturbing atom.

As was mentioned above, condition (11) implies that the boundary between the van der Waals range and the range corresponding to the transition to the Fermi mechanism must shift toward smaller n as the pressure increases. If N = 2.5 relative-density units (r.d.), then the mean distance between atoms is $\bar{R} \approx a_0 n^{*2}$ for n = 9. In [4] the maximum of $\gamma(n)$ was observed for n = 8.9 (see Fig. 1 of the present paper, where $\gamma_{max}$ corresponds to n = 10).

During our work numerous measurements of the shift of cesium lines by krypton at various pressures were published [13]. The results of the measurements are shown in the form of graphs of the dependence of the shift on density $\Delta(N)$ for a specified number of the series. No difference was observed in the behavior of the fine structure components, and this is in agreement with our results (Fig. 3). The case n = 6 for N > 10 r.d. constitutes an exception; here specific effects which we shall not examine here may be substantial.

The functions $\Delta(N)$ for n = 7-10 have the form of two straight lines with sharply differing slopes. For n = 15 only the second of them remains, and for n = 7-12 (~14?) there is a transitional region from one straight region to the other. In accordance with the concepts described above, the first straight region of the dependence of shift on density corresponds to the impact van der Waals range, while the second corresponds to the range where the effects of interaction with an electron and with the atomic core are statistically independent and are both linear in N. According to the theory [2], at high densities deviations from linearity in N are also possible in the second range when $\chi_i \gg 1$, and the broadening connected with the polarization effect is proportional to $N^{4/3}$. This range begins only at N ~ 10 r.d. But even in such a case the deviation from linearity cannot be large, since $\Delta_p$ is considerably less than $\Delta_s$, while the scattering effect is described by the impact mechanism ($\chi_e \ll 1$), i.e. $\Delta_s \sim N$ right up to N of the order of hundreds of r. d.

An attempt was made at a numerical estimation of certain characteristics of the examined dependence of shift on density. The results obtained and the corresponding experimental values from Table 1 of [13] are given in Table 2.

For estimates in the van der Waals range it is necessary to know the magnitudes of the interaction constants $C_6$. They were calculated according to Eqs. (9). The polarizability of krypton $\alpha_2 = 2.49$ Å$^3$ and the polarizability $\alpha_1$ of the excited cesium atom is given in Table 1. The values of $C_6^{2/5}$ obtained from (9) according to the argon broadening case examined above increased by a factor of 2. The results are given in Table 2 in which the shift in cm$^{-1}$/r.d. is given and calculated according to Eqs. (8); the experimental data are given, and the temperature at which the measurements were carried out is indicated in parentheses. The calculated values of the shift also correspond to these temperatures. The agreement between the calculated and experimental quantities is satisfactory. Note that the contribution from many-particle van der Waals and Fermi collisions is not considered in the calculation; this contribution increases with the principal quantum number. Therefore, the calculated values increase with n more slowly than the experimental values. Moreover, the table gives the limiting concentration $N_I$ for which condition (12) $[\chi_{at}=1]$ is satisfied; i.e., many-particle collisions begin to play a basic role, and the first linear region of the function $\Delta(N)$ ends; the experimental values of $N_I$ in accordance with [13] are also given.

For n > 10 condition (10), $\rho_i < a_0 n^{*2}$, is already satisfied, and the function $\Delta(N)$ begins with the transitional range or may be connected with the second straight region corresponding to interaction of the Fermi type.

The transition to the second linear region of the function $\Delta(N)$ for n = 7-10 corresponds to the transition to the Fermi mechanism in accordance with condition (4). Keeping in mind the fact that the shift is predominantly determined by atoms situated at a distance $\bar{R}$, it can be expected that for linearity of $\Delta(N)$ it is sufficient that the inequality be satisfied with a slight margin. Table 2 gives the values of $N_{II}$ calculated from the condition $\bar{R} = a_0 n^{*2}$ (variation of $\bar{R}$ by a factor of 2.2 (see p. 85) changes $N_{II}$ by one order of magnitude), and the experimental values of the density at which the second linear region begins in accordance with [13].

Based on these same concepts, let us examine the variation of the width of the lines with particle density. As mentioned above, the width depends more substantially on the form of the interaction. Regrettably, data on the width of cesium lines are available only for n = 6 and 7 in the case of broadening by means of argon [5] and krypton [13]. The broadening by means of the intrinsic cesium pressure and the apparatus width were not considered in these papers in carrying out the width measurements. Therefore, the results for small N are very approximate. But the limits of the individual regions and the dependence of $\gamma$ on N can be isolated fairly reliably.

The width varies nonmonotonically with density. At small densities $\gamma$ depends weakly on N. In this case binary van der Waals collisions must predominate. In the case of krypton this region must end at the density $N_I$ given in Table 2; for argon (also on the basis of condition (12); for estimating $\rho_{at}$ we took t = 200°C): n = 6 for a density $N_I = 3$ r.d., n = 7-10, which is in good agreement with the results of the papers indicated.

For a further increase in N the width first increases abruptly (approximately $N^2$), and then a second linear region is observed which begins for both Kr and for Ar approximately from $N_{II} = 25$ r.d. for n = 6, and from $N_{II} = 10$ r.d. for n = 7. For n = 7 the rough theoretical value of $N_{II}$ is given in Table 2. In the approximation considered this value does not depend on the type of perturbing gas, which is in agreement with experiments. Thus, the concepts expounded above concerning the interaction of excited alkali atoms with heavy inert gases correspond both to the experimental results of the present work and those described in the literature.

Interaction with helium and with a number of other gases cannot be described by the simple $C_6/r^6$ potential even for the very first terms of the series, which is evidenced by the fact that these lines have a positive (violet) shift. The character of the interaction is more complex in this case (see [1]). The Fermi mechanism remains valid for the high terms of the series, and the sign of the shift is determined by the behavior of the scattering amplitude. Gases for which the Ramsauer effect can be observed yield a negative shift (Ar, Kr, Xe). The shift produced by the effect of gases which do not have the Ramsauer effect (He, Ne) must be positive. Possibly, it is precisely as a consequence of the comparatively large cross sections for the collisions of these gases with an electron at electron velocities corresponding to the first excited state of an alkali metal atom that the repulsion is substantial and the interaction does not go over into the simple $\sim r^{-6}$ law even for a weakly excited alkali atom.

The author is deeply grateful to M. A. Mazing for his constant interest and numerous suggestions, and also to L. P. Presnyakov and I. I. Sobel'man for discussing the work and making valuable comments.

References

1.  Sh. Chen and M. Takeo, Uspekhi Fiz. Nauk, 66:391 (1958).
2.  V. A. Alekseev and I. I. Sobel'man, Zh. Éksperim. i Teor. Fiz., 49:1274 (1965).
3.  M. A. Mazing and N. A. Vrublevskaya, Zh. Éksperim. i Teor. Fiz., 50:343 (1966).
3a. M. A. Mazing and P. D. Serapinas, Preprint of the Physics Institute, Academy of Sciences of the USSR, No. 73 (1968).
4.  C. Füchtbauer and F. Gössler, Z. Phys., 87:89 (1933); F. Gössler and H. E. Kundt, Z. Phys., 89:63 (1934).
5.  S. Y. Chen and R. O. Garrett, Phys. Rev., 144:59 (1966); S. Y. Chen and W. J. Parker, J. Opt. Soc. Am., 45:22 (1965).
6.  H. R. Kratz, Phys. Rev., 75:1844 (1949).
7.  K. B. Eriksson, I. Johansson, and G. Norlen, Arkiv fys., 28:233 (1964); I. Johansson, Arkiv fys., 20:135 (1961); H. Kleiman, J. Opt. Soc. Am., 52:441 (1962); Kh. Kallas, G. Markova, G. Khvostenko, and M. Chaika, Opt. i Spektr., 19:303 (1965).
8.  A. Dalgarno and W. D. Davison, in: Advances in Atomic and Molecular Physics, Vol. 2, D. R. Bates and I. Estermann (Eds.), Academic Press, New York, London (1966), pp. 1-32.
9.  H. R. Griem, Phys. Rev., 128:515 (1962).
10. R. Marrus, D. McColm, and J. Yellin, Phys. Rev., 147:55 (1966).
11. Y. T. Yao, Z. Phys., 77:307 (1932).
12. I. I. Sobel'man, Introduction to the Theory of Atomic Spectra, Fizmatgiz (1963).
13. S. Y. Chen, E. C. Looi, and R. O. Garrett, Phys. Rev., 155:38 (1967).

# INVESTIGATION OF TRANSITIONS BETWEEN EXCITED STATES OF THE NEON ATOM DURING COLLISIONS OF ELECTRONS*

## A. S. Khaikin

Introduction

**1.** In order to understand various phenomena which occur in nonequilibrium plasma it is necessary to know the rates of the processes which determine the concentrations of charged and neutral particles. The concentrations of exced atoms (the populations of the corresponding states of the atoms) are determined basically by the excitation rate due to collisions with electrons and the de-excitation rate due to collisions and spontaneous emission. In relatively cold plasma practically just the ground and first excited states are populated; therefore, the largest contribution to the population of the highly excited states is made principally by excitation from the ground and metastable states. With increasing electron temperature the populations of the highly excited states increase abruptly, and the processes of excitation transfer between these states become very substantial. Such a situation predominates in such frequently encountered systems as a low-pressure gas discharge, gas lasers, the ionosphere, gas nebulae, solar corona, etc. In this connection the investigation of transitions between excited states acquires special importance.

Existing experimental data apply chiefly to transitions from the ground state of atoms (see, for example, [1-3]); however, data on the effective cross sections of transitions between excited levels can be counted on the fingers due to many experimental difficulties [5-12]. Theoretical calculations (a review of the modern state of collision theory can be found in the monographs [1-4], as well as in the papers of the present collection) may be used to find the cross sections of various transitions. However, in view of the sparseness of the experimental material there is no possibility of evaluating the applicability of some calculation method to transitions between excited states of atoms. At the same time the variety of practical problems requires knowledge of a large number of quantities that for the time being could be more realistically found by means of theory. A completely universal method is required for this purpose, which would be applicable to various transitions. Among the approximate methods used in collision theory the only sufficiently universal one is the simplest — the Born method. Other methods, although they allow considerably more exact results to be obtained in particular cases, fail utterly to provide such accuracy for arbitrary transitions. Therefore, the Born approximation is the only suitable one for systematic calculations.

*The abridged text of a dissertation for the degree of Candidate of Physicomathematical Sciences; scientific supervisor — Doctor of Physicomathematical Sciences S. G. Rautian.

In this connection the problem of the applicability of the approximation (with or without consideration of normalization) to the calculation of transitions between excited states of atoms acquires great importance. The point is that the Born approximation is certainly applicable for $e^2/\hbar v \ll 1$ (e, v are the charge and velocity of an electron); i.e., it is applicable when the electron energy $E = mv^2/2$ is very high compared with the ionization energy of a hydrogen atom (13.5 eV). For transitions from the ground state this criterion is practically equivalent to the condition $E/\Delta E \gg 1$, where $\Delta E$ is the threshold excitation energy.

In [10] a hypothesis is stated to the effect that the true criterion for the applicability of the Born method is precisely the latter condition $E/\Delta E \gg 1$. For transitions between excited states this condition is considerably less rigorous than the strict criterion $e^2/\hbar v \ll 1$ since for such transitions the threshold energy $\Delta E$ is considerably lower than the ionization energy of the atom. However, the criterion $E/\Delta E \gg 1$ has not been substantiated theoretically. It is for this reason that experimental investigations of transitions between excited states acquire special importance.

Direct experimental data on transitions between excited states of atoms are practically unavailable. The characteristics of such transitions can be found indirectly from experimental data on the broadening of spectral lines in plasma [10, 11]. The analysis of a number of such data has shown that the theory of line broadening, even in the simplest approximation based on a simplified version of the Born method (with allowance for normalization), does not contradict experiment when E is sufficiently small (several electron-volts), but $E/\Delta \gg 1$.

Under these conditions the problem of the applicability of the Born approximation cannot be considered solved, since data on line broadening can be treated only as an indirect method of solving this problem. Therefore, obtaining direct experimental data is necessary to the extent possible.

**2.** All existing methods of investigating processes involving the excitation of atoms by electron impact [1, 2, 13-16] in principle allow the investigation of transitions between excited states; however, their practical implementation is connected with exceptional difficulties. The point is that in order to investigate such transitions the gas must first be excited. Here it is possible to imagine an experiment with two electron beams of which the one having the higher energy excites the gas up to a specified state, while the other (having an energy insufficient for preliminary excitation) transfers the atoms to a higher excited level. Experimentally, this method is considerably more complex than the method involving a single beam, and therefore experiments of this kind are only in the planning stage so far.

We can, for example, use an electron beam to bombard atoms in a gas discharge. Under these conditions, however, an additional difficulty develops: in a sufficiently hot discharge the excited states are noticeably populated, and it is difficult to determine the levels from which the excitation by the beam occurs. This also applies completely to the investigation of excitation in a gas discharge. Heretofore only one paper [8] is known in which the combined method (an electron beam in a discharge) was used for the direct isolation of the process of exciting the $7^3S$ mercury level from the metastable state $6^3P$ and for measurement of its cross section relative to the cross section for excitation from the ground state $6^1S_0$. In all of the remaining papers on the investigation of transitions between excited states [5-7, 9-12] the corresponding cross sections were determined by indirect methods. Under these conditions either the dependence of the intensity of the spectral lines on the electron concentration [5, 6, 9], or the broadening of the lines [10, 11], or the absorption of light in the corresponding lines [7, 12] were used.

**3.** Thus, if we dwell on the gas discharge method as being the most convenient for obtaining excited atoms, then the following problem arises: how to isolate the original state from the

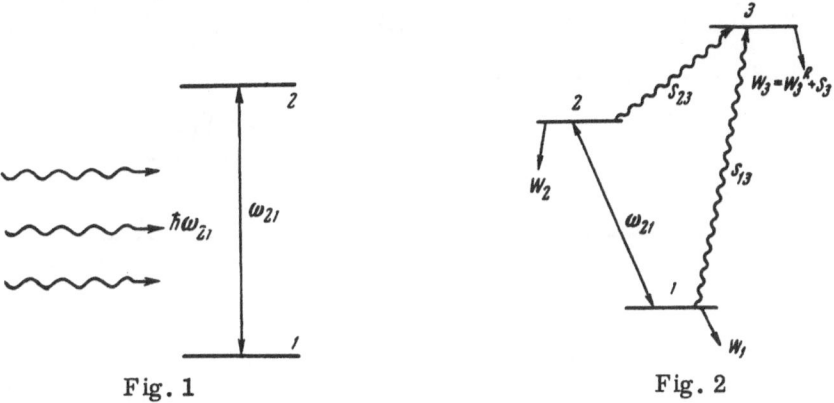

Fig. 1                                    Fig. 2

entire energy spectrum of the atoms? In the present work a method has been developed which seems to us to be promising and consists in the following [18, 20, 21].

If radiation whose frequency coincides with that of the transition between any two excited levels 2 and 1 (Fig. 1) is directed into an excited gas, then the processes of stimulated emission and absorption lead to a redistribution of the original population of levels 2 and 1, the changes in population increasing as the radiation increases in intensity (external irradiation always decreases the difference in the populations of levels 2 and 1; only the case of equal populations constitutes an exception: under these conditions irradiation causes no changes whatsoever).

By superimposing modulation (by periodic interruption) on the external irradiation, we thereby compel the populations $N_2$, $N_1$ to vary periodically with the amplitudes $\Delta N_2$, $\Delta N_1$; under these conditions the phases of these population changes at the levels 2 and 1 will be opposite. Thus, by modulating the populations of levels 2 and 1 we isolate them from the overall energy spectrum of the atoms. At all remaining levels the modulation of populations can occur only as a consequence of the transfer of excitation from levels 2 or 1 during collisions or spontaneous transitions. The phase of the population modulation of such levels must be determined by the phase of the population modulation of that level (1 or 2) whose excitation is more efficient. Measurements of the amplitude of the population modulation and its phase at levels which are not connected with external irradiation allow information to be obtained on the rates of both spontaneous transitions and transitions which accompany collisions with electrons. Below we shall call this method the modulation method throughout.

4. Lasers, for which a concentrated intense directed radiation in one or several atomic lines is characteristic, constitute a very advantageous source of radiation for such experiments. Because of the high intensity and coherence of laser radiation it can be expected that for large saturation the changes of populations (modulation depths) of the irradiated levels will be very noticeable. According to estimates, the modulation depth of the populations can reach 20-40% in a helium-neon laser for generation in the $\lambda = 3.39\,\mu$ line. It is natural that at levels connected by inelastic processes with levels whose population is modulated by irradiation the modulation depth of the populations will be less, but there too a detectable modulation may be expected.

It is clear that the method suggested allows [17-21] measurement of only the rate coefficients $\sigma(v)$ of the processes, and not the cross sections $\langle \sigma v \rangle$. Nevertheless, such results are very valuable both in themselves and in view of the sparseness of other results. In a somewhat different version the modulation method allows measurement of the Einstein coefficients for spontaneous transitions [22]. Let us note here that an analogous method was used to investigate the transition rates for atomic collisions [17, 19].

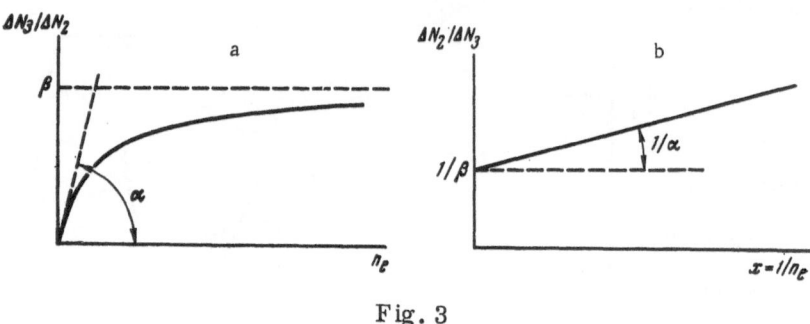

Fig. 3

In the present paper a detailed examination is made of the modulation method, and the results of measurements of the rate coefficients $<\sigma v>$ for $s'-d'$, $p'-d'$, and $s'-p'$ transitions in neon for collisions with electrons are described.

## §I.  Theory of the Modulation Method

**1.** For a quantitative investigation of the method let us turn to Fig. 2. Let us consider how the population of level 3, which is situated above levels 2 and 1 whose populations are modulated by the external irradiation, will behave. We shall assume that our recording equipment responds only to the variable part of the intensity of the spectral lines; therefore, here we shall examine only the amplitudes $\Delta N_i$ of the population modulation of the levels.

The qualitative picture is obtained most easily in the simplest case by assuming that the modulation of the population $N_3$ is caused exclusively by the processes of excitation transfer from level 2 for collisions with electrons. The possibility of exciting level 3 for collisions of electrons with atoms in the state 1 is for the time being excluded from consideration for simplicity.

The rate $S_{ik}$ of the $i-k$ transition for collisions with electrons in a gas discharge is determined by the relationship:

$$S_{ik} = n_e \langle \sigma_{ik} v \rangle , \tag{I.1}$$

where $n_e$ is the electron concentration in the discharge, and $<\sigma_{ik}v>$ is the rate coefficient of the process and consists of the product of the process cross section $\sigma_{ik}$ and the electron velocity v, averaged over the velocities. In our approximation the ratio between the amplitudes of the modulation of the populations at levels 2 and 3, $\Delta N_3/\Delta N_2$ depends on $n_e$ as follows. For small $n_e$, when the contribution of electrons to the de-excitation of level 3 is small, the amplitude ratio $\Delta N_3/\Delta N_2$ is proportional to $n_e$; under these conditions the proportionality coefficient $\alpha$ (Fig. 3a) is determined by the ratio of the rate coefficient $<\sigma_{23}v>$ of the process $2 \rightarrow 3$ to the rate $W_3^R$ of radiative de-excitation of level 3:

$$\alpha = \frac{\langle \sigma_{23}v \rangle}{W_3^R} \tag{I.2}$$

With increasing $n_e$ the role of collisions with electrons in the de-excitation of level 3 increases, and the proportionality between $\Delta N_3/\Delta N_2$ and $n_e$ is violated; then, when the rate $S_3$ of impact de-excitation of the level 3 considerably exceeds the rate $W_3^R$ of radiative de-excitation of this level, the ratio between the modulation amplitudes ceases to depend on $n_e$. The corresponding asymptotic value of $\Delta N_3/\Delta N_2$ is equal to (Fig. 3a):

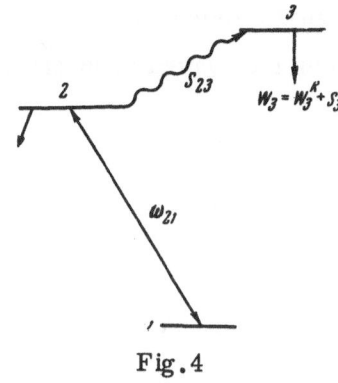

Fig. 4

$$\beta = \frac{S_{23}}{S_3}\frac{\langle \sigma_{23}v\rangle}{\langle \sigma_3 v\rangle}\,. \tag{I.3}$$

Thus, we obtain the dependence illustrated in Fig. 3a:

$$\frac{\Delta N_3}{\Delta N_2} = \frac{n_e\langle \sigma_{23}v\rangle}{W_3^R + n_e\langle \sigma_3 v\rangle} = \frac{n_e}{\frac{1}{\alpha} + n_e\frac{1}{\beta}}\,. \tag{I.3a}$$

The curve which has been found can easily be transformed to a more convenient linear dependence; for this it is sufficient to perform a hyperbolic transformation with respect to both axes and to lay off x = 1/$n_e$ and $\Delta N_2/\Delta N_3$ (Fig. 3b). The slope of this straight line and the point of its intersection with the axis of ordinates are determined by quantities which are the reciprocals of the slope at zero and of the asymptote of the curve shown in Fig. 3a ($\mu = 1/\alpha$, $\nu = 1/\beta$).

Thus, a qualitative examination with allowance for the sole excitation process 2 → 3 leads to the linear function shown in Fig. 3b:

$$\frac{\Delta N_2}{\Delta N_3} = \mu x + \nu = \frac{W_3^R}{\langle \sigma_{23}v\rangle}x + \frac{\langle \sigma_3 v\rangle}{\langle \sigma_{23}v\rangle}\,. \tag{I.4}$$

From this it is evident that by measuring the dependence of the ratio of the modulation amplitude $\Delta N_2/\Delta N_3$ on x = 1/$n_e$, we can find the parameters of the straight line and therefore the rate coefficients of the excitation and de-excitation processes.

The expounded qualitative picture considers only a particular case; more general cases, however, require the exact solution of a system of kinetic equations to which we shall now turn.

**2.** Let us consider the diagram shown in Fig. 4, and let us investigate the behavior of the populations of levels 1, 2, 3 for introduction of irradiation having the frequency $\omega_{21}$. From the total de-excitation rates $W_i$ of the levels let us isolate the rates of the excitation processes $S_{13}$, $S_{23}$ which are of interest to us; all the remaining radiative and nonradiative processes (including the reverse processes 3 → 2, 3 → 1) shall be assumed to be accounted for in $W_i$. The stimulated processes which accompany irradiation shall be characterized by the ratio between the number of stimulated emission (absorption) events $w_2(w_1)$ and the number of events $Q_2(Q_1)$ involving the excitation of the upper (lower) level 2 (1) ($g_2W_2w_2 = g_1W_1w_1$). Then under steady-state conditions the system of kinetic equations for the populations $N_1$, $N_2$, $N_3$ is written as follows in the presence of external irradiation:

$$\begin{aligned}
Q_1 &= N_1W_1 - N_2A_{21} - (Q_2w_2 - Q_1w_1),\\
Q_2 &= N_2W_2 + (Q_2w_2 - Q_1w_1),\\
Q_3 &= -N_1S_{13} - N_2S_{23} + N_3W_3,
\end{aligned} \tag{I.5}$$

where $A_{21}$ is the Einstein coefficient for the spontaneous transition 2 → 1. In the absence of external irradiation we have

$$\begin{aligned}
Q_1 &= N_1^{(0)}W_1 - N_2A_{21},\\
Q_2 &= N_2^{(0)}W_2,\\
Q_3 &= -N_1^{(0)}S_{13} - N_2^{(0)}S_{23} + N_3^{(0)}W_3.
\end{aligned} \tag{I.6}$$

Here it is assumed that the excitation rates do not depend on the external irradiation.*

Subtracting Eqs. (I.6) from the corresponding Eqs. (I.5), we obtain the system of equations for the amplitudes of the modulation of the populations:

$$\Delta N_1 W_1 - \Delta N_2 A_{21} - (Q_2 w_2 - Q_1 w_1) = 0,$$
$$\Delta N_2 W_2 + (Q_2 w_2 - Q_1 w_1) = 0, \tag{I.7}$$
$$-\Delta N_1 S_{13} - \Delta N_2 S_{23} + \Delta N_3 W_3 = 0,$$

We are interested in the ratio between the modulation amplitudes, and therefore we divide all three equations by $\Delta N_2$:

$$\frac{\Delta N_1}{\Delta N_2} W_1 - \frac{1}{\Delta N_2}(Q_2 w_2 - Q_1 w_1) - A_{21} = 0,$$
$$\frac{1}{\Delta N_2}(Q_2 w_2 - Q_1 w_1) + W_2 = 0, \tag{I.8}$$
$$-\frac{\Delta N_1}{\Delta N_2} S_{13} + \frac{\Delta N_3}{\Delta N_2} W_3 - S_{23} = 0.$$

Eliminating $[(Q_2 w_2 - Q_1 w_1)/\Delta N_2]$ from (I.8), we find the system of equations for the amplitude ratios $\Delta N_1/\Delta N_2$ and $\Delta N_3/\Delta N_2$:

$$\frac{\Delta N_1}{\Delta N_2} W_1 + W_2 - A_{21} = 0,$$
$$-\frac{\Delta N_1}{\Delta N_2} S_{13} + \frac{\Delta N_3}{\Delta N_2} W_3 - S_{23} = 0. \tag{I.9}$$

The solutions of the system (I.9) have the form

$$\frac{\Delta N_1}{\Delta N_2} = -\frac{W_2 - A_{21}}{W_1}, \tag{I.10}$$

$$\frac{\Delta N_2}{\Delta N_3} = \frac{W_3/S_{23}}{1 - \frac{S_{13}}{W_1}\frac{W_2}{S_{23}}}. \tag{I.11}$$

From (I.10) it follows that the ratio between the population modulation amplitudes for levels 1 and 2 is determined solely by the total de-excitation rates of these levels and is in no way dependent on the intensity of the external irradiation. Since it can always be assumed that $A_{21} \ll W_2$, it follows that

$$\frac{\Delta N_1}{\Delta N_2} \approx -\frac{W_2}{W_1}. \tag{I.12}$$

From this it is also evident that the amplitudes $\Delta N_1$ and $\Delta N_2$ always have different signs (i.e., that the populations $N_1$ and $N_2$ are always modulated in phase opposition).

**3.** In order to investigate the solutions (I.11) we isolate from $W_3$ the resultant rate $S_3$ of impact de-excitation of level 3 and write the rates $S_3$, $S_{23}$ in the numerator of (I.11) in the form (I.1) by means of the rate coefficient. Then we obtain

---

*This proposition is discussed in detail below.

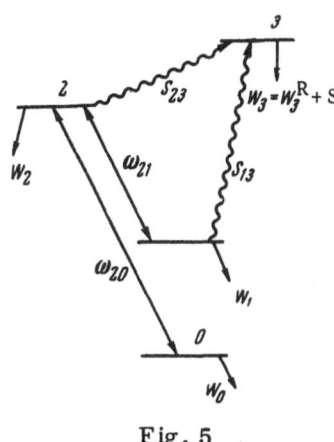

Fig. 5

$$\frac{\Delta N_2}{\Delta N_3} = \frac{\frac{W_3^R}{\langle \sigma_{23} v \rangle} \frac{1}{n_e} + \frac{\langle \sigma_3 v \rangle}{\langle \sigma_{23} v \rangle}}{1 - \frac{S_{13}}{W_1} \frac{W_2}{S_{23}}} = \frac{\mu x + \nu}{1 - \frac{S_{13}}{W_1} \frac{W_2}{S_{23}}} . \qquad (I.13)$$

This equation considers both processes by which the level 3 is excited: $1 \to 3$ and $2 \to 3$.

It is easy to see that for the case in which the contribution of the process $1 \to 3$ to $N_3$ ($S_{13}/W_1 \ll S_{23}/W_2$) can be neglected we obtain the linear Eq. (I.4), which we have already derived by means of qualitative concepts (see § I.1). In such cases the rate coefficients $\langle \sigma_{23} v \rangle$ and $\langle \sigma_3 v \rangle$ are extracted relatively easily from measurements.

The situation is very complex when the process $1 \to 3$ cannot be neglected. Actually, although a linear dependence of $\Delta N_2 / \Delta N_3$ on $x = 1/n_e$ must be obtained experimentally as previously (on the assumption that $W_1$, $W_2$ do not depend or depend weakly on $n_e$), its two parameters (the slope and the point of intersection with the ordinate) now include three unknowns $\langle \sigma_{23} v \rangle$, $\langle \sigma_{13} v \rangle$, $\langle \sigma_3 v \rangle$ for which only two relationships can be obtained by means of (I.13).

It would seem that the situation is hopeless and that the processes $1 \to 3$ and $2 \to 3$ cannot be separated completely. However, the separation of the processes will be possible if we can vary the lower level 1 while keeping the upper level 2 constant. Actually, let us consider the diagram in Fig. 5: assume that it is possible to irradiate the gas with two different spectral lines corresponding to the transitions $2 \to 1$ and $2 \to 0$. For irradiation of the gas with each of these lines we obtain linear functions of the type (I.13) with different parameters $\mu$, $\nu$:

$$\left( \frac{\Delta N_2}{\Delta N_3} \right)_{21} = \frac{\mu_{21} x + \nu_{21}}{1 - \frac{S_{13}}{W_1} \frac{W_2}{S_{23}}} , \qquad (I.14)$$

$$\left( \frac{\Delta N_2}{\Delta N_3} \right)_{20} = \frac{\mu_{20} x + \nu_{20}}{1 - \frac{S_{03}}{W_0} \frac{W_2}{S_{23}}} . \qquad (I.15)$$

From (I.14), (I.15) it is clear that when the 0 level is replaced by the 1 level the parameters $\mu$ and $\nu$ change in identical ratios:

$$k = \frac{\mu_{20}}{\mu_{21}} = \frac{\nu_{20}}{\nu_{21}} = \frac{1 - \frac{S_{13}}{W_1} \frac{W_2}{S_{23}}}{1 - \frac{S_{03}}{W_0} \frac{W_2}{S_{23}}} . \qquad (I.16)$$

The separation of the processes $1 \to 3$ and $2 \to 3$ is carried out most easily if the 0 level is situated considerably farther from 3 than the 1 level. In this case the rate of the process $0 \to 3$ must be considerably lower than the rate of the process $2 \to 3$, since the cross sections and rates decrease rapidly with increasing energy of the transition (the ratios between the excitation rates $S_{ik}$ and the de-excitation rate $W_i$ can be estimated by means of theoretical calculations, since the theory yields relative quantities quite reliably). If this is so, then

$$\frac{S_{03}}{W_0} \ll \frac{S_{23}}{W_2} . \qquad (I.17)$$

Then (I.15) is written approximately as:

$$\left(\frac{\Delta N_2}{\Delta N_3}\right)_{20} \approx \mu_{20}\, x + \nu_{20}\,,\tag{I.18}$$

and the results of the measurements obtained for irradiation of the gas by the $2 \to 0$ can be considered free from the $0 \to 3$ excitation. Under these conditions the rate coefficients $<\sigma_{23}v>$ and $<\sigma_3 v>$ can be found from the parameters $\mu_{20}$ and $\nu_{20}$.

Going over to irradiation of the gas by the $2 \to 1$ line, we obtain the relationship (I.14) which yields us the parameters $\mu_{21}$, $\nu_{21}$. For the condition (I.17) the relationship (I.16) is simplified:

$$k = \frac{\mu_{20}}{\mu_{21}} = \frac{\nu_{20}}{\nu_{21}} \approx 1 - \frac{S_{13}}{W_1}\frac{W_2}{S_{23}}\,.\tag{I.19}$$

The quantity k is determined from the experimental ratios $\mu_{20}/\mu_{21}$, $\nu_{20}/\nu_{21}$; then the rate coefficient $<\sigma_{13}v>$ is found by means of (I.19).

In accordance with the left side of (I.19) both coefficients ($\mu$ and $\nu$) vary in identical ratios for transition from the $2 \to 0$ line to the $2 \to 1$ line. This allows the correctness of the interpretation of the experiment to be monitored. Using arithmetic operations, it is easy to show that the largest possible relative error introduced by neglecting the $0 \to 3$ process is stipulated by the experimental coefficient k. However, the true value of the error is determined by the ratio $(S_{03}/W_0,\ W_2/S_{23})$, which may be much less than unity (see (III.6)).

4. Above we have used the ratios $\Delta N_2/\Delta N_3$ of the population modulation amplitudes throughout. In the experiment these ratios must be measured from the amplitudes of the modulation of the spectral line intensities, beginning with the necessary levels. Therefore, we shall now introduce the corresponding coefficients and write out the equations in final form.

As is well known, the intensity of a spectral line is determined by the relationship

$$I_{ik} = N_i\, A_{ik}\, \hbar\, \omega_{ik}\,,\tag{I.20}$$

where $N_i$ is the population of the upper level i, $A_{ik}$ is the Einstein coefficient; $\hbar\omega_{ik}$ is the energy of a quantum.

In recording a line the signal at the output of the recording system is

$$J_{ik} = H_{ik}\, I_{ik}\,,\tag{I.21}$$

where $H_{ik}$ is the spectral sensitivity of the system.

Thus, for a ratio $y = \Delta J_2/\Delta J_3$ of the measured signals we find the following results by means of (I.20), (I.21) (dropping the second subscripts of the quantities H and A):

$$y = M_{23}\,\frac{\Delta N_2}{\Delta N_3}\,;\qquad M_{23} = \frac{H_2 A_2 \omega_2}{H_3 A_3 \omega_3}\,.\tag{I.22}$$

Here and further on we shall use $M_{ik}$ to denote the coefficient connected with the spectral sensitivity of the apparatus and the Einstein coefficients for the lines, beginning with the i and k levels.

Now introducing the coefficient $M_{23}$ into (I.14) and (I.18), we obtain the expressions for the experimental straight lines $y = f(x)$:

$$y_{21} = \frac{a_{21}x + b_{21}}{1 - \frac{S_{13}}{W_1}\frac{W_2}{S_{23}}} , \tag{I.23}$$

$$y_{20} \approx a_{20}x + b_{20}, \tag{I.24}$$

where the new parameters are denoted by

$$a = M_{23}\mu = M_{23}\frac{W_3^R}{\langle \sigma_{23} v \rangle} ; \tag{I.25}$$

$$b = M_{23}\nu = M_{23}\frac{\langle \sigma_3 v \rangle}{\langle \sigma_{23} v \rangle} . \tag{I.26}$$

For a and b, just as for $\mu$, $\nu$, the relationship (I.19) is similarly valid:

$$k = \frac{a_{20}}{a_{21}} = \frac{b_{20}}{b_{21}} = 1 - \frac{S_{13}}{W_1}\frac{W_2}{S_{23}} . \tag{I.27}$$

Now making use of the relationships (I.23)–(I.27), we write out the equations for calculating $\langle \sigma v \rangle$ from the experimental data:

$$\langle \sigma_{23} v \rangle \approx \frac{A_2}{a_{20}}\frac{W_3^R}{A_3}\frac{H_2\omega_2}{H_3\omega_3} , \tag{I.28}$$

$$\langle \sigma_{13} v \rangle \approx \langle \sigma_{23} v \rangle \frac{W_1}{W_2}(1 - k), \tag{I.29}$$

$$\frac{S_3}{W_0^R} = \frac{b}{a} n_e . \tag{I.30}$$

Let us note that the method allows the ratio (I.30) to be found besides the rate coefficients (I.28), (I.29); this ratio characterizes the role of the electrons in the de-excitation of level 3. It is very essential that the result (I.30) does not depend on the approximation (I.17) and is always valid whether or not the processes $1 \rightarrow 3$, $0 \rightarrow 3$ are considered.

The accuracy of the coefficients $\langle \sigma v \rangle$ obtained by the method described is determined both by the accuracy with which the parameters a and b are determined and by the accuracy of the extraneous data which are included in relationships (I.28), (I.29). Thus, the most realistic way of obtaining the different transition probabilities (Einstein coefficients) is to carry out theoretical calculations; these are generally used in the majority of cases. However, our method allows measurement of certain quantities; the corresponding measurements have been described in [22]. The approximation (I.17), of course, introduces an additional error for which we can obtain only the upper boundary from experiments: the relative error does not exceed the coefficient k from (I.27) and has an order of magnitude equal to the ratio ($S_{03}/W_0$, $W_2/S_{23}$); a final check of the approximation (I.17) can be made only by means of theoretical estimates of this ratio.

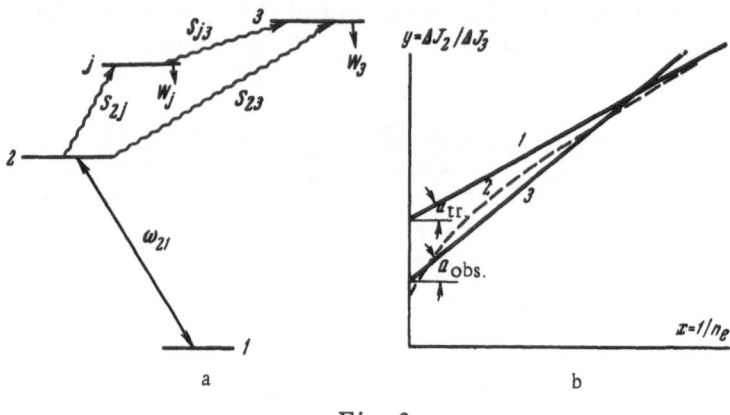

Fig. 6

**5.** Let us recall that the model which we have constructed presupposes the following assumptions: 1) the rates of excitation $Q_i$ from the ground state are considered to be independent of the external irradiation, and 2) all possible transitions which originate during atom–atom collisions are excluded from consideration.

As has already been mentioned in §I.1, the rates of excitation $Q_i$ from the ground (and metastable) states depend weakly on the irradiation. However, a situation is possible in which the irradiation leads to a change of $n_e$ [23] and therefore to change of $Q_i$. Since these changes are usually small, it follows that the kinetic equations can be used to introduce the corresponding correction and to obtain the following relationship for the measured quantity $\tilde{y}$:

$$\tilde{y} = y\,(1 - \delta), \tag{I.31}$$

where y is determined by Eqs. (I.23)-(I.26) and the correction $\delta$ is equal to

$$\delta = \frac{\Delta n_e / n_e}{\Delta N_i / N_i}. \tag{I.32}$$

Measuring both the quantity $\tilde{y}$ and the dependence $\Delta \mathcal{I}/\mathcal{I}$ (i.e., $n_e / \Delta n_e$) the modulation depth $\Delta N_i / N_i$ on current $\mathcal{I}$, it is possible to find the correction $\delta$ and to obtain a linear dependence for the quantity

$$y = \tilde{y}/(1 - \delta). \tag{I.33}$$

Thus, a violation of the first assumption in our model can always be considered and the corresponding correction made.

For a violation of the second assumption (i.e., for cases in which transitions accompanying atom–atom collisions cannot be neglected) this situation is complicated somewhat. However, such processes must reveal an explicit dependence on the concentration of atoms (i.e., on the gas pressure). Therefore, for monitoring purposes it is necessary to carry out measurements at various pressures. Below we shall see how atom–atom collisions of a specific type affect the graphs $y = f(x)$, and what interpretation of the effect is possible in this case. It is clear only that the model should be monitored experimentally in all cases; the criterion of the applicability of the model must be the function $y = f(x)$ obtained experimentally and its dependence on the experimental conditions.

**6.** Of course, the coefficients $<\sigma_{ik}v>$ obtained here consider all $i \rightarrow k$ processes, including the step-by-step processes involving the participation of an actual intermediate level: $i \rightarrow j \rightarrow k$. Let us consider the effect of step-by-step processes in greater detail. For simplicity let us limit the analysis to the simplest case of the $2 \rightarrow 3$ transition. Assume that a level j is situated between levels 2 and 3 (Fig. 6a) and that the excitation of $2 \rightarrow 3$ is possible both directly and via the j level. Introducing the rates $S_{2j}$ and $S_{j3}$ of the corresponding processes, as well as the total de-excitation rate $W_j$, we obtain the kinetic equations for the amplitudes of the population modulation:

$$\Delta N_2 S_{23} + \Delta N_j S_{j3} = \Delta N_3 W_3,$$
$$\Delta N_2 S_{2j} = \Delta N_j W_j. \tag{I.34}$$

From this it is easy to find

$$\frac{\Delta N_3}{\Delta N_2} = \frac{S_{23}}{W_3}(1 + \varkappa), \quad \varkappa = \frac{S_{2j}}{S_{23}}\frac{S_{j3}}{W_j}. \tag{I.35}$$

The quantity $\varkappa$ is the contribution of the step-by-step process to $\Delta N_3$, referred to the contribution of the direct process. If it is assumed that $S_{2j}/S_{23} \sim 1$ (which is quite reasonable, since the energies of the $2 \rightarrow j$ and $2 \rightarrow 3$ transitions are close), then $\varkappa \sim S_{j3}/W_j$. The ratio $S_{j3}/W_j$ is always $\ll 1$, independently of what $W_j$ is determined by — radiative transitions($W_i^R$ for small $n_e$) or collisions with electrons ($S_j$ for large $n_e$), since $W_j$ always includes the sum of the rates of all transitions from the j level. Thus, it can be assumed that the relative contribution of a step-by-step transition is always small compared with the contribution of a direct transition ($\varkappa \ll 1$).

If levels 2 and 3 are spaced far apart, then a considerable number of intermediate j levels may turn out to be situated between them. In this case the quantity $\varkappa$ in (I.35) will be replaced by the sum $\sum_j \varkappa_j$, which may turn out to be considerable compared with unity. Then the experimental graph in hyperbolic coordinates for the intensity modulation amplitude $y = f(x = 1/n_e)$ will differ increasingly from a straight line as $\sum_j \varkappa_j$ approaches unity. For the case in which the contribution of $\sum_j \varkappa_j$ is not very large compared with unity it is possible to find the following relationship for the observed dependence of $y = \Delta J_2/\Delta J_3$ on $x = 1/n_e$ from (I.35) using a hyperbolic transformation:

$$y \approx (ax + b)\left[1 - \left(\sum_j \frac{S_{2j}}{S_{23}}\right)\frac{1}{a_j x + b_j}\right], \tag{I.36}$$

where the parameters a, b are the same as in (I.25), (I.26), and

$$a_j = \sum_j \frac{W_j^R}{\langle \sigma_{j3}v \rangle}, \quad b_j = \sum_j \frac{S_j}{S_{j3}}. \tag{I.37}$$

The quantities $a_j$, $b_j$ are very large, since $W_j^R \gg S_{j3}$ and $S_j \gg S_{j3}$; therefore, the second bracket can be treated as a correction factor to the linear function $ax + b$ for small values of x. The result (I.36) is displayed in Fig. 6b. The straight line 1 would be observed in the absence of step-by-step transitions. The consideration of step-by-step transitions leads to curve 2, which deviates from the straight line 1 in the range of small x (large $n_e$). If the scatter of the measurements is large (for a weak signal), then the curvature may not be detected, and, most probably, we will approximate the curve by the straight line 3 whose slope is larger than the slope of the straight line 1. Since the quantity $<\sigma v>$ is inversely proportional to the slope of the straight line, it follows that fairly effective step-by-step transitions can lead to an understimated value of $<\sigma v>$.

Finally, let us note that the physical model described above does not imperatively require that the radiation of the excited gas be external. All of the conclusions are also fully valid when the irradiating field is generated in the investigated gas proper, as is the case in a laser. The spectral composition of the irradiating line is utterly unimportant, since we consider integrated line intensities. Therefore, all the spectral effects inherent in a laser ("hole burning" of dips in the lines, the existence of many oscillation modes within the limits of the Doppler contour, etc.) may be neglected.

7. The concept expounded above allows the conditions to be written out which must be satisfied by the object investigated. First of all, it is necessary that it be possible in the chosen gas to obtain fairly powerful continous-wave generation, since it is very advantageous to use a laser for irradiation purposes. In order to separate the various processes the possibility must be provided of generation via at least two transitions which begin from the same level; under these conditions the wavelengths of these transitions must differ very noticeably. These requirements are best satisfied by neon, which is a gas we chose as the first object of investigation. The generation lines in neon are very numerous; two of them (having a common upper level) are of greatest interest: $\lambda = 6328\,\text{Å}$ (the transition $5s'\,[^1/_2]^0_1 - 3p'\,[^3/_2]_2$) and $\lambda = 3.39\,\mu$ $(5s'\,[^1/_2]^0_1 - 4p'\,[^3/_2]_2)$, which are completely analogous to one another and whose wavelengths (i.e., transition energies) differ by almost a factor of 5. Since a helium-neon laser is the most common one, its application as a source does not present any special difficulties.

Since our method allows measurement of only the average values of $<\sigma v>$ it follows that the parameter

$$\beta = \frac{\Delta E}{kT_e}$$

acquires essential significance. If $\beta \ll 1$, then the main contribution to $<\sigma v>$ will be made by electrons whose energy considerably exceeds the threshold energy $\Delta E$. In the case $\beta \sim 1$, on the contrary, the main contribution to $<\sigma v>$ is made by the region of the $\sigma(E)$ maximum. For the transitions we have chosen, the quantity $\beta$ varies in the range from 0.014 to 0.1 under the conditions of the experiment.

Preliminary experiments [18] which we performed have already shown that in neon there is a fairly wide ensemble of transitions which are of interest from the point of view of our problem. The spectral lines are situated in the visible and near ultraviolet regions of the spectrum, which simplifies the spectral calibration. (Note that our method does not require measurement of the absolute intensities, and therefore it is necessary to calibrate only the relative spectral intensity.)

From §I.4 it is clear that the work must consist in measuring the relative amplitudes of the intensity modulation y of the spectral lines for various electron concentrations $n_e$ in the discharge. By plotting the graph $y = f(x = 1/n_e)$ it is necessary to find the quantities from which the coefficients $<\sigma v>$ are obtained. In view of the relatively small modulation depths, the apparatus must have a sufficiently narrow pass band for extracting a weak signal from a noise background; moreover, it must be stable and immune from noise. The quantity $n_e$ is of cardinal significance for the results. In accordance with this it is necessary to have available a special technique and apparatus for the continuous measurement of $n_e$ in a gas discharge which allows reliable measurements to be carried out in parallel with the measurements of the intensity modulation amplitudes.

## §II. Experimental Technique and Conditions

**1.** The simplest version of the experiment is to use the "universality" of the model in §I and to use the effect of the field generated by the laser on the populations of the levels of the neon atoms in the laser itself. It is precisely in this way that we proceeded in our preliminary experiments [18]. However, such an experimental scheme is very disadvantageous for our purposes for the following reasons. In order to extract the widest possible information and to monitor the correctness of the model it is necessary to vary the conditions of the discharge in the investigated gas to a considerable degree (pressure, current, etc.). It is well known that the generation power of helium−neon lasers depends to a high degree on the discharge conditions, generation being possible only in a relatively narrow range of pressures and currents. However, the useful signal in our experiments (the amplitude of the modulation intensity of spontaneous lines) is directly related to the irradiation power (i.e., to the generation). Thus, the necessity of maintaining the useful signal (a rather weak one) at the highest possible level sharply restricts the ranges of variation of the discharge conditions.

Therefore, in our basic experiments we used a separate measurement discharge tube containing the investigated gas, which was irradiated by the field of a laser. In order to increase the useful signal we inserted this tube into the laser resonator, since the field in the resonator is considerably stronger than outside it. Because of the fact that the measurement tube has a length which is short compared with the length of the laser tube, its presence in the resonator has practically no effect on the generation operating mode and power. Because of this same short length it is possible to stipulate practically any discharge conditions in the measurement tube without detriment to the useful signal. Finally, it is possible to change the construction and material of the tube without touching the overall laser.

The length of the laser resonator (the distance between mirrors) is L = 164 cm, and the mirrors are spherical with radii of curvature R = 200 cm. The generator tube has a length $l$ = 90 cm of the active discharge section, while the total length of the tube (with flanges for windows at the Brewster angle) is approximately 120 cm. In order to provide as large a gain as possible (for the purpose of reducing the effects of losses on the generation power), a narrow generator tube is used having an inside diameter d = 0.3 cm. With quartz windows generation at two neon lines $\lambda$ = 6328 Å and $\lambda$ = 3.39 $\mu$ is possible in the laser. Without special measures, generation occurs only in the infrared line even with mirrors having maximum reflection for $\lambda$ = 6328 Å. Therefore, in order to switch the generation lines a vessel through which it is possible to pass methane (illumination gas), which absorbs radiation having $\lambda$ = 3.39 $\mu$, is inserted into the resonator. Thus, in order to switch the laser from the infrared line to the visible line it is sufficient to pass methane through the vessel. The resonator mirrors have multilayer dielectric coatings with a reflection peak at $\lambda$ = 6328 Å (for generation in the $\lambda$ = 3.39 $\mu$ line it is sufficient to have Fresnel reflection from the mirrors). Since the extraction of the generated radiation from the resonator is required only to monitor the generation power (conversely, it is advantageous to increase the power inside the resonator), we restricted ourselves to the minimal transmissivity of the mirrors (0 and ~ 0.2%).

The generator tube is made of 3S-5 glass (a molybdenum glass) and has plane polished flanges at the Brewster angle. The windows are attached to the flanges by means of Zapon. The cathode of the tube is an indirectly heated oxide cathode, while the anode is a hollow molybdenum cylinder having a small diameter. The tube is supplied with direct current from a special rectifier which produces a voltage of up to 6 kV.

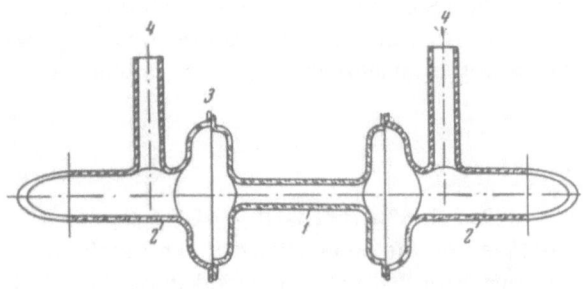

Fig. 7. Construction of the measurement tube.
1) Working section (quartz); 2) flanges (3S-5 glass); 3) joints glued with ÉD-5 epoxy resin; 4) branches to the electrodes.

The construction of the measurement tube is shown in Fig. 7. The length of the active section of the discharge is $l_0 = 5$ cm, and the total tube length is approximately 25 cm. The inside diameter of the working section is $d_0 = 0.35$ cm. The central (working) section of the tube is made of optical quartz — for the purpose of extracting the ultraviolet radiation and for eliminating the effect of the tube on the microwave resonator (see §II.4). The end sections (with the electrodes and flanges for the windows) are made of 3S-5 glass. The end and central sections are glued together with ÉD-5 epoxy resin which is deposited on wide flanges which protect it from heating by the discharge. The cuts for the windows are ground or polished at the Brewster angle, and the windows are attached with picein. The cathode and anode are the same as in the generator tube. The tube can withstand current of up to 200 mA without serious overheating. The tube is supplied from a stabilized rectifier. The electron concentration in the measurement tube is measured by means of a microwave resonator.

Both tubes — the generator tube and the measurement tube — are attached to one vacuum system. During evacuation the tubes are aged by a discharge; special complete annealing is not carried out due to the glued joints. A soft flame from a burner is used to anneal just those parts not endangered by heating (for example, the electrode envelope). After aging, the vacuum system having a glass SDN-1 vapor-oil pump (Hickman pump) provides a vacuum of at worst $10^{-5}$ torr in the tube. The vacuum system has vessels containing gases and taps for filling the tubes; the pressure of the gases is read from thermocouple (up to 0.2 torr) and oil (above 0.2 torr) manometers.

**2.** A DFS-12 spectrometer is used to record the spontaneous emission. This instrument, which has a fairly high aperture ratio and good dispersion, covers a very braod range of the spectrum without a change in gratings — from the infrared region (in the first order) to the ultraviolet region (in the third-fourth orders). All these qualitites provided for the possibility of investigating weak signals, resolving the fairly dense neon spectrum, and joint measurement of lines situated in different regions of the spectrum. The factory cover of the photomultiplier was removed and replaced by a new cover for operation with end-face FEU-38 and FEU-39 photomultipliers. The electronic circuit was similarly replaced by a circuit which records modulated signals and has synchronous detection (see below).

The relative spectral sensitivity of the instrument was calibrated in the visible and ultraviolet ranges of the spectrum. In the visible range (4600-6500 Å second order, FEU-38 with ZhS-12 filter) the calibration was carried out according to a standard filament lamp having a known brightness distribution over the spectrum.* For the ultraviolet range calibration by means of a tungsten lamp was continued down to 3900 Å, and in the 4100-3900 Å range it was "matched" with calibration according to a VSFU-3 hydrogen lamp [24]. In this same range the "matching" of the calibrations in the second and third orders (4000-3000 Å, FEU-39 with UFS-1 and BS-3 filters) was carried out. Note that the hydrogen lamp is not a standard, and therefore the calibration in the ultraviolet can be assumed reliable only in order of magnitude.

---

*The calibration of the brightness temperatures in the 4600-6500 Å range by means of a filament tungsten lamp was carried out by É. A. Lapin in the D. I. Mendeleev All-Union Scientific-Research Institute of Metrology.

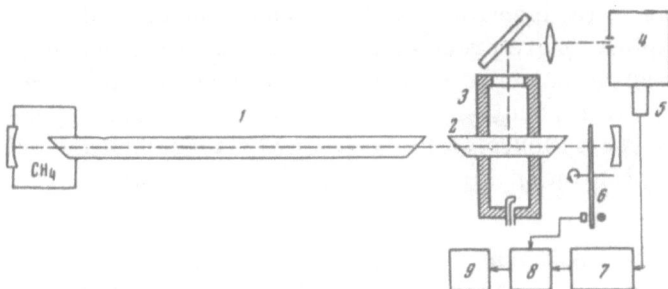

Fig. 8. Block diagram of the measurement apparatus.

The signal from the photomultiplier is recorded by a spectral electronic circuit. In view of the fact that there is no room under the photomultiplier cover for a large narrowband amplifier, a cathode follower is built into the photomultiplier cover; this cathode follower transforms the high output impedance of the photomultiplier to a low impedance, as a result of which the stray effects induced in the cables connecting the cover to the amplifier are reduced practically to zero. In order to eliminate the interference and stray effects the filament of the cathode-follower tube is supplied from a storage battery, while the plate circuit is supplied from a stabilized rectifier. The narrowband amplifier used is of the factory-manufactured U2-6 type. The frequency of the amplifier can be tuned over the range from 18 Hz to 30 kHz and has a sensitivity of several microvolts for a narrow band (attenuation 40 dB per octet). The output signal of the amplifier is applied to a synchronous detector which is synchronized with the shutter. The time constant of the detector is switchable and ranges from 0.25 to 10 sec (the effective passband ranges from 4 to 0.1 Hz). The output stage is differential and is specially chosen for connection to an ÉPP-09 automatic recording potentiometer which records the detected signal.

All of the electronic circuits and supply sources are connected to the line through a ferroresonant stabilizer. The gain drift of the entire system does not exceed 5% per working day, and there is practically no null drift (after warmup). The system is tuned to a frequency of 40 Hz. This frequency is very advantageous from the point of view of suppressing stray line pickups.

The block diagram of the measurement apparatus without the supply source, the apparatus for measuring $n_e$, and the vacuum system is shown in Fig. 8. Here 1 is the generator tube; 2 is the measurement tube; 3 is the microwave resonator for measuring $n_e$; 4 is the DFS-12 spectrometer; 5 is the photomultiplier (FEU-38 or FEU-39) with cathode follower; 6 is the obturator (40 Hz); 7 is the U2-6 amplifier; 8 is the synchronous detector, and 9 is the ÉPP-09 automatic recording potentiometer.

**3.** In view of the fact that the electron concentration in the discharge is of cardinal significance for our work, we used the radio-frequency method for these measurements. All radio-frequency methods of investigating a discharge have a completely different physical basis than probe methods; namely, radio-frequency methods measure the dielectric constant of the plasma at a high frequency, this dielectric constant being uniquely related to the electron concentration. The basic advantage of radio-frequency methods over probe methods lies in the fact that they do not require intrusion into the plasma with metal probes; because of this a weak influence on the plasma is ensured, and additional sources of plasma contamination are eliminated. In the majority of cases radio-frequency methods are in no way connected with the specific form of the electron velocity distribution. Since this distribution frequently deviates greatly from Maxwellian [25] and still more often it is not known at all, it follows that radio-frequency methods have a substantial advantage over probes. Of course, due to the absence of a relation to the velocity distribution of the electrons, radio-frequency methods do not allow measurement of the electron temperature (more precisely, they allow it, but with great difficulty), but in return the electron concentration can be measured directly and considerably more reliably than by probe methods. Therefore, radio-frequency methods more than any are suitable for solving our problem, since for our measurements, as we have already indicated above, it is necessary to know the electron concentration very accurately, while the temperature is of secondary significance.

There are several radio-frequency methods of measuring electron concentration. The resonator method [26, 27], which we chose, is not the simplest of them, but it is most suitable both with respect to the range of measured concentrations and with respect to sensitivity and applicability to narrow tubes. Running ahead, we shall indicate that measurements by the resonator method showed the error of probe measurements under our conditions: the electron concentration measured by probes [20] turned out to be overestimated by a factor of 15–20.

In order to find the relation between the concentration $n_e$ and the shift $\Delta f/f$ of the resonant frequency of the resonator it is common practice to use [26, 27] perturbation theory, within the framework of which the desired relationship turns out to be linear. This approximation, however, is valid only as long as the plasma frequency of the electrons $(\omega_p = (4\pi n_e e^2/m)^{1/2}$, where e, m are the charge and mass of an electron) is small compared with the working frequency $\omega = 2\pi f$. For $\omega_p \sim \omega$ the linearity of $\Delta f/f = \varphi(n_e)$ is violated. Thus, the range of measurable concentrations turns out to be bounded from above by a magnitude of the order of the critical concentration $n_{cr} = m\omega^2/4\pi e^2$ for which $\omega_p = \omega$. Under these conditions it frequently turns out that the linearity of $\Delta f/f = \varphi(n_e)$ is already violated for $n_e < n_{cr}$. In view of the quadratic dependence of $n_{cr}$ on $\omega$, it is difficult to raise the upper boundary of the range of measurable concentrations by increasing the frequency. Figure 9 shows a linear characteristic 1 for the $TM_{020}$ oscillation mode in our resonator, calculated within the framework of perturbation theory according to the equation

$$\frac{\Delta f}{f} = 4.02 \cdot 10^{-3} \, n_e/n_{cr}. \qquad (II.1)$$

Curve 2 in Fig. 9 is the result of an "exact calculation" [27] by means of the boundary conditions on the boundaries in the resonator. As is evident, the more exact calculation already yields a deviation from linearity for $n_e \sim n_{cr}$. However, this calculation does not consider many effects (for example, the presence of the discharge tube, selfscreening of the plasma, etc.), so that it is necessary to use experimental means to obtain that portion of the characteristic corresponding to $n_e$ in order to expand the range of measurement of $n_e \gtrsim n_{cr}$ without raising the working frequency.

For this purpose we used a combination of the resonator method with spectroscopic measurements [28]. The point is that very often the dependence of the intensity of the spectral lines radiated by the discharge on $n_e$ is known. In particular, when some excited states of the atoms are populated by electron impact directly from the ground or metastable states, the intensity of the corresponding spectral lines depends linearly on $n_e$. In such cases it is possible, based on microscopic measurements, experimentally to obtain the nonlinear part of the resonator characteristics $\Delta f/f = \varphi(n_e/n_{cr})$ right up to several $n_{cr}$. Namely, such a situation occurs in the case of a discharge in the helium−neon mixture which we investigated. From experiments [18, 29] it is well known that the intensity of the majority of helium lines is proportional to $n_e$ at moderate discharge currents. Simultaneous measurements of the intensity of helium lines and the shift of the resonant frequency as a function of discharge current allowed the experimental characteristic 3 to be obtained (Fig. 9):

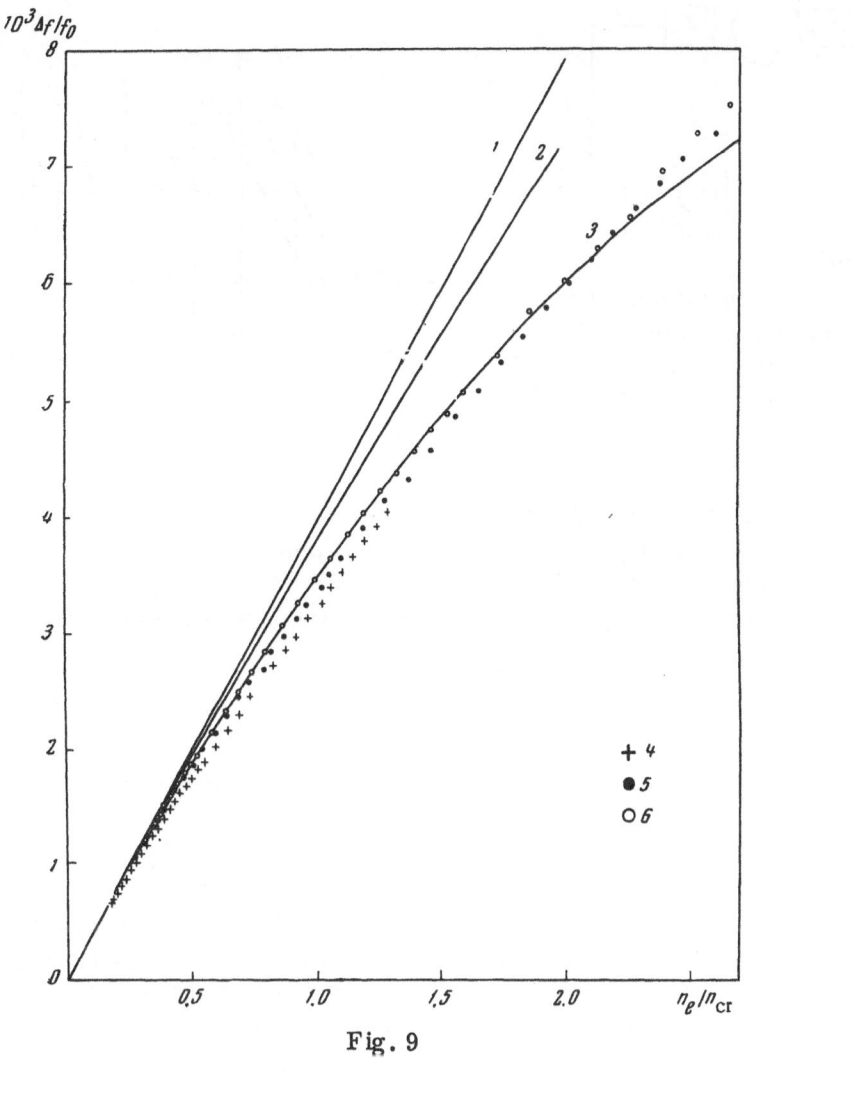

Fig. 9

$$\frac{\Delta f}{f} = 4.02 \cdot 10^{-3} \frac{n_e}{n_{\mathrm{cr}}} \left(1 - 0.12 \frac{n_e}{n_{\mathrm{cr}}}\right), \tag{II.2}$$

whose initial portion, as might be expected, coincides with the linear characteristic 1 calculated in the $\omega > \nu_e$ approximation ($\nu_e$ is the collision frequency of electrons in the plasma). The points in Fig. 9 were obtained at the following pressures (in torr) of the He:Ne mixture (10:1): 4) 0.7; 5) 1.3; 6) 2.0. From Fig. 9 it is evident that the deviations from linearity already begin at $n_e \sim 0.5\, n_{\mathrm{cr}}$ (i.e., most of the characteristic was obtained experimentally).

4. In order to choose the resonator parameters the expected values of concentration $n_e$ and collision frequency $\nu_e$ in our discharge can be estimated approximately on the basis of data available in the literature on the drift velocities of electrons [30] with allowance for the Blane law of mobility in gas mixtures [31].

As is well known, the current density j in a discharge is connected with the mean electron drift velocity $\overline{V}_d$ by the relationship $j = n_e e \overline{V}_d$. Under our conditions the potential gradient E in the discharge is of the order of 20 V/cm for $p \sim 1$ torr, so that [30, 31] $\overline{V}_d \approx 10^7$ m/sec. For a

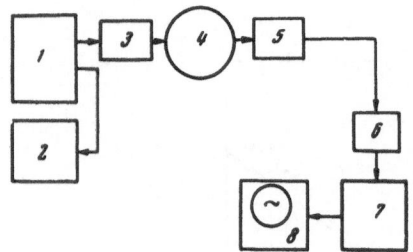

Fig. 10. Block diagram for measurement of the electron concentration. 1) G4-10A oscillator; 2) ShGB-S wavemeter; 3, 5) attenuators; 4) resonator; 6) detector; 7) U2-4 amplifier; 8) S1-19 oscilloscope.

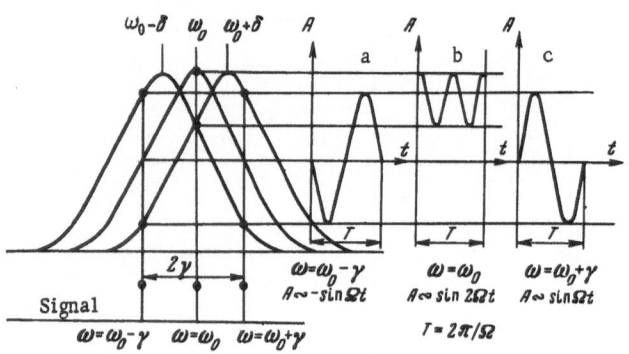

Fig. 11

current of 30 mA ($j \approx 300$ mA/cm$^2$ = $10^9$ CGSE/cm$^2$) we find $n_e \sim 10^{11}$ cm$^{-3}$. In order to estimate $\nu_e$ we make use of its definition [30]: $\nu_e = n_e E / m \bar{V}_d$. Substituting the values of E and $\bar{V}_d$ indicated above into this expression, we find $\nu_e \sim 0.5 \cdot 10^{10}$ sec$^{-1}$ for $p \sim 1$ torr. These estimates allow us to obtain the following criteria for choosing the working frequency:

$$n_{cr} > 10^{11} \text{ cm}^{-3}$$
$$2\pi f_0 = \omega_0 \gg 0.5 \cdot 10^{10} \text{ sec}^{-1}.$$

(II.3)

Hence we find that the working frequency $f_0$ must satisfy the conditions

$$f_0 > 3000 \text{ MHz}.$$

(II.4)

In choosing the resonator and the working oscillation modes we should, based on considerations of small perturbations, strive to make the resonator volume considerably larger than the plasma volume; in this connection it is necessary to choose a working oscillation mode which has the highest possible sensitivity for a stipulated resonator volume.

The cylindrical symmetry of the discharge determines the choice of the cylindrical resonator; in such a resonator the stated conditions are satisfied by the TM$_{020}$ oscillation mode. Among other families the TM$_{0m0}$ family is characterized by the highest concentration sensitivity, whereas the TM$_{020}$ mode has a considerable concentration of the electric field toward the axis (the use of oscillation modes of higher order is inexpedient due to the difficulty of excitation and the possibility of superposition of oscillation modes). The TM$_{0m0}$ family is also convenient in that the resonant frequencies do not depend on the resonator height.

Based on these considerations, a resonator having a radius $a = 5.9$ cm, a height h = 4 cm, and a working oscillation mode TM$_{020}$ ($f_0 = 4464$ MHz, $n_{cr} = 2.4 \cdot 10^{11}$ cm$^{-3}$) was chosen. The resonator was made of cuprite; the sides were soldered to the end covers with silver in a hydrogen atmosphere, and after soldering the resonator was subjected to an electrical polishing. In order to provide for insertion of the discharge tube with the plasma, the resonator was made dismountable; the line of separation is such that it does not intersect the high-frequency currents and thus has no effect on the resonator Q. The apertures for the tube in the end covers have a diameter of 0.8 cm, and the optical radiation from the discharge is extracted from the resonator through a 3.6 × 1.6 cm aperture in a side. All of the apertures have supercritical dimensions for the working frequency and have little effect on the Q. The coupling of the re-

sonator is accomplished with two loops inserted by means of an RK-29 cable through a tube
having a diameter of 0.8 cm.

All of the measurements were carried out in a quartz tube having an inside diameter $d_0 =$
0.35 cm (for a description of the measurement tube (see § II.1). The use of a quartz tube is
very essential, since in conventional glasses the temperature coefficient of the dielectric constant $\varepsilon$ is so high that the variation of $\varepsilon$ of the glass during the heating of the tube by the discharge can completely compensate the frequency shift caused by the plasma.

**5.** The measurements of the resonator frequency shifts were carried out using the standard "transit" scheme (Fig. 10). In order to improve the accuracy of tuning to the resonance
frequency we used the residual pulsations of the discharge current in the tube. The tuning
method is illustrated by Fig. 11. The pulsations of the discharge current (i.e., of $n_e$) cause
pulsations of the resonant frequency of the resonator, $\omega_{res} = \omega_0 (1 + \delta \sin \Omega t)$, where $\Omega$ is the
frequency of the pulsations. If an unmodulated signal is applied to the resonator, then the sig-
nal which has passed through it will be modulated; the frequency and phase of the modulation
are determined by the detuning between the average values of the resonant frequency and the
oscillator frequency. For a detuning of the oscillator to the right $(\omega = \omega_0 + \gamma)$ or to the left
$(\omega = \omega_0 - \gamma)$ from the average value of the resonator frequency, the modulation frequency is
equal to the pulsation frequency $\Omega$, and the phases are opposite for both detunings (Figs. 11c
and 11a). For exact coincidence of the frequencies of the oscillator and the resonator, the
modulation frequency is doubled and is equal to $2\Omega$ (Fig. 11b). Thus, the tuning is carried out
according to an oscilloscope at double the modulation frequency and at the signal minimum.
With the discharge off, tuning is carried out according to the residual modulation of the G4-10A
klystron oscillator. After tuning to resonance, the oscillator frequency is measured by a
ShGV-S heterodyne wavemeter with an accuracy ± 0.05 MHz.

No temperature stabilization of the resonator was used. Experience showed that during
operation with the discharge the resonator proper was heated by no more than 10 deg C, and due to
the relatively low Q ($\leq 2500$) the corresponding frequency change could not be detected. However, the heating of the discharge tube, notwithstanding the use of quartz, turned out to be substantial. The point is that the tube, unlike the resonator which has a fairly large volume and
mass, is heated to 200-300°C. Since the temperature coefficient for quartz is equal to $(d\varepsilon/dT)/$
$\varepsilon \approx 5 \cdot 10^{-5}$ deg$^{-1}$, it follows that heating by 200°C leads to a change in $\varepsilon$ amounting to $\Delta\varepsilon/\varepsilon \sim$
$10^{-2}$; this already yields a fully detectable change in resonator frequency. The correction to
the frequency due to heating of the discharge tube is introduced as follows. Before switching on
the discharge, the resonant frequency of the resonator with a cold tube is measured. Then a
series of standard measurements is carried out at various discharge currents right up to a
certain current $i_{br}$. When the measurements are completed at $i_{br}$ the discharge is switched
off, and the frequency of the resonator with the hot tube is measured immediately. Then the
correction

$$\delta_{br} = f_{cold} - f_{hot}$$

to the frequency is measured for $i_{br}$ (the heating of the tube reduces the resonant frequency)
and then for each value of current $i$ the correction is introduced according to the quadratic
formula

$$\delta_i = \delta_{br} \left( \frac{i}{i_{br}} \right)^2$$

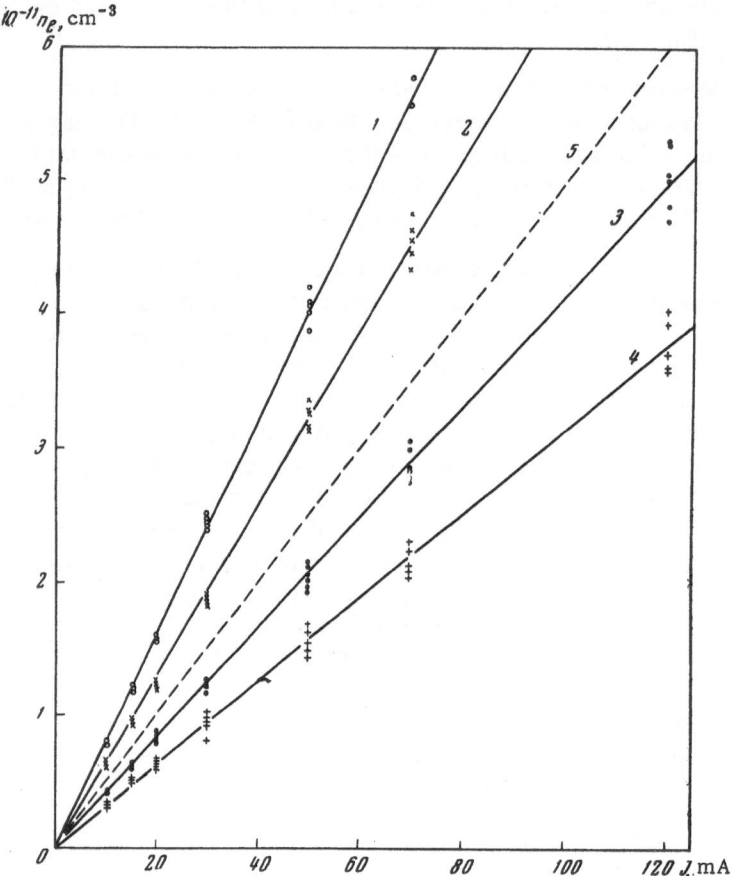

Fig. 12. The results of the measurements of $n_e$. Pressure of
the gas mixture (in torr): 1) 2.7; 2) 2.0; 3) 1.3; 4) 1.0; 5) 1.2
in a tube with $d_0 = 0.3$ cm [32].

in accordance with the fact that the temperature of the tube is approximately proportional to
the power (i.e., to $i^2$). Experiments with various values of $i_{br}$ have shown that the quadratic
correction reflects the actual change of the frequency. Usually, the value of $\delta_{br}$ did not exceed
5% of the maximum frequency shift caused by the plasma; special experiments substantiated
the fact that the values of $\delta_{br}$ are in agreement with those calculated from the temperature coefficient of $\varepsilon$ for quartz with allowance for the temperature of the tube and the shape coefficient
of the system.

Figure 12 shows the result of measurements of $n_e$ for four full pressures of the He − Ne
mixture (10:1) which are typical of our work [28]. The scatter of the experimental points is
caused chiefly by the inaccuracy of reading the pressure from the oil manometer, rather than
by the error in measuring $n_e$. For comparison purposes Fig. 12 also shows the graph obtained
by an analogous method [32] in a narrower tube ($d_0 = 0.30$ cm) and for a different mixture ratio
(5:1). Taking account of the different experimental conditions, the agreement between the results can be considered good. Let us note here that the measured value $n_e = 1 \cdot 10^{11}$ cm$^{-3}$ for
$p = 1$ torr and $\mathscr{I} = 30$ mA is in complete agreement with our preliminary estimates.

Besides measurements of the frequency shift, monitoring measurements were also carried out of the change in the reciprocal $\Delta(1/Q)$ from which estimates of $\nu_e$ were obtained under
our conditions. At a pressure of 1.3 torr a value $\nu_e \sim 0.8 \times 10^{10}$ sec$^{-1}$ was found, which is in

full agreement with the preliminary estimates and indicates good fulfillment of the condition $\omega \gg \nu_e$. Note by the way that fulfillment of this condition was verified from the very beginning by the insignificant deterioration of the Q when the discharge was switched on. If the condition $\omega \gg \nu_e$ is not fulfilled, then the Q drops so rapidly when the discharge is switched on that it is impossible to measure the frequency shift $\Delta f/f$.

**6.** In order to carry out our measurements it turned out to be more advantageous to fill the measurement tube with a mixture of helium and neon rather than with pure neon. The use of the mixture is advantageous because in the presence of helium the electron temperature increases, and the populations of the high neon levels increase. Because of this the intensities of the corresponding lines also increase along with the absolute amplitudes of their modulation. Therefore, all the measurements were carried out in a mixture of helium and neon in the ratio 10:1, and the electron temperature in the mixture corresponded to $kT_e \sim 8\text{--}10$ eV [32].

Let us note here that $T_e$ is not of cardinal significance for our measurements, since the energies of all of the transitions investigated do not exceed 1 eV, which is considerably less than $kT_e$. For this same reason the deviations from a Maxwellian electron velocity distribution, which are observed in the range of high velocities [25], are similarly nonessential for us. From the experimental data of [32], $T_e$ in helium − neon discharge tubes is in thorough accord with the similarity laws; therefore, we did not measure $T_e$, but calculated it from the universal relationships derived by Éngel and Shtenbeck [33]. The accuracy of these calculations is quite sufficient our purposes.

The basic measurement work consisted of measuring the intensity modulation amplitudes of the spontaneous lines, beginning with the required levels (those modulated by the external irradiation, and those modulated due to the transfer of excitations), and the parallel measurement of the electron concentration in the discharge. The measurements were carried out in "runs", each of which yielded 8 points for 8 different values of discharge current (i.e., of $n_e$): 10, 12, 15, 20, 30, 50, 70, and 120 mA. For each value of discharge current, signals which corresponded to the modulation amplitude of each line were recorded on the automatic recording potentiometer, and at the same time the electron concentration $n_e$ was measured. Because of the slight intensity modulation depth (5-20% for levels interacting with the irradiation, and 1-5% for other levels) the illumination of the photomultiplier with unmodulated radiation produced intense noise which often turned out to be larger than the signal. Therefore, the recording of the lines was carried out slowly with a long time constant (the time required to record a line was 60 sec, and the time constant is 3 sec).

The parallel measurements of $n_e$ allowed serious systematic errors connected with the time difference between the measurements to be avoided. A "run" of measurements for eight values of current took one hour. After each "run" the measurement tube (and somewhat more rarely the oscillator tube) was evacuated and filled with a new mixture; then the measurements were repeated. The measurements were carried out for four different mixture pressures: 1, 1.3, 2.0, and 2.7 torr.

## § III.  Results and Discussion

**1.** In the work described in the present paper we investigated the following transitions for collisions of excited neon atoms with electrons (the wavelengths of the lines* for which the

---

*We used the Racah system for denoting the neon levels (see, for example, [4]). The primes in the notation show that the levels belong to the $2p^5\,{}^2P_{1/2}$ state of the original Ne II ion; the notation for the levels belonging to the $2p^5\,{}^2P_{3/2}$ state of the original Ne II ion does not have primes.

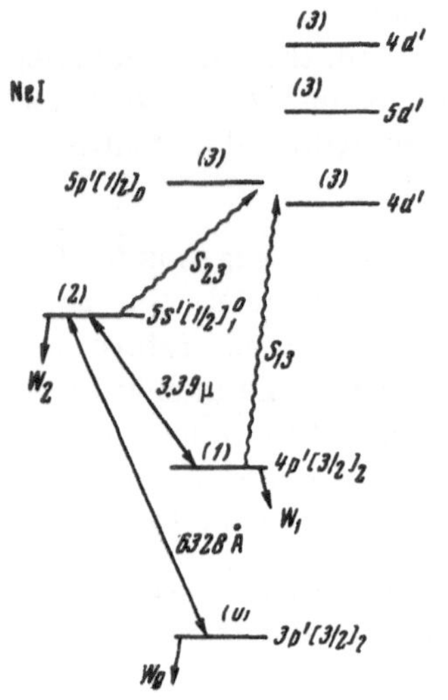

Fig. 13

measurements were carried out are indicated in the parentheses):

$$5s'\left[\frac{1}{2}\right]_1^0 - 4d', \ 5d', \ 6d', \ 5p'\left[\frac{1}{2}\right]_0;$$
$$\text{(6046Å)} \quad \text{(5902Å)} \quad \text{(5145Å)} \quad \text{(4810Å)} \quad \text{(3057Å)}$$

$$4p'\left[\frac{3}{2}\right]_2 - 4d', \ 5d', \ 6d'.$$

The diagram of the neon levels is shown in Fig. 13. The nd' configurations contain four levels each, of which three are very closely spaced (our equipment cannot resolve them), while the fourth levels are spaced at a distance $\gtrsim 10$ cm$^{-1}$ from the first three. The radiation from the separate fourth levels amounts to just 10-15% of the radiation from the closely spaced triplets, so that measurements for the lines indicated above can be assumed valid for the nd' configuration as a whole (with an accuracy up to the radiation from the fourth levels).

The most intense lines were chosen for the purposes of our work, beginning with the nd' levels and ending at the 3p' [$^3/_2$]$_2$ levels. The 5p' [$^1/_2$]$_0$ level was investigated for the most intense of the lines ending at the 3s' levels. In order to observe the original levels (whose populations were modulated by external irradiation), we chose the line 5s' [$^1/_2$]$_1$ – 3p [$^3/_2$]$_1$, $\lambda = 6046$ Å. This line is not the most intense in the "multiplet" beginning with the 5s'[$^1/_2$]$_1^0$ level; however, it was necessary to avoid measurements for the more intense lines $\lambda = 6328$ Å in view of the serious systematic errors caused by the scattering of laser radiation having this wavelength in the walls of the measurement tube. Because of the fact that the population modulation amplitudes at the levels modulated by the external irradiation are rigidly interrelated by the relationships (I.12), it is possible to keep track of all these levels according to the line $\lambda = 6046$ Å.* All of the final levels indicated (with the exception of 5p' [$^1/_2$]$_0$) were investigated for irradiation of the discharge by two laser lines $\lambda = 6328$ Å and $3.39\mu$ in accordance with §I.3. The experimental data are given in Fig. 14 in the form of graphs on which the reciprocal electron concentration on the discharge axis ($x_0 = 1/n_e$) is laid off along the axis of abscissas, while the ratio between the intensity modulation amplitudes for lines beginning with the original and final levels (the ratios between the signals at the output of the recording system) $y = \Delta J_2 / \Delta J_3$ is laid off along the axis of ordinates. The darkened points in Fig. 14 correspond to irradiation of the gas by the $\lambda = 3.39\mu$ line, while the open points correspond to irradiation by $\lambda = 6328$ Å. The method of least squares is used for the experimental points to choose linear functions of the type $y = ax + b$ (1.23-24), and the corresponding parameters a and b (1.25-26) are determined. These parameters are given in Table 1.

For two levels (4d' and 5p') the slope of the straight lines reveals a dependence on the pressure of the mixture. As was already said in §I, our model eliminates transitions which accompany atom – atom collisions from consideration. The dependence of the slope of the straight lines in Figs. 14a and 14b on pressure is evidently connected precisely with the atom – atom collisions causing the 5p' [$^1/_2$]$_0$ – 4d' transition ($\Delta E = -0.1$ eV). However, a detailed investigation of these processes is not included in our problem, and therefore in order to obtain data which are free from the pressure effect we shall extrapolate the slope of the straight line

---

*For operation with the ultraviolet line $\lambda = 3057$ Å (the 5p' [$^1/_2$]$_0$ level) we made use of the $\lambda = 5433$ Å line (5s' [$^1/_2$]$_1^0$ – 3p [$^1/_2$]$_1$) instead of the $\lambda = 6046$ Å line for the purpose of comparing the two lines (3057 and 5433 Å) by means of one FEU-30 photomultiplier.

TABLE 1

| Level | Modulating line | Pressure, torr | Number of points | $a$, $10^{-12}$ | $k$ | $k$ | | $S_t/W_3^R$ for $n_e=3\cdot10^{11}$ cm$^{-3}$ |
|---|---|---|---|---|---|---|---|---|
| | | | | | | $a_{20}/a_{21}$ | $b_{20}/b_{21}$ | |
| 4d′ | 6328A | 1—2.7 | 80 | 0.112±0.002 | 0.14±0.02 | 0.47±0.05 | 0.44±0.09 | 0.38 |
| | | 2.7 | 42 | 0.138±0.003 | 0.31±0.02 | | | |
| | | 2.0 | 46 | 0.164±0.002 | 0.28±0.02 | | | |
| | 3.39μ | 1.3 | 40 | 0.177±0.002 | 0.32±0.02 | | | |
| | | 1.0 | 38 | 0.210±0.004 | 0.38±0.04 | | | |
| | | 0 (extrapolation) | 166 | 0.24±0.02 | 0.32±0.02 | | | |
| 5d′ | 6328A | 1—2.7 | 81 | 0.418±0.006 | 0.76±0.05 | 0.60±0.02 | 0.67±0.07 | 0.54 |
| | 3.39μ | 1—2.7 | 68 | 0.693±0.008 | 1.13±0.05 | | | |
| 6d′ | 6328A | 1—2.7 | 76 | 4.07±0.21 | 12.8±1.0 | 0.68±0.06 | 0.64±0.09 | 0.94 |
| | 3.39μ | 1—2.7 | 58 | 5.96±0.22 | 20.1±1.2 | | | |
| 5p′$[\frac{1}{2}]_0$ | 3.39μ | 2.7 | | 2.14±0.08 | 2.10±0.36 | | | |
| | | 2.0 | | 1.39±0.03 | 2.08±0.18 | | | |
| | | 1.3 | | 0.985±0.028 | 1.64±0.22 | | | |
| | | 1.0 | | 0.820±0.020 | 1.87±0.22 | | | |
| | | 0 (extrapolation) | | 0.35±0.15 | 1.70±0.5 | — | — | 1.45 |

to zero pressure linearly by the method of least squares. A similar extrapolation for 4d′ is possible with an accuracy of better than 5%, while for 5p′ it is possible with a somewhat poorer accuracy; the corresponding results are shown in Table 1.

The slight scatter of the experimental points for the 4d′, 5d′, and 5p′ $[^1/_2]_0$ levels (Figs. 14a, 14b, and 14d) allows straight lines to be drawn through them reliably; if there are departures from linearity, then they are so slight that in these cases we can fully assume the contribution of step-by-step transitions to be negligibly small compared with the contribution of the direct transition (see § I.6). For the 6d′ level (Fig. 14b) the accuracy of the measurements is substantially lower due to the weak signal. It is not excluded that the scatter of the points masks the actual departures from linearity caused by the step-by-step transitions. Thus, we encounter the situation considered above [Eq. (I.36)]; this is all the more probable since between 6d′ and the original states (4p′ and 5s′) there is a set of levels by which step-by-step transitions are possible. In accordance with § I.6 the approximation of the function (I.36) by a linear function must lead to an overestimation of the slope parameter $a$ (i.e., to an underestimation of the observed values $<\sigma v>$). The estimate made by choosing functions of the type (I.36) does not exceed a factor of $\sim1.5$-2, so that the final value of $<\sigma v>$ for transitions into the state 6d′ can be underestimated by a factor $\sim1.5$-2.

It is obvious that the experimental data fit completely within the framework of the model considered in § I. Its validity is also substantiated by the dependence of the parameters of the straight lines on the irradiating lines. As was expected (see Eqs. (I.23) and (I.24)), the replacement of the irradiating line $2\to0$ having a shorter wavelength by the line $2\to1$ having a longer wavelength (for a constant upper level 2) leads to an increase of both the parameters $a$ and $b$, the parameters varying in identical ratios, as is evident from Table 1 (with an accuracy up to experimental errors). The preservation of the phase $\Delta N_3$ (in our experiments it always

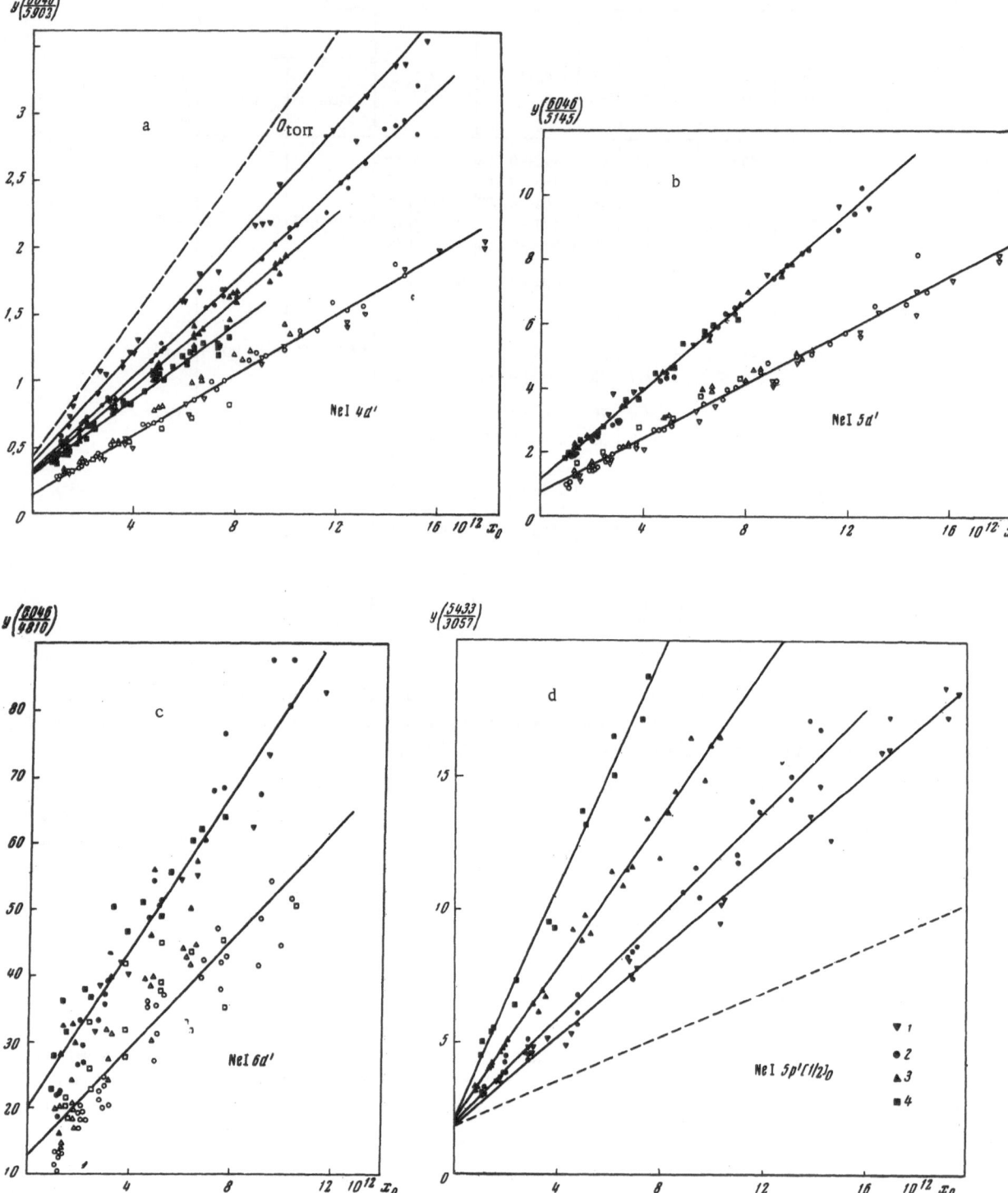

Fig. 14. Results of measurements of the intensity modulation. Pressure of gas mixture: 1)
1.0 torr; 2) 1.3 torr; 3) 2.0 torr; 4) 2.7 torr. The dashed lines in Figs. a and b show the results of extrapolation of the results to zero pressure. The open points were obtained for irradiation of the gas mixture by a 6328 Å line, while the darkened points were obtained for irradiation of the mixture by the 3.39μ line.

coincided with the phase $\Delta N_2$) indicates that the contribution of the $1 \rightarrow 3$ excitation process to $N_3$, although substantial, is nevertheless smaller than the contribution of the $2 \rightarrow 3$ process (see Fig. 13). The data of Table 1 show that the coefficient k — the upper bound of the relative error due to neglect of the $0 \rightarrow 3$ processes (Fig. 13) — does not exceed ~ 0.6 (i.e., ~ 60%). However, the true error, as was indicated in §I.4, must be considerably smaller, which is confirmed by theoretical estimates (see below).

In order to check the influence of external irradiation on the excitation rates $Q_i$ of the levels from the ground state we undertook monitoring measurements of the modulation depths of $n_e$ and $N_i$. They showed [20] that under our conditions the correction $\delta$ from (I.32) has a value ~ 1% on the average and never exceeds ~ 5%. Therefore, it is clear that $Q_i$ is practically independent of irradiation, and the correction $\delta$ can be ignored. Special measurements of the intensity of the spontaneous lines of helium and neon were used to monitor the distribution of the electron concentration $n_e$ over the cross section of the discharge tube. The measurements showed that the intensities, and therefore $n_e$, increase monotonically in the direction from the wall to the tube axis.

**2.** It should be noted that we irradiate the gas with linearly polarized radiation, since the laser has windows at the Brewster angle. Therefore, the changes $\Delta N$ in the populations of levels 1 and 2 are caused exclusively by $\pi$-transitions at $\Delta M = 0$ (M is the magnetic quantum number; here the quantization axis is stipulated by the direction of the electric vector of the laser radiation). This leads to a nonuniform distribution of $\Delta N$ over states with different values of M [34]. At the upper level 2 ($5s'[^1/_2]_1^0$), where all three states $M = 0 \pm 1$ make a contribution to the $\pi$-transitions, the nonuniformity is negligible, $\Delta N_0 : \Delta N_{\pm 1} = 4 : 3$. However, at the lower levels 1 and 0 ($4p'3p'[^3/_2]_2$) there are five states each $M = 0, \pm 1, \pm 2$, of which only three ($M = 0, \pm 1$) participate in the $\pi$-transitions; because of this the changes of the populations $\Delta N_{\pm 2} \sim 0$ and the nonuniformity can be substantial [34]. However, as experience shows [34], collisions between atoms cause transitions between M-states, so that $\Delta N_{\pm 2}$ are nevertheless nonvanishing for a nonzero gas pressure.

The indicated nonuniformity of the distribution of $\Delta N$ over the M-states can lead to a situation in which the spontaneous emission lines for which the measurements are carried out turn out to be partially polarized, so that their radiation pattern in the plane perpendicular to the laser axis will be anisotropic. Actually, a considerable anisotropy of the radiation [34] is observed for transitions from the zero level ($3p'[^3/_2]_2$). For us, however, the nonuniformity of $\Delta N$ and the anisotropy of the radiation are nonessential for the following reasons. First, we trace the original states according to one line $\lambda = 6046$ Å which begins from the upper level 2. Calculation shows that for a nonuniformity $\Delta N_0 : \Delta N_{\pm 1} = 4 : 3$ the anisotropy of the radiation pattern for this line is a negligibly small quantity — of the order of several percent. Second, the possible nonuniformities of the distribution of $\Delta N$ over the M-states, which could be produced at the final levels for excitation by electrons, are averaged by measurement: we observe either the configuration as a whole (4d', 5d', 6d') or the single level $5p'[^1/_2]_0$ with the only possible state $M = 0$. We also add the fact that for excitation by electrons the cross section for excitation from a given M-state decreases as M increases. Therefore, the main contribution to the excitation is made by states with $M = 0, \pm 1$, where the nonuniformity of $\Delta N$ is small, while the absence of excitation from the state $M = \pm 2$ has little effect on the final results.

Thus, it is clear that we are quite able to use the model of §I for finding the coefficients of the processes of interest to us.

**3.** Let us turn once more to Eqs. (I.28)–(I.30) from which we find the rate coefficients from the experimental data. Let us first of all indicate Eq. (I.30) which does not require any extraneous data. The quantity $S_3/W_3^R$ which is given by this equation characterizes the role of impact de-excitation of level 3 by electrons. It is very essential that this quantity be determined solely from experimental data.

Making use of relationship (I.28), we can calculate the rate coefficients $<\sigma_{23}v>$ for processes $2 \rightarrow 3$ ($5s' - nd'$), and then we can use (I.29) to calculate the coefficients $<\sigma_{13}v>$ for the processes $1 \rightarrow 3$ ($4p' \rightarrow nd'$) also. For this purpose, however, it is necessary to have additional data on the Einstein coefficients for the lines for which the measurements are carried out, as well as to know the resultant rate of radiative de-excitation of the levels. Certain of the quantities which we require are measured experimentally, while the others can be found quite reliably by theoretical calculations using a computer (according to the method in [35]). The various ratios of the transition rates can be calculated especially reliably.

The quantity $A_2$ in (I.28) — the Einstein coefficient for the line $\lambda = 6046$ Å — was measured experimentally by us [22] and is equal to $A_2 = 0.44 \cdot 10^6$ sec$^{-1}$. The "branching ratio" $W_3^R/A_3$ for the 4d' configuration was obtained by a combination of calculations and experiments [20]. Radiative transitions to the 2p (ground), 3p, and 4p levels are possible from the 4d' levels. Therefore,

$$\frac{W_3^R}{A_3} = \frac{1}{A_3} \left( A_{4d'-2p} + \sum_{3p} A_{4d'-3p} + \sum_{4p} A_{4d'-4p} \right).$$ (III.1)

The transitions 4d' − 3p, 3p' are situated in an accessible range of the spectrum, so that we were able to measure the ratio

$$\xi = \frac{1}{A_3} \sum_{3p} A_{4d'-3p}$$ (III.2)

experimentally ($A_3$ in this case refers to the line $\lambda = 5902$ Å) and to obtain the value $\xi = 4.3 \pm 0.5$ for it. The further refinement of the factor $W_3^R/A_3$ can be achieved by using the results of semiempirical calculations of the Einstein coefficients according to the method given in [35]. The results of the calculations allow us to obtain the value 6.5 for $\xi$, which is in satisfactory agreement with the experimental value $\xi = 4.3 \pm 0.5$. Thus, it is possible to adopt $\xi \approx 5$ and to consider the theoretical calculations of the relative Einstein coefficients to be reliable within the limits of ~30%. To our regret, the two other terms in (III.1) cannot be determined experimentally, and we can only assume that the corresponding calculations are fairly reliable.

The calculated values for the three terms in (III.1) are respectively equal to $5 \cdot 10^9$, $2 \cdot 10^7$, and $0.5 \cdot 10^7$ sec$^{-1}$. The Einstein coefficient for the transition to the ground state 2p greatly exceeds the remaining terms. However, the imprisonment of radiation plays an essential role for this transition, and we must use the effective value of the Einstein coefficient [36]

$$A_{\text{eff}} = AT(R),$$ (III.3)

where

$$T(R) = \frac{1}{\sqrt{\pi}} \int_{-\infty}^{\infty} \exp\left[-(x^2 + k_0 R e^{-x^2})\right] dx = \begin{cases} 1 - \dfrac{k_0 R}{\sqrt{2}}, & k_0 R \ll 1, \\[2mm] \dfrac{1}{k_0 R \sqrt{\pi \ln k_0 R}}, & k_0 R \gg 1. \end{cases}$$ (III.4)

is the imprisonment factor which represents the probability of photon transit through a distance R without absorption [36]. In (III.4) $k_0$ is the absorption coefficient in the center of the Doppler contour of the line. For our discharge tube $2R = d_0 = 0.35$ cm, so that its optical thickness is equal to $\tau = k_0 R \approx 650$. For $\tau \gg 1$, making use of the corresponding expression for T(R) from (III.4), we find that under our conditions $A_{eff} \approx 0.2 \cdot 10^7$ sec$^{-1}$. Thus, the main contribution to $W_3^R/A_3$ is made by the 4d' → 3p, 3p' transitions whose role is estimated experimentally. The remaining transitions from the nd' levels lead to a comparatively small correction ~ 30%. Based on all this we took $W_{4d'}^R/A_{5902} \approx 7$. In similar fashion we found $W_{5d'}^R/A_{5145} \approx 12$, for the 5d' configuration. The values of the "branching ratios" for 6d' and 5p' $[^1/_2]_0$ were obtained by pure calculation and were found to equal $W_{6d'}^R/A_{4810} \approx 20$, $W_{5p'}^R/A_{3057} \approx 4$.

Let us note here that the least quantity of additional data is required to find the coefficients $\langle \sigma_{23}v \rangle$ for the 2 → 3 (5s' − nd') transitions. However, the coefficients $\langle \sigma_{13}v \rangle$ for the 1 → 3 (4p' − nd') transitions are calculated from the $\langle \sigma_{23}v \rangle$ which have already been found by means of the ratio of the resultant de-excitation rates $W_1$ and $W_2$ (see Eq. (I.29)) of the 1 (4p' $[^3/_2]_2$) and 2 (5s' $[^1/_2]_1^0$) levels. Thus, this ratio is very essential from the point of view of the reliability of the results.

Regrettably, the ratio $W_1/W_2$ was not measured directly by experiment, but it would have been possible to use separate measurements of $W_2$ [37] and $W_1$ [38, 39]. However, the measurements in [38, 39] yielded an exceptionally large value ~$0.7 \cdot 10^9$ sec$^{-1}$ for $W_1$, which is all the more unexpected since there are very few transitions from the 1 (4p' $[^3/_2]_2$) level to lower states, the transition to the ground state, which could have such a large Einstein coefficient (only in the absence of radiation imprisonment), not being among them. From the level 4p' $[^3/_2]_2$ radiative transitions to the states 3s, 4s, 3d are possible, from which it is obvious that the main contribution to $W_1$ must be made by the transitions 4p' − 3s, 4s. As is well known [4], the approximate rule of $f$-sums says that the sum of the oscillator forces over all possible transitions from the state n is $\sum_{n'} f_{nn'} \sim 1$, the main contribution being given by transitions from n' = n for which $f_{nn'} \sim 1$. The contribution of the remaining transitions is considerably less than unity. From these concepts it is clear that the transitions 4p' − 4s, 4s' must correspond to $f \sim 1$, while the transitions 4p' → 3s, 3s' must correspond to $f \ll 1$. The Einstein coefficient is associated with the oscillator force by the relationship.

$$g_i A_{ik} = 0.67\, g_k f_{ki}\, \widetilde{\nu}^2, \tag{III.5}$$

where $g_k$, $g_i$ are the statistical weights of the lower and upper states, while $\widetilde{\nu}$ is the transition frequency in cm$^{-1}$. The squares of the transition frequencies differ approximately by a factor of 40: $\widetilde{\nu}^2_{4'p3s} / \widetilde{\nu}^2_{4p'4s} \sim 40$, while the oscillator forces differ by approximately an order of magnitude; therefore it can be assumed that the main contribution to $W_1$ is made by the transition 4p' − 3s, 3s'. Then, substituting the experimental value $W_1 \approx 0.7 \cdot 10^9$ sec$^{-1}$ and the statistical weight g (4p' $[^3/_2]_2$) = 5, $g_{3s} = 1$ (the statistical weight of the configuration with $l = 0$) into (III.5), we find

$$f_{3s4p'} \approx \frac{g_{4p'}}{g_{3s}} \frac{W_{4p'}}{0.67\widetilde{\nu}^2\,_{4p'3s}} \approx 6.5, \tag{III.6}$$

which is incompatible with $f_{3s4p''} \ll 1$. Therefore, it is obvious that the experimental value of $W_1$ [38, 39] is overestimated. The cause of this may lie in elastic collisions between atoms, which do not de-excite the states but yield a contribution to the quantity $W_1$ that is determined according to the methods of [38, 39]. Since the cross sections of such collisions can be very

TABLE 2

| Transition | $\Delta E$, eV | $\beta = \dfrac{\Delta E}{kT_e}$ for $kT_e \approx 10$ eV | $10^8 \langle \sigma v \rangle$ exp, cm³/sec | $10^8 \langle \sigma v \rangle$ theor, cm³/sec | |
|---|---|---|---|---|---|
| | | | | 1 | 2 |
| $5s'\left[\frac{1}{2}\right]_1^0$ —4d' | 0.14 | 0.014 | 2200 | 500 | — |
| —5d' | 0.45 | 0.045 | 370 | 41 | 260 (5p') |
| —6d' | 0.62 | 0.062 | 39 | 13 | 56 (5p') <br> 44 (6p') |
| $4p'\left[\frac{3}{2}\right]_2$ —4d' | 0.51 | 0.051 | 1230 | 1730 | — |
| —5d' | 0.82 | 0.082 | 140 | 178 | — |
| —6d' | 0.98 | 0.098 | 14 | 54 | — |
| $5s'\left[\frac{1}{2}\right]_1^0$ — $5p'\left[\frac{1}{2}\right]_0$ | 0.24 | 0.024 | ~300 | 320 | — |

Note: 1) First Born approximation; 2) second Born approximation.

large: $\sim 10^{13}$ cm⁻² [40], it follows that the overestimation of the quantity $W_1$ is quite probable. In this case finding $W_1/W_2$ according to the data of separate measurements [37-39] carried out under different conditions using different methods could lead to inaccuracies, and we consider the calculated value $W_1/W_2 \approx 1.0$ to be more reliable. This value was found from calculations of the Einstein coefficient using the method of [35] while considering radiation imprisonment for the 5s' — 2p transition according to the method described above.

Thus, the reliability of the experimental results given below for $\langle \sigma v \rangle$ is estimated as follows. For the forbidden transition $2 \rightarrow 3$ (5s'—nd') the error does not exceed approximately $\pm 50\%$. The error for the allowed transitions $1 \rightarrow 3$ (4p'—nd') contains an additional contribution both from the neglected process $0 - 3$ and from the possible errors in determining $W_1/W_2$. Finally, the result for the 5s' — 5p' transition can be considered reliable only within the limits of the factor $\sim 2$, since 1) absolutization of this result is inaccurate due to unreliable calibration of the spectral sensitivity in the ultraviolet range, and 2) due to the lack of sensitivity this transition was investigated for irradiation by only 1 line $\lambda = 3.39 \mu$. It should similarly be kept in mind that the values of $\langle \sigma v \rangle$ for the transition to the 6d' state can be underestimated by a factor of $\sim 1.5 - 2$ (see § III.1).

4. The rate coefficients $\langle \sigma v \rangle_{exp}$ found from the experimental data according to the method of § III.3 are displayed in Table 2 along with the parameters of the transitions ($\Delta E$ is the threshold energy of the transitions) [21]. The values of $\langle \sigma v \rangle$ published in our paper [20] for the transitions 5s' — 4d', 5d' differ from the corresponding data in Table 2. The point is that in [20] the electron concentration $n_e$ was measured by the probe method and, as was subsequently clarified, it was overestimated by a factor of 15-20 as against the true value. Here we present the results obtained after a serious improvement of the entire method; the decisive role here was played by the use of the resonator method of measuring $n_e$.

The experimental values of $\langle \sigma v \rangle$ in Table 2 are several orders of magnitude in excess of the values of $\langle \sigma v \rangle$ which are usual for transitions from the ground state of atoms ($\sim 10^{-8}$ cm³/sec). In our case the original states are strongly excited (n = 4 and 5), and the purely geometric cross section of the excited atoms is hundreds of times ($\sigma \sim n^4$) larger than the geometric cross section of an atom which is in the ground state. These general physical concepts find direct experimental substantiation in our results.

It is obvious that because of the large values of $\langle \sigma v \rangle$ the role of the electrons in the de-excitation of high levels of the atoms must be considerable. Actually, as the results of the last column of Table 1 show, the ratio $S_3/W_3^R$ for the states nd' are close to unity; in particular,

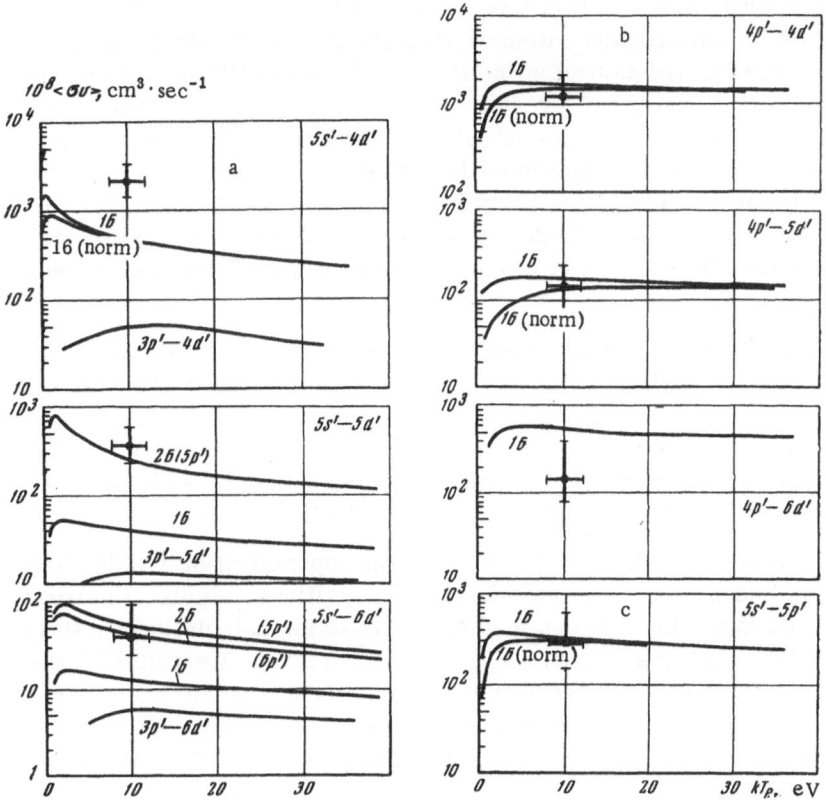

Fig. 15. Comparison of experimental and theoretical data for $<\sigma v>$.

for 6d' the rate $S_{6d'}$ of impact de-excitation is almost equal to the rate $W_{6d'}^{R}$ of radiative de-excitation. Making use of the data of Table 2, one can calculate the rates of the 5s'−nd' transitions which accompany collisions with electrons and can obtain the following sum over three

terms for $n_e \approx 3 \cdot 10^{11}$ cm$^{-3}$: $\sum_{n=4}^{6} S_{5s'-nd'} \approx 10^7$ sec$^{-1}$. Since the total sum $S_{5s'}$ is certainly

larger than this quantity, we see that the de-excitation of the 5s' $[^1/_2]^0_1$ level is determined to a considerable degree by collisions with electrons.

Thus, the results obtained here show the substantial role played by electrons in transitions between excited states of neon and can serve as a basis for estimates of the various transition rates.

**5.** As has already been mentioned in the introduction, obtaining a large quantity of data on the rates of transitions between excited states is more realistic using theoretical calculations (in particular, using the Born method as the most universal one). However, the applicability of the Born approximation to such transitions has not been proved. Therefore, the comparison of our results with calculations according to the Born method acquires special interest.

In this connection calculations of the coefficients were carried out by a semiempirical method in the generalized Born approximation [41] on an M-20 electronic computer. The results of the calculations are given in the last two columns of Table 2 (for the values of $\beta = \Delta E/kT_e$ indicated in the table); however, Fig. 15 allows a more detailed comparison between theory and experiment. Here the curves for the dependence of the calculated values of $<\sigma v>$ on $kT_e$ are given; the experimental data are plotted in the form of points with error intervals. For all of the transitions investigated the coefficients $<\sigma v>$ were calculated in the first Born approximation while neglecting normalization (see Table 2, Fig. 15). For certain transitions calculations were carried out with consideration of normalization using the R-matrix method [42], and the corresponding curves are plotted in the figure. As is evident, under our conditions consideration of normalization changes the result only slightly. For the forbidden transitions 5s'−5d', 6d' the quantities $<\sigma v>$ were also calculated in the second Born approximation with allowance for just one virtual intermediate level [43] (this level is indicated in parentheses in Table 2 and next to the corresponding curves in Fig. 15). The absence of a result from the second approximation for 5s'−4d' is due to the fact that the calculation program in its modern form is inapplicable to those cases in which the final level (4d') lies below the intermediate level (5p').

From a comparison of the theoretical and experimental data the following is clear. The first Born approximation satisfactorily describes the allowed transitions 4p'−4d', 5d' and 5s'−5p'. In the case of the 4p'−6d' transition a noticeable discrepancy in the data is observed, notwithstanding consideration of the possible systematic error connected with step-by-step transitions. It is not excluded that this discrepancy is associated with the subtler effects of the step-by-step transitions, perhaps with their interference.

For the forbidden transitions 5s'−nd' the first approximation yields sharply underestimated results and evidently is unsuitable for such transitions. Figure 15 shows that satisfactory agreement with experiment is obtained in this case for calculations in the second Born approximation which considers one virtual intermediate level. Regrettably, we were not able to calculate $<\sigma v>$ for the 5s'−4d' transition in such an approximation, but the position of the experimental point on the graph allows the assumption, by analogy with the 5s'−5d', 6d' transitions, that the result of the calculation in the second approximation would be close to the experimental results.

Thus, the comparison of the data evidently substantiates the applicability of the Born method to the calculation of the rates of transitions between excited states, at least in the $kT_e \gtrsim 10\,\Delta E$ range.

**6.** Here we note still another result of our experiments. All of the transitions considered occur between shifted neon states belonging to the $2p^5\,{}^2P_{1/2}$ state of the Ne II atomic core. The experiments showed that transitions between shifted and unshifted (the $2p^5\,{}^2P_{3/2}$ state of the Ne II core) systems of levels are not observed during electron collisions. At the same time spontaneous transitions between these systems, although weaker than those which occur within the limits of one system, nevertheless take place. This implies that evidently excitation by electrons can be treated as an essentially one-electron transition without a change of the state of the atomic core. However, for spontaneous emission two-electron transitions are clearly more probable.

The fact that the theory yields reasonable results for the rates of transitions between excited states allows us a posteriori to prove the validity of neglecting the $0 \to 3$ excitation process in Eq. (I.16). Actually, as calculations showed, the rates of the $0 \to 3$ (3p'−nd') processes are an average of 1.5 orders of magnitude lower than the rates of the corresponding $2 \to 3$ (5s'−nd') processes. For the sake of clarity the calculated values of $<\sigma v>$ are noted on the corresponding graph in Fig. 15. Since the condition for neglecting (I.17) resides in the fact that

$$S_{03}/S_{23} \ll W_0/W_2,$$

and the ratio $W_0/W_2$ obtained from calculated and experimental data [12, 43] is close to unity, it follows that the inequality can be considered satisfied. Under these conditions the error of neglecting (I.17) will be approximately 1.5 orders of magnitude smaller than its upper bound indicated in § III.1; i.e., it constitutes a negligible value ~ 5%. Therefore, the additional error in $< \sigma v >$ for allowed $1 \rightarrow 3$ (4p' − nd') transitions is connected solely with the possible inaccuracy of $W_1/W_2$.

## References

1.  J. Hasted, Physics of Atomic Collisions [Russian translation], Izd. Mir (1965).
2.  I. McDaniels, Collision Processes in Ionized Gases [Russian translation], Izd. Mir (1967).
3.  Atomic and Molecular Processes, D. Bates (ed.), Academic, New York (1962).
4.  I. I. Sobel'man, Introduction to the Theory of Atomic Spectra, Fizmatgiz (1963).
5.  F. L. Mohler, NBS J. Res., 9:493 (1932).
6.  V. A. Fabrikant, F. A. Butaeva, and I. P. Tsirg, Dokl. Akad. Nauk SSSR, 14:423 (1937); V. A. Fabrikant and I. P. Tsirg, Dokl. Akad. Nauk SSSR, 16:271 (1937).
7.  A. V. Phelps and J. P. Molnar, Phys. Rev., 89:1203 (1953).
8.  K. I. Rozgachev, Opt. i Spektr., 4:549 (1958).
9.  S. É. Frisch and V. F. Revald, Opt. i Spektr., 15:726 (1963).
10. I. I. Sobel'man, Zh. Éksperim. i Teor. Fiz., 48:965 (1965).
11. L. A. Vainshtein, M. A. Mazing, and P. D. Serapinas, Preprint of the Physics Institute, Academy of Sciences of the USSR, No. 54 (1967).
12. V. P. Chebotaev, Dissertation, Institute of Semiconductor Physics, Siberian Branch, Academy of Sciences of the USSR, Novosibirsk (1965).
13. H. Massey and E. Barhope, Electronic and Ionic Impact Phenomena, Oxford (1962).
14. H. S. W. Massey, Handbuch der Physik, Vol. 36, Berlin (1956), p. 307
15. J. D. Craggs and H. S. W. Massey, Handbuch der Physik, Vol. 37, Berlin (1959), p. 314.
16. W. Fite, in: Atomic and Molecular Processes (D. Bates, ed.), Academic (1962), Chap. 12.
17. J. H. Parks, A. Szöke, and A. Javan, Bull. Am. Phys. Soc., 9:490 (1964).
18. V. M. Kaslin, G. G. Petrash and A. S. Khaikin, Opt. i Spektr., 23:28 (1967). (Report to the Third All-Union Conference on the Physics of Electronics and Atomic Collisions, Khar'kov, 1965); Preprint of the Physics Institute, Academy of Sciences of the USSR, No. 26 (1966).
19. J. H. Parks and A. Javan, Phys. Rev., 139:A1351 (1965).
20. A. S. Khaikin, Zh. Éksperim. i Teor. Fiz., 51:38 (1966).
21. A. S. Khaikin, Zh. Éksperim. i Teor. Fiz., 54:52 (1968).
22. T. V. Bychkova, V. G. Kirpilenko, S. G. Rautian, and A. S. Khaikin, Opt. i Spektr. 22:679 (1967).
23. G. Schiffner and F. Seifert, Proc. IEEE, 53:1657 (1965); A. Carcadden and S. L. Adams, Proc. IEEE, 54:427 (1966); Ch. Vauge and J. -F. Delpech, Compt. Rend., 263B:443 (1966).
24. M. N. Smolkin and N. B. Berdnikov, Opt. i Spektr., 14:414 (1963).
25. Yu. M. Kagan, Beitr. Plasma-Physik, 5:479 (1965); J. Y. Wada and H. Heil, IEEE J. Quant. Electr., 1:327 (1965).
26. V. E. Golant, Zh. Tekh. Fiz., 30:1265 (1960).
27. S. J. Buchsbaum, L. Mower, and S. C. Brown, Phys. Fluids, 3:806 (1960).
28. A. S. Khaikin, Zh. Tekh. Fiz. (in Press)
29. A. D. White and E. I. Gordon, Appl. Phys. Letters, 3:197 (1963).
30. S. Brown, Elementary Processes in Gas-Discharge Plasma [Russian translation], Atomizdat (1961).
31. M. A. Biondi and L. M. Chanin, Phys. Rev., 122:843 (1961).
32. E. F. Labuda and E. I. Gordon, J. Appl. Phys., 35:1647 (1964).
33. A. Éngel', Ionized Gases, Fizmatgiz (1959).
34. Th. Hänsch and P. Toschek, Phys. Letters, 22:150 (1966).

35.  L. A. Vainshtein, Opt. i Spektr., 3:313 (1957); L. Vainshtein and L. Minaeva, Preprint of the Physics Institute, Academy of Sciences of the USSR, No. 23 (1967).
36.  T. Holstein, Phys. Rev., 72:1212 (1947); 83:1159 (1951).
37.  L. S. Vasilenko and V. P. Chebotaev, Zh. Priklad. Spektr., 6:536 (1967).
38.  W. R. Bennett Jr. and V. P. Chebotayev [Chebotaev], Report to the Fifth International Conference on the Physics of Electronic and Atomic Collisions [Russian translation], Leningrad (1967).
39.  M. P. Chaika et al., Report to the Symposium on Theoretical Spectroscopy, Erevan (1966).
40.  R. H. Cordover, J. Parks, A. Szöke, and A. Javan, Physics of Quantum Electronics, McGraw-Hill (1966), p. 591.
41.  L. A. Vainshtein, this volume, p. 1; L. A. Vainshtein and I. I. Sobel'man, Preprint of the Physics Institute, Academy of Sciences of the USSR, No. 66, Part 1 (1967).
42.  M. Seaton, Proc. Phys. Soc., London, A77:174 (1961).
43.  L. A. Vainshtein and L. P. Presnyakov, this volume, p.37.

# EXPERIMENTAL DETERMINATION OF
# THE MATRIX ELEMENTS OF THE DIPOLE MOMENT OF
# AN ELECTRONIC TRANSITION IN THE SYSTEM OF
# SWAN BANDS OF THE $C_2$ MOLECULE*

## A. G. Sviridov

### INTRODUCTION

The problems of radiative heat exchange in hot gases have acquired special significance in recent years in connection with the extensive development of rocket engineering and the necessity of clarifying the operating conditions of the frontal portion of satellites and spacecraft during re-entry into the dense atmospheric layers of the earth and other planets. The contribution of radiative heating to the overall thermal effect during re-entry of a spacecraft into the dense layers of the atmosphere at cosmic velocities can be very large.

The conditions for re-entry of spacecraft into the atmosphere of other planets differ from the conditions of their interaction with the earth's atmosphere due to the difference in the atmospheric composition. Measurements transmitted from the Soviet automatic station "Venera-4" showed that the atmosphere of Venus consists almost completely of $CO_2$ [1]. Therefore, it is of interest to know the processes which occur in $CO_2$ at temperatures up to 10,000°K (i.e., at temperatures which develop during the flight of bodies in the atmosphere at cosmic velocities). At such temperatures the basic contribution to radiation and absorption will be made by the molecular components: $C_2$, $C_3$ and CO. The absorptive and radiative capacities of the Venusian atmosphere depends on the characteristics of the radiation and absorption of these molecular components. The intensity of the radiation or absorption of a molecule is determined by the square of the matrix element of the corresponding dipole transition or by the oscillator force.

Recently, great successes have been achieved in determining the oscillator forces of atoms by experiment and by calculation. The most enormous amount of experimental material is available on the oscillator forces of many atoms. The oscillator forces of the simplest atoms can be calculated with a fair reliability which is supported by a comparison of the results of theory and experiment for a large number of examples.

---

*Dissertation for the degree of Candidate of Physicomathematical Sciences. Scientific supervisor — Doctor of Physicomathematical Sciences, Professor N. N. Sobolev.

The situation with the determination of oscillator forces from molecules is considerably worse. Just five years ago data on the oscillator forces of electronic transitions were unavailable even for the molecules which are most important from the applied point of view. Theoretical calculations of the oscillator forces of electronic transitions remain unreliable to this day. Modern approximate methods of calculation do not contain an internal accuracy criterion, and therefore they cannot provide for satisfactory accuracy. Experimental measurements of oscillator forces at the present time are urgently needed for the development and improvement of quantum-mechanical methods of calculation, since only by comparing calculated values with measured ones can one establish the degree of reliability of the theoretical data.

The present paper is included in a series of investigations on determining the strength of oscillators of electronic transitions in $O_2$, $CN$, $NO$, and $N_2$ molecules, which were carried out in the Low-Temperature Plasma Optics Laboratory of the Physics Institute, Academy of Sciences of the USSR [2-4]. The papers devoted to the experimental determination of the oscillator force of the $d^3\Pi_g - a^3\Pi_u$ electronic transition of the $C_2$ molecule. The Swan bands situated in the visible range of the spectrum and having a considerable intensity correspond to this transition.

The interest in the $C_2$ molecule is connected not only with the demands of rocket engineering. Reliable values of oscillator forces for electronic transitions are necessary for astrophysics and the study of low-temperature plasma. Moreover, special interest is evoked by the study of the process of carbon black formation which is important from the scientific and applied points of view. The molecules $C_2$ and $C_3$ are evidently transitional radicals which accompany the formation of the chain of the polyatomic molecule $C_n$ (where n is several hundred) from monatomic carbon; this chain then becomes the "center" of the carbon particle [5]. A knowledge of the radiative properties of the $C_2$ molecule provides the basis for evaluating the concentration and concentration variations of $C_2$ molecules during the process of polymerization and formation of carbon particles. The $C_2$ molecule is often present in various engineering flames (rocket motors, internal combustion engines, etc.). Spectroscopic investigations of such flames will be successful and can yield substantial data for the understanding of the combustion process only for the case in which the radiative properties of the radicals formed during the combustion process are known. We carried out measurements of the oscillator force in a shock tube by two methods: the radiation method and the absorption method. The shock tube is a comparatively new source for the excitation of the spectrum, which has a series of significant advantages over widely known and well-investigated spectroscopic sources such as a flame, an arc, and a spark.

During the past ten years the shock tube has been studied intensively as a source of excitation of a spectrum. One of the basic trends in experimental research was the study of the state of inert gases (argon, krypton, xenon) behind a shock wave [6-10]. The experimental study of the state of inert gases is of interest from two points of view. First, after a shock wave having an intensity which can easily be obtained under laboratory conditions has passed through an inert gas, the gas is heated to comparatively high temperatures — 10,000-15,000°K. Second, gasdynamic calculations of the state of inert gases are comparatively simple and do not require the number of additional data which are required in calculating the state of more complex molecular gases. This allows calculated data to be obtained and compared with experiment rather simply, and thereby the validity of the premises on which the calculations are based can be established. It was also found experimentally that a gas heated by a reflected shock wave is in a state of semiequilibrium. Improved methods of calculating the composition of plasma and the development of methods for measuring temperature allowed the shock tube to be converted into a convenient instrument for studying radiation and absorption processes in gases at high temperatures.

CHAPTER I

## THE RADIATIVE AND ABSORPTIVE PROPERTIES OF THE $C_2$ MOLECULE

### §1.  The Radiation Spectrum of the $C_2$ Molecule

The system of energy levels of the $C_2$ molecule are shown in Fig. 1. There are two groups of electronic states of the $C_2$ molecule: triplet and singlet. Allowed transitions are transitions which occur only between levels having an identical multiplicity. Intercombinational transitions (i.e., transitions between levels having different multiplicities) were not observed in the spectrum of $C_2$. Until recently, the lower electronic level of the $C_2$ molecule was assumed to be the $^3\Pi_u$ level, and only comparatively recently [11] was it established that the ground energy level is the $^1\Sigma_g^+$ level which lies $610 \pm 20$ cm$^{-1}$ below the $^3\Pi_u$.

The $C_2$ molecule emits in a very broad frequency range — from the infrared range to the ultraviolet range. The most intense, and therefore the best known, are the Swan bands of the $C_2$ molecule. The bands are shaded for a shorter wavelength and form easily distinguished se-

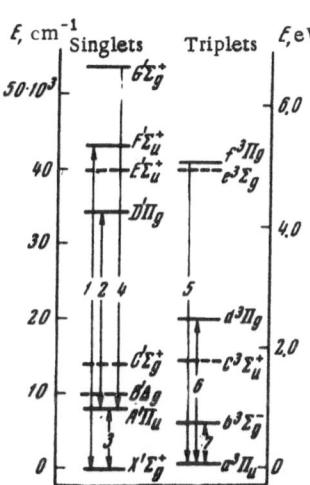

Fig. 1. Diagram of the energy levels of the $C_2$ molecule. The horizontal solid lines denote the experimentally observed electronic states; the dashed lines denote the theoretically calculated states. The electronic transitions which can be observed in the $C_2$ spectrum are denoted by arrows, where 1 is the Mulliken system; 2 is the Delendre — Asambuch system; 3 is the Phillips system; 4 is the Freimark system; 5 is the Fox — Herzberg system; 6 is the Swan system; 7 is the Ballik — Ramsay system.

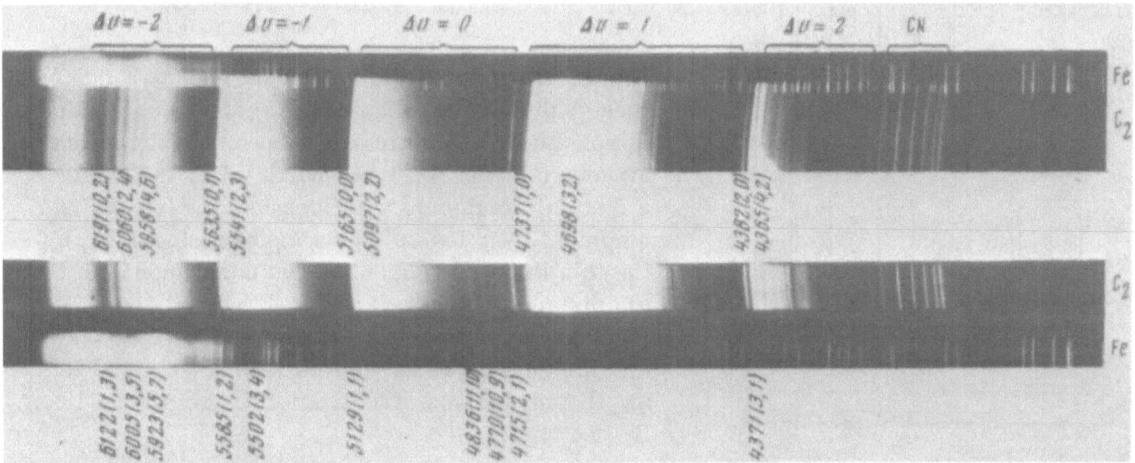

Fig. 2. The Swan bands of the $C_2$ molecule. The spectrum was obtained by means of an ISP-51 spectograph (the camera and collimator are respectively equal to 270, 300 mm; the input width is 0.02 mm) behind a reflected shock wave in a mixture of 70% Ar−30% CO at an initial pressure $P_1 = 5.0$ mm Hg. The pressure corresponding to rupture of the diaphragm is $P_4 = 7$ atm.

quences (Fig. 2). The sequences begin with the (2,5), (0,2), (0,1), (0,0), (1,0), (2,0) bands, whose edges are located at $\lambda = 6677.3$; 6191.2; 5635.5; 5165.2; 4737.1; 4382.5 Å.

Table 1 presents the wavelengths of all the edges [12] which were observed experimentally, where v' is the quantum vibration number of the upper level; v" is the quantum vibration number of the lower level. Swan bands have long been well known. However, the nature of the spectrum carrier remained unclear for a long time. Many attempts were made at interpreting this spectrum. The investigations of a discharge in carbon monoxide and pentane, and the spectra of a carbon arc in an air atmosphere [13] showed unambiguously that the Swan bands belong to the $C_2$ molecule.

TABLE 1. Edges of the Bands of the Swan System of the $C_2$ Molecule

| $\lambda$ | $\Delta v$ | $v'$ | $v''$ | $\lambda$ | $\Delta v$ | $v'$ | $v''$ |
|---|---|---|---|---|---|---|---|
| 6677.3 | | 2 | 5 | 5501.9 | | 3 | 4 |
| 6599.1 | | 3 | 6 | 5470.3 | | 4 | 5 |
| 6533.7 | −3 | 4 | 7 | 5165.2 | | 0 | 0 |
| 6580.5 | | 5 | 8 | 5129.3 | 0 | 1 | 1 |
| 6442.3 | | 6 | 9 | 5096.7 | | 2 | 2 |
| 6191.2 | | 0 | 2 | 4737.1 | | 1 | 0 |
| 6122.1 | | 1 | 3 | 4715.2 | | 2 | 1 |
| 6059.7 | −2 | 2 | 4 | 4697.6 | 1 | 3 | 2 |
| 6004.9 | | 3 | 5 | 4684.8 | | 4 | 3 |
| 5958.4 | | 4 | 6 | 4678.6 | | 5 | 4 |
| 5923.4 | | 5 | 7 | 4382.5 | | 3 | 0 |
| 5635.5 | | 0 | 1 | 4371.4 | 2 | 3 | 1 |
| 5585.5 | | 1 | 2 | 4365.2 | | 4 | 2 |
| 5540.7 | −1 | 2 | 3 | | | | |

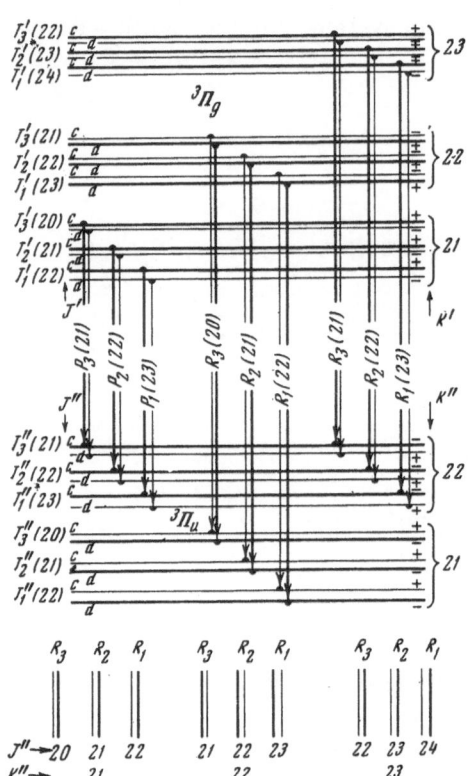

Fig. 3. Diagram of the rotational energy bands for large values of K (when case (c) according to Hund is realized), which clarifies the formation of the structure of the $^3\Pi_g - ^3\Pi_u$ band. For the case of a diatomic molecule with identical nuclei (for example, $C_2$) the levels and transitions indicated by the thin lines are absent. As a consequence of this the distance between triplets with K = 21 and 22 (see the diagram at the bottom) will be less than the distance between triplets with K = 22 and 23 for the unresolved structure of the rotational lines $R_1$, $R_2$, and $R_3$.

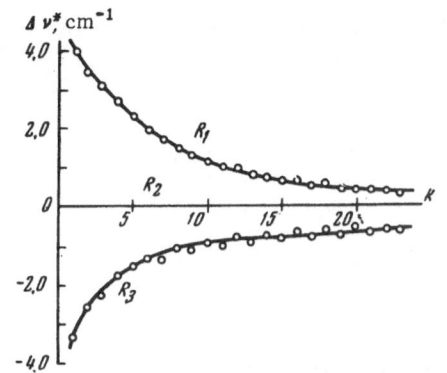

Fig. 4. Variation of the multiplet splitting $\Delta\nu$ of the branches with increasing K.

The system of Swan bands is connected with the $d^3\Pi_g - a^3\Pi_u$ transitions (see Fig. 1). The investigation of the fine structure of Swan bands on an instrument having a high resolving power shows that the bands consist of R- and P- branches, while the lines which form the fine structure of the bands are triplet lines. Each rotational line in turn must be a doublet by virtue of $\Lambda$-doubling. But one of the two components of the rotational line of the $C_2$ molecule is a characteristic feature of the spectrum of diatomic molecules consisting of identical atoms [14]. Furthermore, the $^3\Pi_g$ and $^3\Pi_u$ levels are inverted. This leads to the observed alternation of distances between unresolved triplets (Fig. 3). This effect is called "vibration." The observation of this effect is facilitated by the fact that with an increasing rotational quantum number K the distance between the triplet lines $R_1, R_2$, and $R_3$ decreases (Fig. 4), and the $\Lambda$-doubling of each level increases in proportion to the square of K.

An analysis of the vibration bands of the Swan system is contained in [15], while in [16] the constants $B_e$, $B_0$, $\alpha$, $D$, $r_e$, etc., of the lower levels were refined; on the basis of this an equation was given for the edges of the bands:

$$\nu = 19373,87 - 1773,42\,v' - 19,35\,v'^2 - (1629,88\,v'' - 11,71\,v''^2). \tag{1}$$

A detailed analysis of the rotational structure of the (0,0), (1,1), (1,0), and (1,2) bands is given in [17], and the intensity distribution over the rotational lines was considered theoretically in [18].

Besides the bands displayed in Table 1, new weaker bands with emission maxima at $\lambda =$ 4770, 4836, 4911, 4997, 4395, 4734 Å were found in [19]. The 4395 and 4734 Å bands have no edges, while the remaining bands are shaded toward the red range. A detailed analysis showed that the emission of these bands belongs to the $C_2$ molecule and corresponds to $d^3\Pi_g - a^3\Pi_u$ transitions from high vibrational levels. The 4770 and 4836 Å bands appear in the luminescence of the shock tube, and they can be distinguished in Fig. 2. Besides these bands, which are frequently called "tail" bands, "high pressure" bands similarly corresponding to the $d^3\Pi_g - a^3\Pi_u$ transition from the $v'' = 0, 1, 2, \ldots, 11$ levels to the $v'' = 6$ level are also well known [20, 21].

Several other bands connected with the $C_2$ emission are observed in the spectra of discharge tubes and flames in addition to the intense system of Swan bands. The Fox–Herzberg [16] system of bands, which has a common lower state $a^3\Pi_u$ with a system of Swan bands and an upper state $f^3\Pi_g$, was similarly detected in the spectrum of hydrocarbon flames at atmospheric pressures [22-24]. The edges are shaded toward long waves, and the most intense of them are situated at $\lambda$ 2855 and 2987 Å (the (0,3) and (0,4) bands). The wavelengths of the edges of the Fox–Herzberg system of bands are displayed in Table 2.

The luminescence spectrum of the $C_2$ molecule is sometimes accompanied by a group of diffuse bands near 4050 Å. This group of bands was detected for the first time in the spectra of the heads of comets, and then was obtained in the spectrum of a discharge in a stream of hydrocarbon vapors, in the spectrum of diffuse hydrocarbon flames burning with fluorine [25], and also in the emission spectrum of the inner cone and envelope of rich acetylene–oxygen flames [26]. It has been established [27] that these bands do not reveal an isotopic shift when the hydrogen is replaced by deuterium. At present it has been convincingly shown [28-31] that the emitter of the bands is the $C_3$ molecule. Since these bands frequently develop with the appearance of nucleating hydrogen particles, it follows that they can be of definite interest in connection

TABLE 2. Edges of the Bands of the Fox — Herzberg System for the
C₂ Molecule

| λ | $v'$ | $v''$ | λ | $v'$ | $v''$ | λ | $v'$ | $v''$ |
|---|---|---|---|---|---|---|---|---|
| 2378.2 | 4 | 1 | 2656.3 | 1 | 2 | 2987 | 0 | 4 |
| 2429.3 | 3 | 1 | 2698.8 | 2 | 3 | 2996.4 | 1 | 4 |
| 2486.3 | 2 | 1 | 2731.5 | 0 | 2 | 3129.0 | 0 | 5 |
| 2527.9 | 3 | 2 | 2772.1 | 1 | 3 | | | |
| 2589.0 | 2 | 2 | 2855 | 0 | 3 | 3283 | 0 | 6 |

TABLE 3. The Relative Probabilities of the Vibrational Transitions
of the Swan Bands of the C₂ Molecule (Experimental Values)

| $v', v''$ | λ, Å | Johnson and Tawde [34] (average of five measurements) | Patel [36] | Patel [37] (average of five measurements) | Tawde and Laud [38] | King [39] | Phillips [40] | Phillips [41] |
|---|---|---|---|---|---|---|---|---|
| 2.0 | 4382 | 0.04 | 0.09 | 0.11 | 0.04 | 0.044 | | |
| 3.1 | 4371 | 0.09 | | | | 0.111 | | |
| 4.2 | 4365 | 0.17 | | | | 0.252 | | |
| 1.0 | 4737 | 0.44 | 0.65 | 0.57 | 0.38 | 0.358 | 0.362 | 0.298 |
| 2.1 | 4715 | 0.56 | 0.65 | 0.61 | 0.55 | 0.525 | 0.51 | 0.407 |
| 3.2 | 4697 | 0.59 | | | | 0.718 | 0.76 | |
| 4.3 | 4684 | 0.66 | | | | 0.952 | | |
| 5.4 | 4673 | 0.85 | | | | 1.56 | | |
| 0.0 | 5165 | 1.00 | 1.00 | 1.00 | 1.00 | 1.000 | 1.000 | 1.000 |
| 1.1 | 5129 | 0.42 | 0.43 | 0.46 | 0.48 | 0.520 | 0.64 | |
| 2.2 | 5097 | 0.12 | 0.22 | 0.21 | 0.20 | 0.256 | | |
| 3.3 | 5070 | 0.05 | | | | | | |
| 0.1 | 5635 | 0.46 | 0.59 | 0.46 | 0.35 | 0.244 | 0.263 | 0.413 |
| 1.2 | 5585 | 0.42 | 0.48 | 0.39 | 0.43 | 0.470 | 0.37 | 0.529 |
| 2.3 | 5540 | 0.44 | 0.43 | 0.43 | 0.46 | 0.497 | | |
| 3.4 | 5501 | 0.45 | | | | 0.802 | | |
| 4.5 | | 0.33 | | | | | | |
| 0.2 | 6191 | 0.32 | 0.24 | 0.18 | 0.10 | 0.043 | | |
| 1.3 | 6122 | 0.49 | 0.28 | 0.20 | 0.16 | 0.123 | | |
| 2.4 | 6059 | 0.63 | 0.48 | 0.26 | 0.19 | 0.188 | | |

with the study of the mechanism by which carbon black is formed. Notwithstanding the fact that the carrier of the bands does not contain hydrogen, the emission from the group of bands at 4050 Å was usually observed solely in the spectra of sources in which hydrogen is present. But in [32] these bands were obtained in a King furnace (i.e., in the absence of hydrogen). The maxima of the C₃ bands are situated at the following wavelengths [33]: 4075.0; 4073.5; 4069.3; 4064.3; 4054.2; 4051.5; 4043.5; 4042.1; 4039.1; 4033.2; 4019.2; 4013.2; 4002.2; 3992.6; 3987.2 Å.

§2. The Relative Probabilities of the Electron-Vibrational

Transitions of the Swan Bands of the C₂ Molecule

There are comparatively few experimental papers devoted to the determination of the relative probabilities of electron-vibrational transitions. In an experimental investigation of the Swan bands one should consider the fact that pronounced superposition exists between neighboring bands of the sequence (see Fig. 2). Therefore, only a portion of each of the bands (for example, the head of the band) can be measured, and then the intensities of the entire band can be determined by calculation. Both the portion of the overall intensity which is contained in the head of the band and the degree of superposition vary from band to band and depend on the temperature of the excitation source. In the paper by Johnson and Tawde [34] the Swan bands were investigated for five spectrum sources: in Bunsen and oxygen — carbon flames, in a carbon arc and hydrogen, in a discharge tube with argon containing traces of carbon at a pressure of 30 mm Hg, and in a condensed discharge in glycerine. The radiation was recorded photographically by a spectrograph having a moderate dispersion. The intensity of the bands was estimated according to the maximum brightness of the edges; under these conditions it was assumed that the brightness of the edge radiation was proportional to the intensity of the band emission. The values of relative probabilities of the electron-vibrational transitions obtained in various sources differ from each other by a factor of 1.5-2.0.

The values averaged over all five light sources are shown in the third column of Table 3. Laud [35], Patel [36, 37], and Tawde and Laud [38] conducted a series of analogous measurements in various ethyl alcohol and oxygen – carbon flames.

In the paper by King [39] the Swan bands were excited in a carbon furnace heated by electric current. The measurements were based on the following assumptions: 1) the distribution of $C_2$ molecules over the high vibration levels corresponds to the Boltzmann formula with an equilibrium temperature equal to the temperature of the walls of the graphite tube of the furnace; 2) the number of molecules is proportional to the pressure of the $C_2$ vapors at the temperatures used; 3) the total intensity of the bands is proportional to the intensity peaks in the edges; 4) the self-absorption in the bands from the ground vibrational (v" = 0) level is insignificant. The values of the relative probabilities obtained by the radiation method are also displayed in Table 3.

Phillips [40, 41] measured the relative intensities of the Swan bands of the $C_2$ molecule in the emission spectrum of a carbon furnace by scanning the spectrum with a photoelectron multiplier. In the first paper [40] an investigation was carried out of the intensity distribution in a band using a wide slit (i.e., without resolution of the rotational structure of the spectrum): in the second [41] an investigation was made of the relative integrated intensities of neighboring rotational lines conforming to different bands. In the second paper the relative probabilities for the (1,0), (2,1) bands were found to be lower, while for the (0,1), (1,2) bands they were found to be higher (see Table 3).

The methods of calculating the relative probabilities of the vibrational transitions of diatomic molecules or the Franck – Condon factors, as well as the difficulties which are encountered under these conditions, have been elucidated in a number of monographs [14, 42-44] and journal papers, references to which can be found in [45, 46]. Here we shall limit ourselves to a brief review of papers of a theoretical character which have been devoted to the $C_2$ molecule.

In order to calculate the Franck – Condon factors

$$q = \left| \int \psi_{n'v'} \psi_{n''v''} \, dr \right|^2 \tag{2}$$

it is necessary to know the vibrational wave functions $\psi_{n'v'}$ and $\psi_{n''v''}$ of the molecules for the levels v' and v" of the upper and lower electron states n' and n" between which the transition occurs. The wave functions are determined by solving the wave equation which describes the motion of the nuclei of the diatomic molecules. The accuracy with which the wave functions are determined depends on the extent to which the approximate potential curves used correspond to the true potential of the molecule, and on those approximations which are allowed in solving the wave equations. Under these conditions, however, one is obliged not only to solve the problem of finding the wave functions $\psi_{n'v'}$ and $\psi_{n''v''}$ most accurately, but also to consider the fact that the analytical form of the wave functions obtained should be suitable for integration.

The first calculations of the superposition integral $\int \psi_{n'v'} \psi_{n''v''} \, dr$ for symmetrical molecules and small quantum numbers were carried out by Hutchison [47] by means of the potential function of a harmonic oscillator. Wurm [48] used the Hutchison method for estimating the relative intensities of the Swan bands of the $C_2$ molecule. He showed that the greatest error in the calculations is caused by the error in determining the equilibrium internuclear distance

TABLE 4.  The Relative Probabilities of Vibrational Transitions of the
Swan Bands of the $C_2$ Molecule (Calculated)

| $v'$, $v''$ | $\lambda$, A | McKeller, Buscomb [49] | Pillow [53] | Pillow [52] | Wyler [59] | Wyler [60] | Tawde, Laud [32] | Fraser, Jarmain, Nichols [63] |
|---|---|---|---|---|---|---|---|---|
| 2.0 | 4382 | 0.067 | 0.008 | 0.04 | 0.030 | 0.029 | 0.05 | 0.033 |
| 3.1 | 4371 | 0.149 | 0.20 | 0.11 | 0.115 | 0.111 | | 0.082 |
| 4.2 | 4365 | 0.215 | | 0.19 | 0 212 | 0.200 | | 0.133 |
| 1.0 | 4737 | 0.324 | 0.46 | 0.32 | 0.345 | 0.336 | 0.46 | 0.324 |
| 2.1 | 4715 | 0.422 | 0.65 | 0.49 | 0.560 | 0.528 | 0.65 | 0.487 |
| 3.2 | 4697 | 0.392 | 0.67 | 0.59 | 0.622 | 0.565 | | 0.554 |
| 4.3 | 4684 | 0.334 | | 0.65 | 0.631 | 0.546 | | 0.584 |
| 5.4 | 4673 | | | | | | | |
| 0.0 | 5165 | 1.000 | 1.00 | 1.00 | 1.000 | 1.000 | 1.00 | 1.000 |
| 1.1 | 5129 | 0.436 | 0.48 | 0.51 | 0.473 | 0.453 | 0.48 | 0.496 |
| 2.2 | 5097 | 0.144 | 0.28 | 0.24 | 0.174 | 0.160 | 0.23 | 0.222 |
| 3.3 | 5070 | 0.022 | 0.13 | 0.10 | 0.026 | 0.023 | | 0.078 |
| 0.1 | 5635 | 0.354 | 0.18 | 0.31 | 0.313 | 0.310 | 0.15 | 0.289 |
| 1.2 | 5585 | 0.511 | 0.25 | 0.40 | 0.448 | 0.421 | 0.18 | 0.383 |
| 2.3 | 5540 | 0.544 | 0.29 | 0.38 | 0.415 | 0.376 | 0.17 | 0.383 |
| 3.4 | 5501 | 0.498 | 0.24 | 0.29 | 0.315 | 0.275 | | 0 360 |
| 4.5 | | | | 0.27 | | | | |
| 0.2 | 6191 | 0.048 | 0.24 | 0.05 | 0.060 | 0.059 | 0.36 | 0.057 |
| 1.3 | 6122 | 0.121 | 0.05 | 0.13 | 0 152 | 0.146 | 0.07 | 0.120 |
| 2.4 | 6059 | 0.203 | 0.10 | 0.21 | 0.227 | 0.214 | 0.10 | 0.170 |

$\Delta r_e = |r_e' - r_e''|$ during an electronic transition.  The values which he calculated for the relative intensities of certain Swan bands differ almost by a factor of 2 for the extremal values $\Delta r_e = 0.046$ and $0.043$ Å.  McKellar and Buscomb [49] used refined values of the molecular constants to obtain values of the Franck – Condon factors according to the Hutchison method which differ by 5-10% from the Wurm values.  Tawde and Patel [50] used the same method to find values of the Franck – Condon factors which differ from the McKellar – Buscomb values.  Later McKellar and Tawde [51] established the fact that the difference can be explained completely by the difference between the values adopted for the internuclear distance in the calculations.

Pillow [52, 53] perfected the method adopted previously [54] of distorting the wave functions of the simple harmonic oscillator for a diatomic molecule.  Having used three different forms of graphical distortion, she obtained wave functions close to the wave functions obtained from the Morse potential.  Ta-jou Wu [55] indicated that the wave functions obtained by this method for different vibrational levels are not orthogonal.  The absence of orthogonality leads to a noticeable error in calculating the superposition integral.  He suggested a semianalytical approximate method of determining the superposition integral in the case of the Morse potential, and the Wentzel – Kramers – Brillouin method [56-58] for solving the Schrödinger equation for the purpose of obtaining the vibrational wave functions of an anharmonic oscillator.  This method of calculation was used by Wyler [59, 60] who used the more exact Hulbert – Hirschfeld potentials [61] under these conditions.  He showed that the Pillow calculations are in better agreement with the experimental values obtained by King [39], whereas the values obtained by the Wentzel – Kramers – Brillouin method are closer to the experimental values obtained by Johnson and Tawde [34].

Tawde and Laud [52] used the Hutchison method to calculate the relative transition probabilities, and also obtained the same "distortion" method independently of Pillow [52, 53].  The data obtained by them are in better agreement with the experimental data obtained by King than with the data obtained by the distortion method (see Tables 3 and 4).  Fraser, Jarmain, and Nichols [63], having applied the method of "simplified" Morse potentials which they worked out and the Pillow $r_e$-shift method, achieved the best accuracy compared with other approximate methods in calculating the superposition integrals for the Swan bands.

Ortenberg [64] calculated the Franck–Condon factors of the $C_2$ molecule on the basis of the Morse potential. For penetration into the range of higher quantum numbers the calculation was carried out on an M-20 computer having the capacity of carrying out calculations with rank 20. Because of the fact that each operation was carried out with 18 significant figures it was possible to obtain results over wider ranges of quantum numbers than was possible earlier. The results of these calculations for the Swan bands are shown in Table 5. Unlike Tables 3 and 4, the sum of the values of $q_{v'v''}$ over each original level is normalized to unity.

The three-parameter Morse curve and the five-parameter Hulbert–Hirschfeld curve are good but insufficiently accurate representations of the true molecular potentials, especially for high quantum numbers. Jarmain [65] calculated the values of the Klein–Dunham potential curve on an electronic computer for several diatomic molecules, including the $C_2$ molecule. Singh and Jain [66], Read and Wanderslice [67] constructed the true potential curves for the $C_2$ molecule using the Rydberg–Klein–Rhee potentials. The calculation of the vibrational wave functions by means of the true potential curves plotted on the basis of spectroscopic data yields better results than calculations using the Morse and Hulbert–Hirschfeld potential.

The experimental values of the relative transition probabilities obtained by various investigators using various methods are displayed in the summary Table 3, while the calculated values are displayed in Tables 4 and 5. In compiling these data it is evidently impossible to avoid a certain quota of subjectivism. First, the transition probability for the (0,0) bands is not always taken as the comparison unit. Second, several very good experimental investigations apply only to a limited number of transitions, and they are difficult to include in the overall averaging scheme; this is all the more true if it is necessary to carry out a scaling of the normalization data to unity probability of transitions from some level. Third, the accuracy of determination for various bands in one and the same system is not identical. Experimentally, the transition probabilities for the more intense bands, as a rule, will be measured with greater accuracy than those for the less intense ones. Relative probabilities for bands with small vibrational quantum numbers $v'$ and $v''$ are determined more accurately by computation. The method of determining the "smooth" values which was used in [68] decreases the role of the subjective factor to a known degree in estimating the relative transition probabilities, but it does not eliminate it entirely.

TABLE 5. The Franck–Condon Factors for the System of Swan Bands of the $C_2$ Molecule [64]

| $v'$ / $v''$ | 0 | 1 | 2 | 3 | 4 | 5 | 6 | 7 | 8 | 9 |
|---|---|---|---|---|---|---|---|---|---|---|
| 0 | 0.73520 | 0.21299 | 0.04312 | 0.00737 | 0.00113 | 0.00016 | 0.00002 | 0.00000 | 0.00001 | 0.00038 |
| 1 | 0.23963 | 0.36359 | 0.27355 | 0.09158 | 0.02168 | 0.00422 | 0.00063 | 0.00000 | 0.00271 | |
| 2 | 0.02445 | 0.36141 | 0.16196 | 0.27065 | 0.12927 | 0.03872 | 0.00639 | 0.00317 | | |
| 3 | 0.00071 | 0.05933 | 0.41396 | 0.06010 | 0.22896 | 0.13629 | 0.01216 | | | |
| 4 | 0.00000 | 0.00218 | 0.09832 | 0.43053 | 0.01099 | 0.08880 | | | | |
| 5 | 0.00000 | 0.00000 | 0.00430 | 0.12739 | 0.56152 | | | | | |
| 6 | 0.00000 | 0.00000 | 0.00003 | 0.01875 | 0.00341 | | | | | |
| 7 | 0.00000 | 0.00000 | 0.00045 | 0.06226 | | | | | | |
| 8 | 0.00000 | 0.00003 | 0.06833 | | | | | | | |
| 9 | 0.00000 | 0.00037 | | | | | | | | |

With increasing accuracy of the experimental determinations of the relative intensities of the vibrational transitions of diatomic molecules and the perfection of calculation methods, it became obvious that the interaction of electronic and vibrational motions in a molecule cannot be neglected. The usual notion of the approximate constancy of the electronic angular momentun $R_e(r)$ of the transition for the entire system of bands is unsuitable in the majority of cases. For certain diatomic molecules the quantity $R_e(r)$ changes by a factor of 2 with a change in internuclear distance r. Therefore, the relative transition probability can be written as follows [69]:

$$p_{v'v''} = \left| \int \psi_{v'} R_e(r) \psi_{v''} \right|^2 dr = R_e^2(r_{v'v''}) \left| \int \psi_{v'} \psi_{v''} dr \right|^2 = R_e^2(r_{v'v''}) q_{v'v''},$$  (3)

where $q_{v'v''}$ is the Franck – Condon factor (see (2)), and

$$r_{v'v''} = \int \psi_{v'} r \psi_{v''} dr \Big/ \int \psi_{v'} \psi_{v''} dr$$  (4)

is a unique value of the internuclear distance which is characteristic of the vibrational bands. The quantity $\bar{r}_{v'v''}$ is called an r-centroid.

The values of the r-centroids for the Swan bands were calculated by Nichols and his associates [68] and by Jain [70]. The latter also found the dependence of $|R_e|^2$ on the vibrational quantum numbers [71], and this dependence turned out to be very insignificant. Mentall and Nichols [71] recently reanalyzed the experimental data [39, 41] and found, unlike [70], a strong dependence of $R_e$ on r for the Swan bands. For a final solution of the problem of the dependence of $|R_e|^2$ on the vibrational quantun number new additional experimental data are required. Therefore, in measuring $|R_e|^2$ by the radiation method we assumed in our work that it is possible to postulate that the dependence of $R_e$ on r is insignificant for the Swan bands.

§ 3.  The Absolute Probabilities of the Electron-Vibrational

Transitions of the $C_2$ Molecule

At present the calculation of the absolute transition probabilities of diatomic molecules cannot be carried out with any reliability. In constructing the total electronic wave function of the molecule it is necessary to make a number of simplifying assumptions which are not actually realized. The validity of any particular approximation can be evaluated according to the extent to which the calculated data obtained approximates the experimental data. Thus, the experimental values of the oscillator forces of the electronic transitions for molecules are the basic starting point for improving methods of theoretical calculation.

It is from this point of view that the brief review of theoretical papers dealing with this trend should be considered.

Lyddane, Rogers, and Roach [72], having used atomic and molecular orbitals for an approximate representation of the wave functions of the molecules, determined the ratio of the oscillator forces for CN and $C_2$. Then, using the obtained ratio as the value of the electronic oscillator force for CN which is known from the White experiments [73], they calculated the oscillator force for the Swan bands of $C_2$, which turned out to equal 0.024.

Stevenson [74] calculated the oscillator force for the Swan bands and the Delendre – Asambuch bands of the $C_2$ molecule, having approximately expressed the wave functions 6u2s and 6g2p of the molecular orbitals of the appropriate linear combination of atomic wave functions,

TABLE 6. Calculated [79] Values of the Oscillator Forces of the $C_2$ Molecule

| Transition | System | $v_0$, cm$^{-1}$ | $f_e$ |
|---|---|---|---|
| $d^3\Pi_g - a^3\Pi_u$ | Swan | 19306.26 | 0.0485 |
| $D'\Pi_g - A^1\Pi_u$ | Delendre – Asambuch | 25870.24 | 0.0650 |
| $F^1\sum_g^+ - X^1\sum_g^+$ | Mulliken | 43240.23 | 0.1025 |
| $f^3\Pi_g - a^3\Pi_u$ | Fox – Herzberg | 40080.41 | 0.8184 |
| $A^1\Pi_u - X^1\sum_g^+$ | Phillips | 8391.66 | 0.0027 |
| $b^3\sum_g^- - a^3\Pi_u$ | Ballik – Ramsay | 5656 | 0.0066 |
| $E^1\sum_u^+ - X^1\sum_g^+$ | | (40341) | 0.7520 |
| $G^1\sum_g^+ - A^1\Pi_u$ | | 8875 | 0.0101 |
| $B^1\Delta_g - A^1\Pi_u$ | | 1614 | 0.0018 |

just as for the hydrogen ion. The oscillator force and the lifetime for the Swan bands, according to his calculations, are respectively equal to 0.029 and $1.47 \cdot 10^{-7}$ sec.

Shull [75-77] calculated the values of $f_e$ for the Swan bands using the dipole length method and obtained the value $f_e = 0.13$. Then he used the dipole velocity method, which yielded the value $f_e = 0.18$.

Coulson and Lester [78] obtained the higher value $f_e = 0.24$ as a result of complex and multistep calculations, having used the orthogonalization of the atomic wave functions and molecular orbitals and having employed hybridization.

The oscillator forces for nine electronic systems of bands of the $C_2$ molecule were calculated by Clementi [79], who used the dipole moment operator and wave functions of the Slater type with $Z = 3.25$. In order to determine the degree of hybridization he used the semiempirical method of estimating the energy of the electrons in the inner shell. The values of the oscillator forces obtained by Clementi are displayed in Table 6.

From this table it is evident that the Fox – Herzberg bands, the $E^1\sum_u^+ - X^1\sum_g^+$ bands, the Mulliken bands, the Delendre – Asambuch bands, and, finally, the Swan bands have the largest values of the oscillator force. The $E^1\sum_u^+ - X^1\sum_g^+$ bands, notwithstanding the large calculated value of the oscillator force, evidently are not very intense, since heretofore they have not been detected experimentally. The upper levels of the Mulliken and Delendre – Asambuch singlet bands are comparatively high (see Fig. 1), and therefore at temperatures up to 10,000°K they cannot have a substantial influence on the radiative and absorptive properties of the $C_2$ molecules. Thus, the radiative and absorptive properties of the $C_2$ molecule, according to the values given for the oscillator forces, will basically be determined by the Swan bands and the Fox – Herzberg bands.

At the time we began our investigations the sole experimental work [80] on the determination of the absolute values of the oscillator forces of the Swan band had been carried out. Regrettably, this investigation has still not been published by Hicks, and we do not know either the method of measurement or the excitation source of the spectrum. Based on indirect indications from other literature sources it can only be conjectured that these measurements were evidently carried out by the radiation method using a heated carbon furnace.

In 1963 a short communication [81] was published on the measurement of the oscillator forces of Swan bands by radiation and absorption methods, while in 1965 the paper [82], which was devoted to these same bands and was based on measurements of the lifetime of the electron-vibrational levels during the excitation of the spectrum of Swan bands by heating a graphite block with short pulses from a ruby laser. Regrettably, many essential details of the experiment and the method of interpreting the results obtained are absent from the latter publication, so that it is impossible to form a correct concept of this work. It is also not known whether the values of the oscillator forces $f_e = 0.005 \pm 0.003$ [81], $f_e = 0.0043$ [82] apply to an electronic or a vibrational transition.

We know of no experimental papers on the determination of the electronic oscillator force of the Fox – Herzberg bands.

CHAPTER II

ORIGINAL DATA FOR DETERMINING $|R_e|^2$ BY MEANS OF A SHOCK TUBE

## §1. General Foundations of the Radiation Method
## and Derivation of the Equations for the Work

As is well known, the mean radiation energy (in erg/sec) of a harmonic oscillator having a charge e and a frequency $\nu$ is determined from the expression

$$\mathscr{I} = \frac{(2\pi\nu)^4}{3c^3} \left| e\mathbf{r}_0 \right|^2 = \frac{(2\pi\nu)^4}{3c^3} e^2 (x_0^2 + y_0^2 + z_0^2), \tag{5}$$

where the vector $\mathbf{r}_0$ having the components $x_0$, $y_0$, $z_0$ is the amplitude of the vibration. In accordance with quantum mechanics the energy radiated by a molecule during a transition from the quantum state m to the state n is calculated analogously. Replacing $|er_0|^2$ by the expression $|2\bar{R}^{mn}|^2$, where $|\bar{R}^{mn}|$ is the dipole moment corresponding to the m − n transition, we obtain

$$\mathscr{I}_{m \to n} = \frac{64\pi^4}{3c^3} \nu_{mn}^4 |\bar{R}^{mn}|^2. \tag{6}$$

Assume $N_{n',v',J'}$ is the population of the upper electronic vibrational − rotational state, while v', v", J', J" are the vibrational and rotational numbers in the upper and lower electronic states. Then the total radiation energy of a solitary rotational line will be equal to

$$\int I_\nu \, d\nu = \frac{64\pi^4}{3c^3} \nu^4_{n',v',J,'\,n'',v'',J''} \frac{N_{n',v',J'}}{g'} \left| \bar{R}_{n',v',J',n'',v'',J''} \right|^2, \tag{7}$$

the integration being carried out over the spectral interval in which the intensity of the line is nonvanishing. Here $|\bar{R}_{n',v',J',n'',v'',J''}|^2$ determines the intensity of the line and is called the strength of the line or the strength of the transition [83]; g' is the degree of degeneration of the original level. The calculation of $|R_{n',v',J',n'',v'',J''}|^2$ for certain simplifying assumptions leads to

$$\left| \bar{R}_{n',v',J',n'',v'',J''} \right|^2 = |\bar{R}_e|^2 \left| \int \psi_{v'} \psi_{v''} \, dr \right| \delta_{J'J''} = |\bar{R}_e|^2 \cdot g_{v'v''} \, S_{J'J''}, \tag{8}$$

where $|R|^2$ is the matrix element of the electronic transition; $q_{v'v''} = |\int \psi_{v'} \psi_{v''} dr|^2$ is the Franck−Condon factor, and $S_{J'J''}$ is the Höhnli−London intensity factor.

The factors $q_{v'v''}$ and $S_{J'J''}$ determine the distribution of the intensities in the vibrational and rotational structure, respectively, and the square of the dipole moment of the electronic transition $|R_e|^2$ determines the absolute radiation intensity of the line. At thermal equilibrium the population of the electronic vibrational − rotational level which is characterized by the quantum numbers n', v' and J' (or K'), is equal to

$$N_{n',v',J'} = \frac{N}{Z} g' \exp\left[ -\frac{T_0' + G_0'(v) + F_v'(J)}{kT} \right], \tag{9}$$

where N is the number of molecules in $cm^3$; Z is the sum over the electronic, vibrational, and rotational states of the molecules, and $T_0'$, $G_0'(v)$, $F_{v'}(J)$ are the values of the electronic, vibrational, and rotational terms of the molecule. Substituting (9) into (7) and considering (8), we obtain the expression for the radiation energy of one rotational line in a solid angle equal to $4\pi$:

$$\int I_\nu d\nu = \frac{64\pi^4}{3c^3} \nu_{av}^4 \frac{Nl}{Z} \exp\left[-\frac{T_0' + G_0'(v)}{kT}\right] q_{v'v''} |\bar{R}_e|^2 \sum_{P,Q,R} \sum_{J'_{min}}^{J'_{max}} S_{J'J''} \exp\left[-\frac{F_v'(J)}{kT}\right], \quad (10)$$

where $l$ is the thickness of the plasma layer.

For the radiation energy of the entire electron-vibrational band we have the following results in accordance with the relationship (10):

$$\sum_{P,Q,R} \sum_{J'_{min}}^{J'_{max}} \int I_\nu d\nu = \frac{64\pi^4}{3c^3} \nu_{av} \frac{Nl}{Z} \exp\left[-\frac{T_0' + G_0'(v)}{kT}\right] q_{v'v''} |R_e|^2 \sum_{P,Q,R} S_{J'J''} \exp\left[-\frac{F_v'(J)}{kT}\right], \quad (11)$$

the summation being carried out over all P-, Q-, and R-branches and over all rotational lines of the bands. Since for a diatomic molecule of the C$_2$ type $\Sigma S_{J'J''}$ is equal to $3(2J'+1)$ over all P-, Q-, and R-branches for a stipulated J', and

$$\sum_{J'_{min}}^{J'_{max}} (2J+1) \exp\left[-\frac{F_v'(J)}{kT}\right] = Z_r , \quad (12)$$

while the sum over the states Z can be written approximately in the form of the product $Z = Z_n Z_v Z_r$, where $Z_n$, $Z_v$, $Z_r$ are respectively the sums over the electronic, vibrational, and rotational levels, then

$$\sum_{P,Q,R} \sum_{J'_{min}}^{J'_{max}} \int I_\nu d\nu = \frac{64\pi^4}{c^3} \nu_{av}^4 \frac{Nl}{Z_n Z_v} |\bar{R}_e|^2 \exp\left[-\frac{T_0' + G_0'(v)}{kT}\right] q_{v'v''}. \quad (13)$$

If the radiation energy of an entire sequence of bands is measured, for example, (0,0), (1,1), (2,2), (3,3), etc., then it is additionally necessary to sum over the sequence of bands. Thus,

$$\sum_{(0,0),(1,1),\ldots} \sum_{P,Q,R} \sum_{J'_{min}}^{J'_{max}} \int I_\nu d\nu = \frac{64\pi^4}{c^3} \nu_{av}^4 \frac{Nl}{Z_n Z_v} |\bar{R}_e|^2 \exp\left(-\frac{T_0'}{kT}\right) \sum_{v_{min}}^{v_{max}} q_{v'v''} \exp\left[-\frac{G_0'(v)}{kT}\right]. \quad (14)$$

Therefore,

$$|\bar{R}_e|^2 = \frac{c^3}{64\pi^4 \nu_{av}^4} \frac{Z_n Z_v}{Nl \exp\left(-\frac{T_0}{kT}\right)} \frac{\sum\sum\int I_\nu d\nu}{\sum_{v_{min}}^{v_{max}} q_{v'v''} \exp\left[-\frac{G_0'(v)}{kT}\right]}. \quad (15)$$

Thus, in order to determine $|R_e|^2$ it is necessary to know in addition to the experimentally determined quantity $\Sigma\Sigma\Sigma \int I_\nu d\nu$ the temperature T, the number of molecules N in cm$^3$, the sum of the states $Z_n Z_v$, the factors $q_{v'v''}$, and the energy levels of the C$_2$ molecule.

If it is assumed that $|R_e|^2$ for the Swan bands does not depend on the internuclear distance, then the electronic force of the oscillator

$$f_e = \frac{8\pi^2 mc}{3he^2}\overset{*}{v}_{av}|\bar{R}_e|^2 \tag{16}$$

can be determined; under these conditions the frequency for the edge of the (0,0) band can be used as $\nu^*_{av}$ (in cm$^{-1}$).

## §2. General Foundations of the Absorption Method

The intensity of radiation transmitted through a gas is expressed by means of the Beer law [14]:

$$I_\nu = I_\nu^0 \exp(-k_\nu \Delta x), \tag{17}$$

where $I_\nu$ and $I_\nu^0$ are the intensities of the transmitted and incident radiation; $k_\nu$ is the absorption coefficient. The total intensity absorbed by the line, which corresponds to the transition from the state n to the state m, is equal to

$$I^{nm} = \int(I_\nu^0 - I_\nu)\,d\nu = I_\nu^0 \Delta x \int k_\nu d\nu. \tag{18}$$

On the other hand $I^{nm}$ depends on the intensity of the incident radiation $I_\nu^0$, the number of absorptive particles in the lower state $N_m$, the magnitude of the absorbed quanta $h\nu_{nm}$, and the thickness $\Delta x$ of the absorptive layer; i.e.,

$$I^{nm} = I_\nu^0 h\nu_{nm} \Delta x B_{nm} N_n. \tag{19}$$

Here $B_{nm}$ is the Einstein coefficient for the absorption, which determines the absorption probability. In writing Eqs. (18) and (19) it was assumed that the intensity of the incident radiation was constant in a frequency interval exceeding the width of the line.

Comparing (18) and (19), we obtain the following expression for the integrated absorption factor:

$$\int k_\nu\, d\nu = N_n B_{nm} h\nu_{nm}. \tag{20}$$

Expressing the coefficient $B_{nm}$ in terms of the matrix element of the dipole moment of the transition

$$B_{nm} = \frac{8\pi^3}{3h^2c}|\bar{R}^{mn}|^2 \tag{21}$$

and substituting it into (20), we find

$$\int k_\nu d\nu = \frac{8\pi^3\nu_{nm}}{3hc} N_n |\bar{R}^{mn}|^2. \tag{22}$$

This equation is applicable only in the case of transitions between nondegenerate levels. However, in the case of transitions from a level with degeneration $g_n$ it is necessary to replace $|\bar{R}^{mn}|^2$ in $\frac{1}{g_n}\sum_{i,k}|\bar{R}^{mn}|^2$, where i and k are the number of degenerate sublevels in the lower (n) and upper (m) states. Taking account of relationship (8) and having determined the population of the lower electronic vibrational — rotational state according to Eq. (9), we finally obtain

$$\int k_\nu d\nu = \frac{8\pi^3\nu}{3hc}\frac{Nl}{Z}|\bar{R}_e|^2 \exp\left[-\frac{T_0'' + G_0''(v) + F_v''(J)}{kT}\right] q_{v'v''}S_{J'J''},\tag{23}$$

whence

$$|\bar{R}_e|^2 = \frac{3hc}{8\pi^3\nu}\frac{Z\int k_\nu d\nu}{Nlq_{v'v''}S_{J'J''}}\exp\left[\frac{T_0'' + G_0''(v) + F_v''(J)}{kT}\right].\tag{24}$$

Equation (24) allows calculation of the square of the matrix element $|R_e|^2$ of the dipole moment of the electronic transition from the experimentally determined integrated absorption coefficient $\int k_\nu\,d\nu$. For this purpose it is also necessary to know the system of energy levels of the molecule, the sum over the states Z, the Franck — Condon factors $q_{v'v''}$ and the Hanley — London factors $S_{J'J''}$, as well as the temperature T of the absorbative gas and the number of molecules N per unit volume.

## §3.  Choice of the Gas

Preliminary experiments on a shock tube in comparatively low operating modes were used to establish the fact that luminescence of the Swan bands behind reflected shock wave is observed in $C_2H_2 + Ar$, $CO_2 + Ar$, and $CO + Ar$ mixtures if the accelerating gas is hydrogen.

At an initial pressure $P_1 = 5$ mm Hg of a 10% $C_2H_2$ + 90% Ar mixture and for diaphragms which open at a pressure $P_4 = 7$-10 atm, sharp Swan bands were obtained by means of an ISP-51 spectrograph with an intermediate chamber ($f = 270$ mm). For the same pressure differential but in pure ethylene a continuous luminescence spectrum was recorded, while lamp black was detected on the walls of the tubes after the experiment. Evidently, the decay of $C_2H_2$ molecules occurred at comparatively low (T < 3500°K) temperatures behind the shock wave, which facilitated the formation of sooty particles. For operation with a $CO_2$ + Ar mixture high temperatures are unattainable for small pressure differentials due to the large dissociation losses. Therefore, in order to conduct the experiments we chose the CO + Ar mixture, which yielded the most intense Swan bands. Luminescence was observed over a wide a wide range of ratios between the concentrations of CO and Ar. Thus, the spectrum whose photograph is reproduced in Fig. 2 was obtained behind the shock wave for lateral observation at a distance of 3 mm from the end of the tube using a 70% Ar + 30% CO mixture at an initial pressure $P_1 = 5.0$ mm Hg and a hydrogen pressure $P_4 = 7$ atm. Spectra in CO + Ar mixtures were recorded over a very wide range of variation of CO concentration — from 4 to 95% in comparatively low operating modes. We chose a 50% CO + 50% Ar mixture for our work, and the calculation is carried out for it.

## §4.  Gasdynamic Calculation

It is assumed that the gas mixture is in thermal equilibrium behind the front of the shock wave. In order to determine the equilibrium composition of the mixture it is necessary to solve the system of chemical equilibrium equations, which consist of the equation for the Dalton law (25), the material balance equations (26), (27), the equations for the law of effective masses (28)-(37), and the charge conservation equation (38):

$$X_{CO} + X_{Ar} + X_C + X_O + X_{C_2} + X_{e^-} + X_{C^+} + X_{O_2} + X_{CO} + X_{Ar^+} + X_{CO_2} + X_{O^+} + X_{C_3} + X_{O_2^+} = 1, \quad (25)$$

$$\frac{2X_{CO_2} + X_{CO} + 2X_{O_2} + X_O + X_{CO^+} + 2X_{O_2} + X_{O^+}}{X_{CO} + X_C + 2X_{C_2} + X_{C^+} + X_{CO^+} + X_{CO_2} + 3X_{C_3}} = 1, \quad (26)$$

$$\frac{X_{Ar} + X_{Ar^+}}{X_{CO} + X_C + 2X_{C_2} + X_{C^+} + X_{CO} + X_{CO_2} + 3X_{C_3}} = 1. \quad (27)$$

$$\frac{X_C X_O^2}{X_{CO_2}} = \frac{K_p(CO_2)}{P}, \quad (28) \qquad\qquad \frac{X_{CO} X_{e^-}}{X_{CO}} = \frac{K_p(CO)}{P}, \quad (33)$$

$$\frac{X_O^2}{X_{O_2}} = \frac{K_p(O_2)}{P}, \quad (29) \qquad\qquad \frac{X_{C^+} X_O}{X_{CO^+}} = \frac{K_p(CO^+)}{P}, \quad (34)$$

$$\frac{X_C^2}{X_{C_2}} = \frac{K_p(C_2)}{P}, \quad (30) \qquad\qquad \frac{X_{O^+} X_{e^-}}{X_O} = \frac{K_p(O)}{P}, \quad (35)$$

$$\frac{X_C X_O}{X_{CO}} = \frac{K_p(CO)}{P}, \quad (31) \qquad\qquad \frac{X_{O_2^+} X_{e^-}}{X_{O_2}} = \frac{K_p(O_2)}{P}, \quad (36)$$

$$\frac{X_C^3}{X_{C_3}} = \frac{K_p(C_3)}{P}, \quad (32) \qquad\qquad \frac{X_{C^+} X_{e^-}}{X_C} = \frac{K_p(C)}{P}, \quad (37)$$

$$X_{e^-} + X_C + X_{Ar} + X_{O_2^+} + X_{O^+} + X_{CO^+} = 0, \quad (38)$$

In writing these equations the following notation was used: P is the pressure in atm; X are the molar fractions of the components; $K_p$ are the equilibrium components of the corresponding reactions.

The system of Eqs. (25)-(38) allows the composition of the mixture to be determined as a function of temperature and pressure. Knowing the composition, one can find the molecular weight $\mu$ and the enthalpy h of the mixture as functions of temperature and pressure:

$$\mu(P, T) = \sum_i \mu_i X_i(P, T), \quad (39)$$

$$h(P, T) = \frac{\sum_i H_i(T) X_i(P, T)}{\mu(P, T)}, \quad (40)$$

where $\mu_i$, $H_i(T)$ are the molecular weight and molar enthalpy of components i.

In order to determine the state of the gas mixture behind the fronts of the incident and reflected shock waves it is necessary to use the equations of the conservation laws and the equations of state. The equations of the conservation laws are written as follows for the case of the one-dimensional problem in the system associated with the fronts of the incident and reflected shock waves.

For the incident wave:

$$\rho_1 v_1 = \rho_2 v_2 \quad \text{(the continuity equation)}, \tag{41}$$

$$h_1(P_1, T_1) + \frac{v_1^2}{2J_\tau} = h_2(P_2, T_2) + \frac{v_2^2}{2J} \quad \text{(the energy conservation equation)}, \tag{42}$$

$$P_1 + \rho_1 v_1 = P_2 + \rho_2 v_2 \quad \text{(the momentum conservation equation)}. \tag{43}$$

For the reflected wave:

$$\rho_2(v_1 - v_2 + v_5) = \rho_5 v_5 \quad \text{(the continuity equation)}, \tag{44}$$

$$h_2(P_2 T_2) + \frac{(v_1 - v_2 + v_5)^2}{2J} = h_5(P_5, T_5) + \frac{v_5^2}{2J} \quad \text{(the energy conservation equation)}, \tag{45}$$

$$P_2 + \rho_2(v_1 - v_2 + v_5)^2 = P_5 + \rho_5 v_5^2 \quad \text{(the momentum conservation equation)}. \tag{46}$$

In Eqs. (41)-(46) the subscript 1 refers to the state in front of the incident wave; the subscript 2 refers to the state behind the front of the incident wave; the subscript 5 refers to the state behind the front of the reflected wave. Under these conditions v and $\rho$ are the velocity and density of the corresponding fluxes; J is the mechanical heat equivalent.

To the equations of the conservation laws it is necessary to add the equations of the states

$$P_1 = \frac{R}{\mu_1(P_1, T_1)} \rho_1 T_1 \quad \text{(the initial state)}, \tag{47}$$

$$P_2 = \frac{R}{\mu_2(P_2, T_2)} \rho_2 T_2 \quad \text{(the state behind the front of the incident wave)}, \tag{48}$$

$$P_5 = \frac{R}{\mu_5(P_5, T_5)} \rho_5 T_5 \quad \text{(the state behind the front of the reflected shock wave)}. \tag{49}$$

For Eqs. (41)-(43) for the incident wave and the equations of state (47), (48) one can obtain the equations:

$$\frac{T_2}{\mu_2 P_2} P_1^2 + \left(\frac{h_2 - h_1}{2RJ} - \frac{T_2}{\mu_2} + \frac{T_1}{\mu_1}\right) P_1 - \frac{T_1}{\mu_1} P_2 = 0, \tag{50}$$

$$v_1^2 = \frac{RT_1(P_2 - P_1)}{\rho_1 \mu_1 \left(1 - \frac{\rho_1 \mu_1 T_2}{\rho_2 \mu_2 T_1}\right)}. \tag{51}$$

From Eqs. (44)-(46) and the equations of state (47)-(49) one can obtain [84] the equations

$$\frac{T_5}{\mu_5 P_5} P_2^2 + \left(\frac{h_5 - h_2}{2RJ} - \frac{T_5}{\mu_5} + \frac{T_2}{\mu_2}\right) P_2 - \frac{T_2}{\mu_2} P_5 = 0, \tag{52}$$

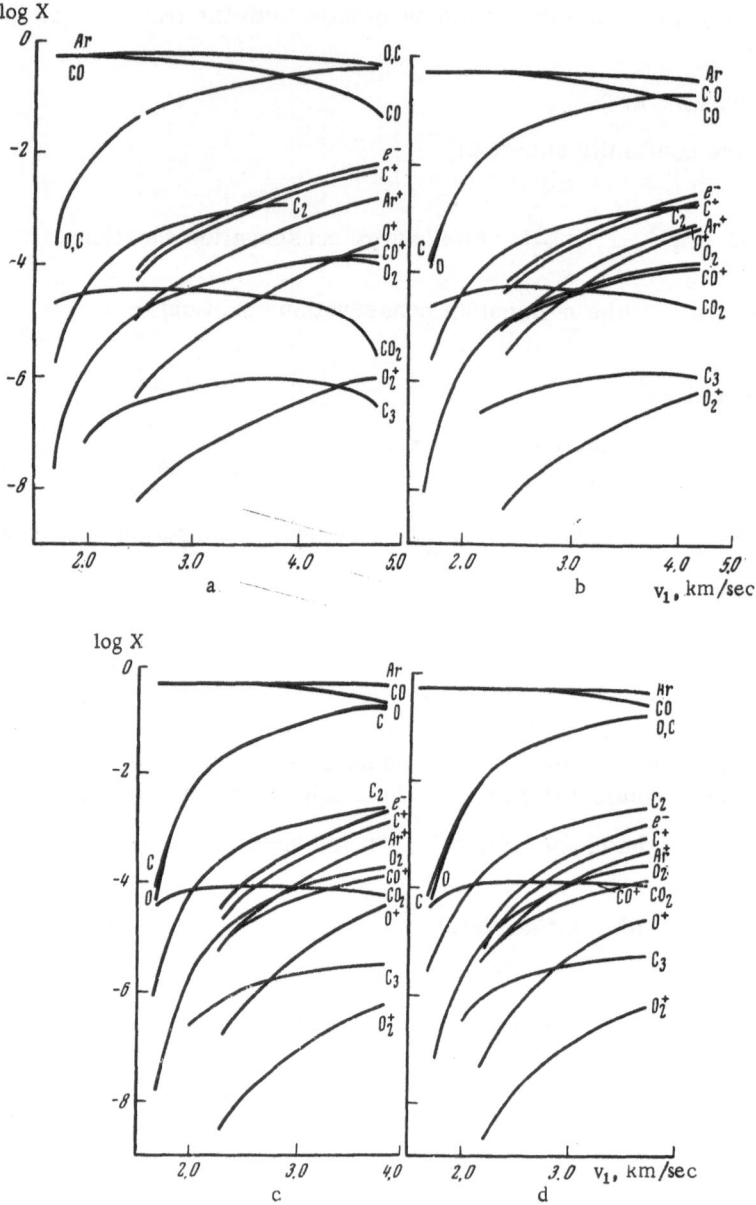

Fig. 5. Dependence of the molar fractions of the components on
the velocity $v_1$ of the incident wave at $P_1 = 5$ (a), 10 (b), 25 (c),
and 50 mm Hg (d). The initial composition was 50% CO + 50% Ar.

$$v_1^2 = \frac{2J\,(h_5 - h_2)\left(\dfrac{T_2}{\rho_2\mu_2} - \dfrac{T_5}{\rho_5\mu_5}\right)}{\left(1 - \dfrac{T_2}{\rho_2\mu_2}\dfrac{\rho_1\mu_1}{T_1}\right)^2\left(\dfrac{T_2}{\rho_2\mu_2} + \dfrac{T_5}{\rho_5\mu_5}\right)}\,. \tag{53}$$

Stipulating the temperature behind the incident shock wave $T_2$ for a certain pressure $P_2$,
we find $h_2$ and $\mu_2$ by means of the system of equations (25)-(40). Assume that the pressure

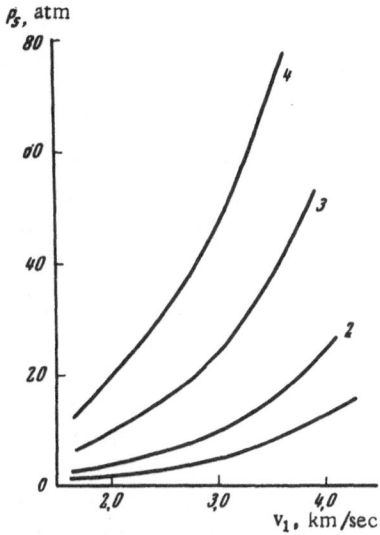

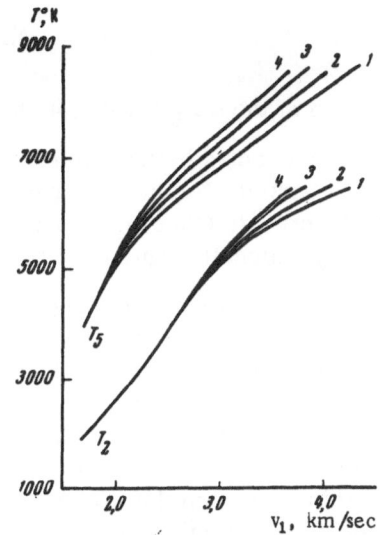

Fig. 6. Dependence of the pressure behind the front of the reflected shock wave on the velocity $v_1$ of the incident wave. The initial pressure was $P_1$ = 5 (1), 10 (2), 25 (3), and 50 mm Hg (4). The initial composition was 50% CO + 50% Ar.

Fig. 7. Dependence of the temperature $T_2$ behind the front of the shock wave and the temperature $T_5$ behind the front of the reflected shock wave on the velocity $v_1$ of the incident shock wave. The initial pressure was $P_1$ = 5 (1), 10 (2), 25 (3), and 50 mm Hg (4). The initial composition was 50% CO + 50% Ar.

behind the shock wave will be $P_1^0$. Then for a stipulated $T_2$ we find a pair of values $P_1^{(1)}$ and $P_2^{(2)}$ for several values of $P_2$ using relationships (50), for which the condition

$$P_2^{(1)} < P_1^{(0)} < P_1^{(2)}$$

will be fulfilled. By choosing $P_2$ it is possible to ensure an arbitrarily small difference $(P_1^{(1)} - P_1^{(2)}$ (i.e., a high accuracy in determining $P_1$). In this manner $P_2$, $T_2$, $h_2$, $\mu_2$ (and therefore the composition also) are determined as a function of the initial pressure $P_1^{(0)}$. The velocity $v_1$ is related to $T_2$ and $P_2$ by relationship (51). Proceeding analogously with Eqs. (52) and (53), we obtain the same quantities behind the front of the reflected shock wave.

The solution of the system of Eqs. (52), (53) was obtained on the M-20 electronic computer by I. S. Berestovoi and A. T. Matachun. The relative fractions of the following components were calculated: CO, Ar, C, O, C$_2$, e$^-$, C$^+$, O$_2$, CO$^+$, Ar$^+$, CO$_2$, O$^+$, C$_3$, O$_2^+$, as well as the pressure and temperature behind the fronts of the incident and reflected shock waves. The calculation was carried out for four initial pressures in the low-pressure chamber: $P_1$ = 5, 10, 25, and 50 mm Hg for the 50% CO + 50% Ar mixture.* The results of the calculation are displayed in the form of graphs (Figs. 5-7).

The graphs allow one to obtain a clear idea of the chemical composition of the gas for various shock wave velocities. Using the graphs, it is possible to choose the operating mode in such a way that the optimal conditions are more completely fulfilled. The radiation intensity of the Swan bands is proportional to the concentration of C$_2$ molecules:

―――

*Later [85] an additional calculation was carried out at our request for this same mixture at a pressure $P_1$ = 100 mm Hg and for pure carbon monoxide CO.

$$N_{C_2} = \frac{P_5}{kT_5} X_{C_2} \tag{54}$$

and the population of the $d^3\Pi_g$ level, which increase with increasing temperature.

In experiments on radiation one should not strive for a high concentration $N_{C_2}$, since the $a^3\Pi_u$ level which participates in the formation of the Swan bands lies above the ground level $X^1\Sigma_g^+$ by just 610 cm$^{-1}$ (0.075 eV). For a high concentration of $C_2$ molecules the reabsorption of radiation by $C_2$ molecules can affect the results of measuring the intensity of the bands.

An argument in favor of choosing high operating modes lies in the fact that the curves for the dependence of log $X_{C_2}$ on the wave velocity has a variable slope. For velocities of 1.7 to 2.3 km/sec the slope is considerable, for higher velocities the slope is slight, and for a velocity of 3.5 km/sec the quantity log $X_{C_2}$ depends very weakly on velocity. It is advantageous to operate on the flat portion of the log $X_{C_2} = f(v_1)$ curve, since the error in determining the velocity of the incident shock wave (which is equal to 2.5%) will lead to the minimum error in determining $X_{C_2}$ in this case; thereby the error of the method will be lowered. In order to estimate the error which develops due to the error in measuring the velocity we assumed that the accuracy of measuring the velocity was equal to 2.5% and considered four operating modes corresponding to the velocities $v_1$ = 2.00, 2.30, 2.65, and 2.89 km/sec. Under these conditions the error in determining $X_{C_2}$ turned out to be equal to 38, 20, 18, and 16%, respectively. Calculations showed that the error in determining $N_{C_2}$ will drop respectively by 58, 31, 23, and 19% under these conditions.

The accuracy of calculating the composition of the plasma is determined basically by the accuracy with which the equilibrium constants $K_p$ are determined. In the calculations we used $K_p$ from the handbook [86]. The main or definitive process affecting the concentration of $C_2$ molecules is the process of dissociation of CO and $C_2$ molecules. The values of $K_p$ for $C_2$ were determined by the authors [86] on the basis of refined data on the heat of formation of $C_2$: $\Delta H^0_{f_0} = 196 \pm 3$ kcal/mole, which corresponds to a dissociation energy $D_0(C_2)$ = 143.17 $\pm$ 3 kcal/mole. This value is in agreement with the results of [87]. The dependence of the equilibrium constant $K_p$ on the reduced thermodynamic potential $\Delta\Phi^*$ and the heat of formation $\Delta H^0_0$ is determined from the Eqs. [86]

$$\log K_p = \frac{1}{4.57584}\left( \Delta\Phi^* - \frac{\Delta H^0_0}{T} \right), \tag{55}$$

whence

$$\delta \log K_p = \frac{\delta\Delta H^0_0}{T}\frac{1}{4.57584}, \tag{56}$$

where $\Delta H^0_0$ is the thermal effect corresponding to the reaction at 0°K and is equal to the dissociation energy or the ionization energy.

Knowing the error in determining the dissociation energy, it is possible to calculate the error $\delta \log K_p$ in determining the equilibrium constant. At a temperature of 6000°K, $\delta \log K_p$ ($C_2$) = 0.11; for a temperature 7000°K, $\delta \log K_p$ ($C_2$) = 0.09. The dissociation energy $D_0$ (CO) = 255.79 $\pm$ 0.43 kcal/mole was determined more exactly: $\delta \log K_p$ (CO) = 0.016 for T = 6000°K, and $\delta \log K_p$ (CO) = 0.013 for T = 7000°K. Elementary calculations showed that the maximum error in determining the concentration of $C_2$ molecules is equal to $\pm$30% for T = 6000°K and $\pm$27% for T = 7000°K. In other words, in using calculated data on the concentration of $C_2$ molecules it should be remembered that these data can deviate by 30% in either direction, and thereby a systematic error can be introduced into the results of the measurements. In our reasoning we omitted, as a consequence of its insignificance, the effect of the error of the equilibrium constant on such plasma characteristics as temperature, pressure, and shock wave velocity.

## § 5. Calculating the Sum of the States of the $C_2$ Molecule

In quantum statistics [88] the sum of the states (or the Z-function) is defined as follows:

$$Z = \sum_i g_i \exp\left(-\frac{E_i}{kT}\right), \tag{57}$$

where $g_i$ is the statistical weight of the level having the energy $E_i$; k is Boltzmann's constant; T is temperature.

The decisive contribution to the overall sum of the states of a molecule is made by the lower electronic states, whereas the high electronic levels yield only adjusting corrections. If the molecule is represented by means of the model consisting of a rigid rotator and a harmonic oscillator, then the sum of the states can be expressed approximately by the product of four independent quantities

$$Z = Z_n Z_v Z_r Z_s, \tag{58}$$

where $Z_n$, $Z_v$, $Z_r$ respectively represent the sums of the electronic, vibrational, and rotational states, while $Z_s$ is the degeneration due to nuclear spin. Each of the factors can be determined relatively simply by summation (or integration) if the energy levels of the molecule are known:

$$Z_n = \xi(\Lambda)(2\sigma + 1)\exp\left(-\frac{T_0 hc}{kT}\right), \tag{59}$$

$$Z_v = \sum_{v=0}^{v_{max}} g_v \exp\left[-\frac{G_0(v)}{kT}\right] \approx \sum_{v=0}^{v=\infty} \exp\left(-\frac{\omega_0 v hc}{kT}\right) = \frac{1}{1 - \exp\left(-\frac{\omega_0 hc}{kT}\right)}, \tag{60}$$

$$Z_r = \sum_{J_{min}}^{J_{max}} g_r \exp\left[-\frac{F_v'(J)}{kT}\right] = \sum_{J_{min}}^{J_{max}} (2J + 1)\exp\left[-\frac{J(J+1)B_v hc}{kT}\right] \approx$$

$$\approx \int (2J + 1)\exp\left[-\frac{J(J+1)B_v hc}{kT}\right] dJ \approx \frac{kT}{B_v hc}, \tag{61}$$

$$Z_s = (2s_1 + 1)(2s_2 + 1) \quad \text{(for molecules consisting of different atoms)}, \tag{62}$$

$$Z_s = \frac{1}{2}(2s + 1)^2 \quad \text{(for molecules consisting of identical atoms)}. \tag{63}$$

The notation in Eqs. (59)-(63) is conventional: $\xi(\Lambda)$ is a coefficient which considers $\Lambda$-doubling and is equal to unity for $\Sigma$-states, while for $\Pi$- and $\Delta$-states we have $\xi(\Lambda) = 2$; $\sigma$ is the resultant spin: for the singlet state $\sigma = 0$, for the doublet state $\sigma = \frac{1}{2}$, and for the triplet state $\sigma = 1$. Thus, the term $2\sigma + 1$ considers the multiplicity of the state, which is equal to 1, 2, or 3, respectively, for the singlet, doublet, and triplet states. The number s represents the nuclear

TABLE 7. Values Adopted for the Molecular Constants of $C_2$

| Level | $T_0$ | $\omega_e$ | $\omega_e x_e$ | $B_e$ | $\alpha$ |
|---|---|---|---|---|---|
| $X^1 \sum_g^+$ | 0 | 1855.18 | 13.71 | 1.82052 | 0.01832 |
| $a^3\Pi_u$ | 610.5 | 1641.328 | 11.625 | 1.6326 | 0.01683 |
| $b^3\sum_g^-$ | 6243.5 | 1470.433 | 11.166 | 1.49852 | 0.01634 |
| $A^3\Pi_u$ | 8268.33 | 1608.284 | 12.047 | 1.61700 | 0.01720 |
| $B^1\Delta_g$ | 10 000 | 1481 | 11.14 | 1.486 | 0.015 |
| $C^1\sum_g^+$ | 14 000 | 1470 | 10.06 | 1.475 | 0.013 |
| $c^3\sum_u^+$ | 14 300 | 1891 | 14.23 | 1.897 | 0.018 |
| $d^3\Pi_g$ | 19916.26 | 1788.22 | 16.44 | 1.7527 | 0.01608 |

spin corresponding to one atom. For example, for the carbon atoms which form the $C_2$ molecule, s = 0, for nitrogen atoms s = 1, for oxygen atoms s = 0, for hydrogen atoms s = $1/2$, etc.

Since the model based on a rigid rotator and a harmonic oscillator is approximate, it follows that this method yields approximate values for the molecules; however, this is sufficient in a number of practical applications. For the exact calculation of the sum over the states it is necessary to consider the anharmonicity of the vibrations and the fact that the vibrational and rotational states in the molecule are independent of each other. In this case

$$Z = Z_s \sum \sum \sum g_e \exp\left(-\frac{T_0 hc}{kT}\right) g_r \exp\left[-\frac{F_v(J)hc}{kT}\right] g_v \exp\left[-\frac{G_0(v)hc}{kT}\right] =$$
$$= Z_s\left(g_{n_0} Z_{r,v}^0 + g_{n_1} \exp\left(-\frac{T_0^{(1)} hc}{kT}\right) Z_{r,v}^{(1)} + g_{n_2} \exp\left(-\frac{T_0^{(2)} hc}{kT}\right) Z_{r,v}^{(2)} + \ldots\right),$$
(64)

where

$$Z_{r,v}^{(i)} = \sum_J \sum_v g_r \exp\left[-\frac{F_v(J)hc}{kT}\right]$$
(65)

for the i-th electronic level.

The rotational terms $F_v(J)$ are determined according to the equation

$$F_v(J) = B_v J(J+1) + D_v J^2 (J+1)^2,$$
(66)

where J is the rotational quantum number, and the rotational constant $B_v$ is equal to

$$B_v = B_e - \alpha\left(v + \frac{1}{2}\right),$$
(67)

$$B_e = \frac{h}{8\pi^2 c\theta}.$$
(68)

Here v is the vibrational quantum number; $\theta$ is the moment of inertia of the molecule; $\alpha$ is a correction coefficient whose values are given in Table 7. The term $D_v J^2 (J+1)^2$ considers the effect of the centrifugal rotation force of the molecule (a nonrigid rotator), $D_v$ depending on the vibrational state of the molecule and being small in comparison with $B_e$. The statistical weight of the rotational level is $g_r = 2J + 1$.

TABLE 8. Sum over the Electronic and Vibrational States
of the $C_2$ Molecule

| T, °K | Level | | | | $Z_n Z_v Z_s$ |
|---|---|---|---|---|---|
| | $X^1\sum_g^+$ | $a^3\Pi_u$ | $b^3\sum_g^-$ | $A^1\Pi_u$ | |
| 4000 | 1.051 | 5.513 | 0.395 | 0.122 | 7.086 |
| 4500 | 1.145 | 6.225 | 0.557 | 0.180 | 8.118 |
| 5000 | 1.242 | 6.829 | 0.744 | 0.250 | 9.090 |
| 5500 | 1.339 | 7.500 | 0.949 | 0.340 | 10.172 |
| 6000 | 1.437 | 8.184 | 1.173 | 0.433 | 11.301 |
| 6500 | 1.537 | 8.882 | 1.417 | 0.540 | 12.493 |
| 7000 | 1.629 | 9.566 | 1.670 | 0.675 | 13.709 |
| 7500 | 1.724 | 10.279 | 1.937 | 0.800 | 14.977 |
| 8000 | 1.826 | 11.137 | 2.215 | 0.900 | 16.398 |

For the vibrational term of an anharmonic oscillator quantum statistics yields the following expression:

$$G_0(v) = \omega_0 v - \omega_0 x_0 v^2 + \omega_0 y_0 v^3 + \ldots, \qquad (69)$$

where

$$\omega_0 = \omega_e - \omega_e x_e + 0.75\omega_e y_e,$$
$$\omega_0 x_0 = \omega_e x_e - 1.5\omega_e y_e.$$

The statistical weight of the vibrational level is $g_v = 1$.

Available data in the literature [89, 90] on the sum of the states of the $C_2$ molecule are not accurate. This can be explained by the fact that a) until recently no data were available on certain comparatively low excited levels of the $C_2$ molecule and the vibrational and rotational constants were not known sufficiently accurately, and b) the ground level of the molecule was incorrectly assumed to be the $^3\Pi_u$ level. Table 7 displays the current accepted values of the molecular constants of $C_2$. Based on the refined data, a calculation was carried out in [91], but only according to an approximate equation and only for one value of temperature.

In order to calculate the radiation energy of the entire band according to Eq. (9) or of the entire sequence of bands (Eqs. (10) and (11)) it is necessary to know $Z_n Z_v$ — the sum over the electronic and vibrational levels of the $C_2$ molecule. We determined the sum of the vibrational states of the i-th electronic level according to the equation

$$Z_i Z_v Z_s = \xi(\Lambda) \frac{2\sigma + 1}{2} \exp\left(-\frac{T_0}{kT}\right) \sum_{v_{min}}^{v_{max}} \exp\left[-\frac{G_0(v)}{kT}\right], \qquad (70)$$

and then carried out summation over all electronic levels of the $C_2$ molecule which yield a contribution. The number 2 in the denominator of the fraction reflects the fact that one half of the levels in diatomic molecules having identical nuclei with spin zero is missing. The values of $T_0$ and the molecular constants of $C_2$ required for the calculation were taken from [86] and are displayed in Table 7. The results of the calculation are shown in Table 8.

The contribution to the sum over the states from the $B^1\Delta_g$, $C^1\Sigma_g^+$, and $c^3\Sigma_u^+$ levels, whose existence has not yet been verified experimentally, have not been considered. The consideration

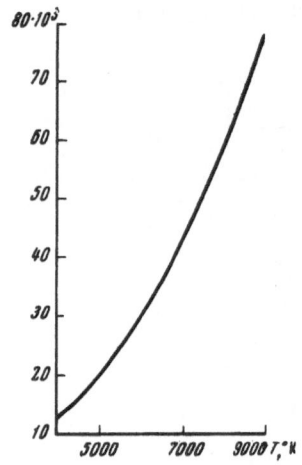

Fig. 8. Vibration of the sum over the states of the $C_2$ molecule with temperature.

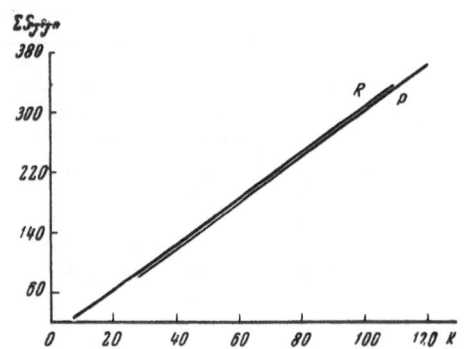

Fig. 9. Dependence of the resultant value of the Höhnli – London $\Sigma S_{J'J''}$ -factors for the triplet P- and R-bands of the $^3\Pi - ^3\Pi$ transitions of the $C_2$ molecule on the rotational quantum number K.

of these levels would increase the overall sum Z over the states (this sum is shown in the graphs) by 1.4% at 4000°K, by 2.9% at 5000°K, by 4.6% at 6000°K, by 6.2% at 7000°K, and by 7.7% at 8000°K according to our estimates. In the calculation the high electronic levels $d^3\Pi_g$, $D^1\Pi_g$ and others are similarly not considered in view of the fact that the contribution from them is small. Thus, for example, the contribution from the $d^3\Pi_g$ level at 7000°K is about 1%.

Having calculated the sum $Z_r$ of the rotational states according to the approximate Eq. (61)

$$Z_r \approx \frac{kT}{B_v hc} = \frac{0.695T}{1.6} = 0.428T,$$

one can calculate the total sum of the states from the data in Table 8 using Eq. (58). Below we give the sum over the electronic, vibrational, and rotational states of the $C_2$ molecule, which constitutes the results of the calculation:

| $T$, °К . . . | 4000 | 4500 | 5000 | 5500 | 6000 | 6500 | 7000 | 7500 | 8000 |
|---|---|---|---|---|---|---|---|---|---|
| $Z \cdot 10^{-3}$ . . . | 12.2 | 15.6 | 19.4 | 23.9 | 29.0 | 34.7 | 41.7 | 47.9 | 56.1 |

More exact values of the sum Z over the states of the molecule $C_2$ were provided to us by L. V. Gurvich, I. V. Veits, and N. I. Ptishcheva. These data were not given in [86], since they were used only as an intermediate stage in calculating the thermodynamic properties of the $C_2$ molecule. These values of the sum Z over the states of the $C_2$ molecule are displayed in Fig. 8 as a function of temperature.

A comparison of the data obtained from the approximate calculation (see above) with the data obtained by the more accurate calculation (see Eq. (64)) shows that the accurate values of Z are 2-3% above the approximate values for temperatures 4000-5000°K, whereas this difference reaches 8% at a temperature 8000°K.

TABLE 9. Formulas for Determining the Höhnli–London Factors $S_{J'J''}$ for the $^3\Pi - {}^3\Pi$ Transitions (Case (c) according to Hund)

| Branch | $S_{J'J''}$-factors | Branch | $S_{J'J''}$-factors |
|---|---|---|---|
| $P_1(J)$ | $(J-2)(2J+1)/(J-1(2J-1)$ | $^QP_{12}(J)$ | 0 |
| $Q_1(J)$ | $(J+1)(2J+1)/J^3$ | $^PQ_{13}(J)$ | $(J-1)(J+1)/J^3$ |
| $R_1(J)$ | $(J-1)(J+1)(2J+3)/(2J+1)$ | $^QR_{12}(J)$ | $(2J+3)/J(J+1)^3$ |
| $^QP_{21}(J)$ | $(2J+1)/(J-1)J^3$ | $P_2(J)$ | $(J-1)^3(J+1)^2/J^3$ |
| $^RQ_{12}(J)$ | $(J-1)(J+1)/J^3$ | $Q_2(J)$ | $(2J+1)(J^2+J-1)^2/J^3(J+1)$ |
| $^SR_{21}(J)$ | 0 | $R_2(J)$ | $J^2(J+2)^2/(J+1)^3$ |
| $^RP_{31}(J)$ | $(J-1)(J+1)/(2J-1)(2J+1)J^3$ | $^QP_{32}(J)$ | $(2J-1)/J^3(J+1)$ |
| $^SQ_{31}(J)$ | 0 | $^RQ_{32}(J)$ | $J(J+2)/(J+1)^3$ |
| $^TR_{31}(J)$ | 0 | $^SR_{32}(J)$ | 0 |

| Branch | $S_{J'J''}$-factors | Branch | $S_{J'J''}$-factors |
|---|---|---|---|
| $^NP_{13}(J)$ | 0 | $R_{23}(J)$ | $(2J+1)/(J+1)^3(J+2)$ |
| $^OQ_{13}(J)$ | 0 | $P_3(J)$ | $J(J+2)(2J-1)/(J+1)(2J+1)$ |
| $^PR_{13}(J)$ | $J(J+2)/(J+1)^3(2J+1)(2J+3)$ | $Q_3(J)$ | $J(2J-1)/(J+1)^3$ |
| $^OP_{23}(J)$ | 0 | $R_3(J)$ | $(J+1)(J+3)(2J+1)/(J+2)(2J+5)$ |
| $^QQ_{23}(J)$ | $J(J+2)/(J+1)^3$ | | |

## §6. The Distribution of the Intensity over the Rotational Lines in the Triplet $^3\Pi - {}^3\Pi$ Transitions of the $C_2$ Molecule

The problem of the intensity distribution in multiplet bands of diatomic molecules corresponding to cases (a), (c), and the intermediate case following Hund was solved in principle in [92]. The difficulty of calculating the intensity distribution over the rotational line (i.e., of calculating the Hönli-London factors for triplet bands) resides in the fact that no simple expressions exist for the energy of the $^3\Pi$ states, which could be valid with sufficient accuracy for all values of the coupling constants $Y = A/B$. The author of [18] succeeded in deriving more or less simple relationships for the energy of the terms, on the basis of which formulas were obtained for calculating the $S_{J'J''}$ factors for the $^3\Pi - {}^3\Pi$ transitions for all 27 branches. Table 9 displays the formulas for determining the $S_{J'J''}$ factors for case (c) which is realized in the $C_2$ molecule for fairly large K numbers $(K = J, K = J - 1, K = J + 1)$.

The calculation of $S_{J'J''}$ factors according to these equations was carried out* on an M-20 electronic computer. As might be expected, for all values except $R_1$, $R_2$, $R_3$ and $P_1$, $P_2$, $P_3$, the factors turned out to be negligibly small. The resultant values of the factors over each R and P triplet are laid off on the graph shown in Fig. 9. Notwithstanding the cumbersome nature of the calculation formulas, the dependence of $\Sigma S_{J'J''}$ on K is almost linear. Attention should be paid to the fact that in these formulas the sum of the factors over all P-, Q-, and R-branches has been normalized to $3(2J' + 1)$.

---

*The programming was carried out by P. P. Lazarev, associate of the Laboratory of Low-Temperature Plasma Optics, Physics Institute, Academy of Sciences of the USSR.

CHAPTER III

## THE RADIATION METHOD

### §1.  Description of the Apparatus and Experimental Procedure

The apparatus on which the experiments were carried out consisted of a shock tube equipped with two spectrographs and two systems for measuring velocity (the velocities of the incident and reflected shock waves).  The arrangement of the individual parts of the apparatus is shown schematically in Fig. 10.

A glass tube 1.65 m long with an inside diameter 3.5 cm and a wall thickness of 5 mm, which had been subjected to preliminary testing at a pressure of 16 atm, was used as the low-pressure chamber.  The high-pressure chamber was made of brass and was constructed in two versions: 15 and 45 cm long.  At comparatively low diaphragm-rupture pressures (up to 15 atm) the long 45 cm chamber was used; this allowed a normal thickness of the plug to be obtained behind the front of the reflected shock wave, since the rarefaction wave arrives at the end of the tube with a large delay.

The short chamber was used in high operating modes.  The residual pressure for operation with the short chamber did not exceed the residual pressure in low operating modes and was under 5 atm, which eliminated the possibility of rupture of the shock tube.  According to our calculations, which we shall not present here, the rarefied wave did not have time to destroy the plug of luminescent gas behind the front of the reflected shock wave in this case either.

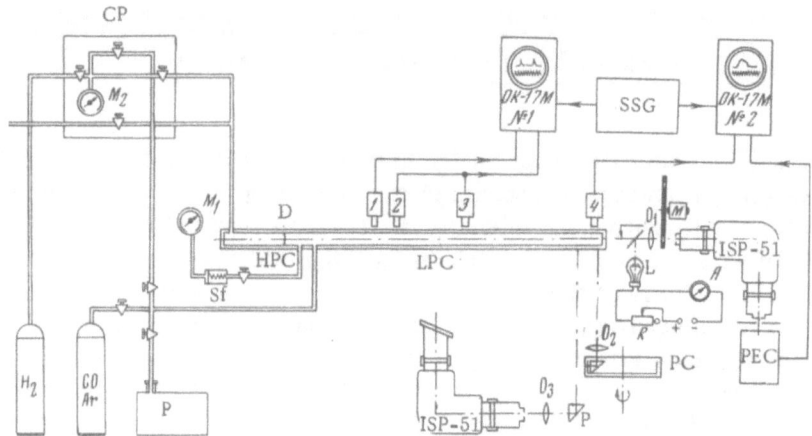

Fig. 10.  Diagram of the experimental apparatus.  HPC is the high pressure chamber; LPC is the low pressure chamber; CP is the control panel; $M_2$ is a manometer for measuring the low pressure; Sf is the safety valve; $M_1$ is the manometer for measuring the rupture pressure of the diaphragm D; $H_2$ is the hydrogen container; CO, Ar is a batcher containing a mixture of carbon monoxide and argon; P is a preevacuation pump; 1, 2, 3, 4 are photomultipliers; PC is a camera with a rotating drum; OK-17M No. 1 is the oscilloscope for recording the velocity of the incident wave; OK-17M No. 2 is the oscilloscope for recording the luminescence brightness.  The diagram also shows the placement of the ISP-51 spectrographs, the temperature lamp L along with its equipment, the modulator M, and the GSS-6 standard signal generator SSG.

The cylindrical surface of the glass tube was covered with black paper in which slits were cut for the photomultipliers (1 and 4) which trigger the oscilloscope and measure the velocity (2, 3), as well as for streak photography at the end of the tube.

The high- and low-pressure chambers were separated by celluloid diaphragms D having thicknesses ranging from 0.15 to 0.45 mm. In order to achieve higher operating modes two-three diaphragms which were joined together were used. Rupture of the diaphragms occurred at pressures from 10 to 45 atm. Under these conditions the velocities of the incident shock waves varied in the range $(2.0-2.9) \times 10^5$ cm/sec, while the temperature behind the reflected shock wave varied from 5000 to 6650°K and the pressure varied from 2.0 to 4.5 atm. The mixture of CO and Ar in the volume proportion 1:1 was used as the accelerating gas. The mixture was first prepared in a small (3000 cm³) container which served as a batcher. Before the container was filled, it was evacuated to a residual pressure of $10^{-3}$ mm Hg. The batcher was connected to the system 40-50 h after it had been filled. This delay was necessary to ensure a more complete mixing of the mixture. Hydrogen was the accelerating gas.

The entire system (including the inlet main) was subjected to preliminary evacuation down to $10^{-3}$ mm Hg. The low-pressure chamber was scavenged twice with a CO + Ar mixture, after which the working pressure of 5 mm Hg was established. Such scavenging with evacuation turns out to be necessary in order to avoid impurities of foreign gases in the tube, which could produce an undesirable continuum background; this would reduce the accuracy of the experiment. The use of a glass tube instead of a metallic one has the same purpose, although it restricts the possibility of raising the operating modes of the tube, since the shock wave velocity which is obtained (and therefore the temperature in the "plug" as well) depends on the pressure $P_4$ of diaphragm rupture (more precisely, on the ratio $P_4/P_1$).

The degree of purity of the carbon monoxide CO was specially checked. Volume chemical analysis yielded the following results: 98.5% CO, 0% $CO_2$, 0.1% $O_2$, 0.4% $H_2$, and 1% $N_2$. The degree of purity of argon was not specially monitored, since according to the factory specifications for the chemically pure argon the percentage content of impurities in it was small (below 0.5%).

In order to estimate the degree of purity of the working mixture immediately before the experiment, it is necessary to know the rate of in-leakage and the time which elapses from the instant at which the working pressure is established until the instant of diaphragm rupture (~ 2 min). The in-leakage was checked in our experiments and amounted to less than $10\mu$ Hg/min. Thus, approximately $20\mu$ Hg leaks in during 2 min; this is about 0.4% of the working pressure.

In order to determine the matrix element it is necessary to know the integrated intensity of the band and the number of radiating centers whose luminescence is recorded. In order to obtain sufficiently intense radiation it is necessary to have as high an equilibrium temperature as possible, which is best achieved behind the front of the reflected shock wave. Therefore, all measurements were carried out behind the front of the reflected shock wave.

The velocity of the incident shock wave was measured by means of an OK-17M pulse oscilloscope to which pulses were successively applied from two photomultipliers (2 and 3, Fig. 10) spaced 19.3 cm apart. The oscilloscope screen was photographed (Fig. 11). The oscilloscope was triggered by means of photomultiplier 1. After the signals of the photomultipliers had been recorded, the signal from the GSS-6 signal generator having a known frequency was applied to the same beam as a time marker. The accuracy of determining the velocity was 2.5%.

The maximum length of the "plug" behind the front of the reflected shock wave was determined according to the photograph obtained in the streak camera, which yielded the time sweep

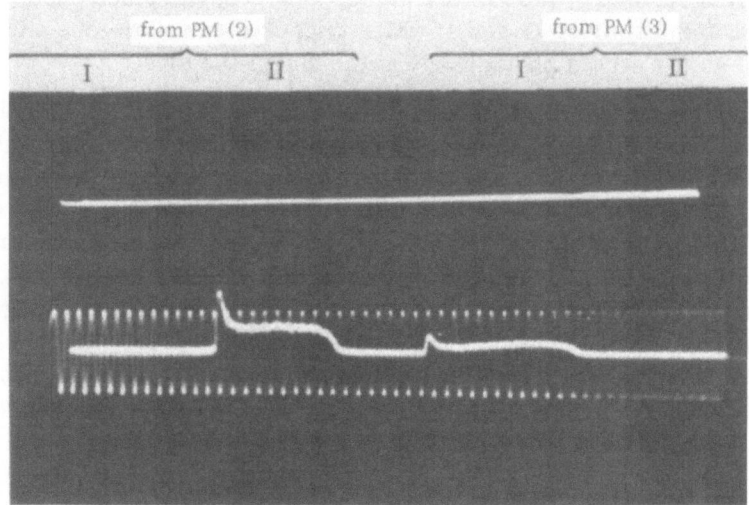

Fig. 11. Oscillograms for determining the velocity of the incident wave. The frequency of the time markers is 200.0
kHz. The distance between two maxima of the sinusoid corresponds to 5 $\mu$ sec. The velocity of the incident shock wave
is equal to 2.30 km/sec according to the oscillogram. I corresponds to the forward front; II corresponds to the contact
surface.

Fig. 12. Photograph of the luminescence of the "plug," which was obtained on the
streak camera (the front of the reflected shock wave (A) was observed).

of the luminescence distribution behind the front of the reflected shock wave. The thickness of
the plug was estimated according to markers marked on the tube at a known distance from each
other, which were also registered on the photograph. The streak camera PC consisted of a
drum connected to an AOL–21–4 motor running at 1400 rpm via a multiplier. The photographic
film was attached to the inside surface of the drum. The camera was equipped with an objective lens and a rotatable prism, which allowed the image of the portion of the tube adjoining its
ends to be focused onto the surface of the photographic film. From the slope angle of the sweep

Fig. 13. Photograph of the luminescence of the "plug," which was obtained on the streak camera (in addition to the reflected shock wave (A), the front of the incident shock wave (B) can be clearly seen).

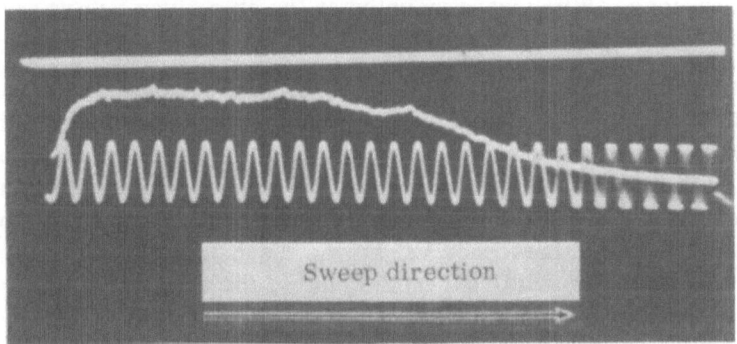

Fig. 14. Oscillogram of the luminescence of the gas behind the front of the reflected shock wave.

of the shock wave front (Figs. 12, 13) and the linear velocity of the photographic film one could determine the velocity of the reflected wave (Fig. 12), while in the case of a sufficiently intense luminescence of the front it was also possible to determine the velocity of the incident wave (Fig. 13).

The measurement of the luminescence intensity of the bands behind the front of the reflected shock wave was carried out by means of an ISP-51 spectograph aimed at the end window of the tube. The window was a circular plate 4 mm thick made of used quartz. A photomultiplier was mounted at the output of the spectograph; the signal from it was then passed through a cathode follower to the OK-17M oscilloscope equipped with a camera. A slit with a micrometer screw was placed in the focal plane of the spectograph in front of the photomultiplier. The setting of the width of the slit and the isolation of the required spectral region was carried out according to the spectra of neon, iron, and sometimes zinc, first visually by means of a microscope, and then photographically. A piece of motion picture film was placed behind the sides of the output slit for purposes of photographic checking; after exposure and development of the peak it was possible to evaluate both the magnitude of the spectral region transmitted by the output slit and the focusing. The procedure described allows a certain spectral region to be extracted with an accuracy up to 1 Å. The oscilloscope was triggered by photomultiplier 4 mounted directly at the end of the tube. After a "shot," sinusoidal pulses from the standard signal generator GSS-6 were applied to the oscilloscope in the capacity of time markers.

The determination of the radiation intensity of the bands was carried out by comparison with the radiation from an SI-16 tungsten temperature lamp.

The temperature lamp was mounted in such a way that the length of the optical path from the lamp filament to the spectrograph slit was equal to the length of the optical path from the end of the tube to the slit. A mirror with a high reflectivity in the visible range of the spectrum was adapted for this purpose and was removed during the "shot." The radiation from the lamp was modulated at a frequency of 5 kHz, which allows the same oscilloscope as that used for recording the shock wave radiation to be used during calibration. The calibration of the equipment according to the temperature lamp was carried out twice — before and after the experiment.

The radiation from the temperature lamp was recorded for different heating currents from 13 to 18 A, which corresponded to brightness temperatures ranging from 2000 to 2560°K. The dependence of the brightness temperature of the lamp on the current strength was calculated for each spectral range according to the brightness temperature at $\lambda = 0.65\mu$ (see Appendix I). The lamp was first calibrated in the metrological laboratory with an accuracy of $\pm 5°$.

If the brightness of the luminescence behind the shock wave front exceeded the maximum radiation intensity of the lamp, then neutral NS-2, NS-7, NS-8, and NS-9 filters were used to attenuate the radiation; these filters were mounted at the objectives before the "shot." The objective $O_1$ was irised down to a relative aperture of 1:10.

The luminescence of the reflected shock wave was also recorded by a second three-prism ISP-51 spectrograph (slitless). An optical system P (Fig. 10), which rotated the image of the tube slit through 90° so that this slit was focused vertically in the focal plane of the ISP-51 collimator, was mounted in front of the spectrograph. The spectrograms obtained on this spectrograph allowed the purity and intensity of the luminescence of the Swan bands to be evaluated. One could also determine the luminescence brightness distribution along the tube from them. In particular, they were used for additional monitoring of the estimate of the maximum thickness of the luminescence plug behind the front of the shock wave.

§2.  Results of the Radiation Experiments

a.  Study of the Time Characteristics of the Luminescence behind the Reflected Shock Wave. The shape of the pulses of gas luminescence behind the forward shock wave in the 50% CO + 50% Ar mixture for observation in the direction perpendicular to the direction of the velocity (see Fig. 11) showed that the front of the shock wave is characterized by a sharp luminescence peak. Behind the front of the shock wave the luminescence is more or less constant. Several explanations exist for this fact which has been repeatedly observed by many experimentalists. The majority of investigators considers that this peak is connected with the de-excitation of atoms at high (nonequilibrium) temperature in the shock wave front. There is no single opinion as to which atom or molecule makes the main contribution to the luminescence of the shock wave front. In [93] the opinion is stated that the luminescence peak in the forward shock wave in argon in the presence of small impurities of carbon dioxide is due to the $C_2$ molecule. In certain of our experiments this peak was obtained in the forward shock wave even when Swan bands were not observed behind the reflected shock wave, and only sodium D-lines appeared on the spectrogram. Evidently, the appearance of this peak is connected with the presence of traces of alkali and alkali-earth elements which are present in the low-pressure chamber before the experiment.

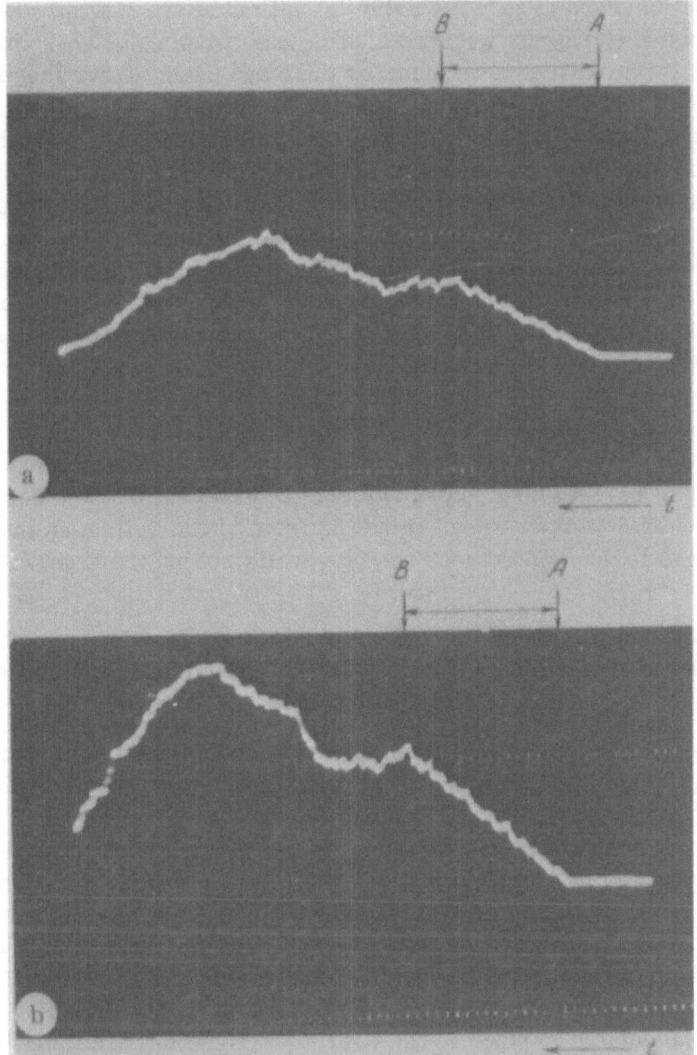

Fig. 15. Oscillogram of the luminescence of the gas behind the front of the reflected shock wave. a) $\tau =$ 55 $\mu$/sec; b) $\tau = 75\,\mu$/sec. A corresponds to the beginning of the luminescence; B corresponds to the time of the encounter with the contact surface; AB is the linearity region.

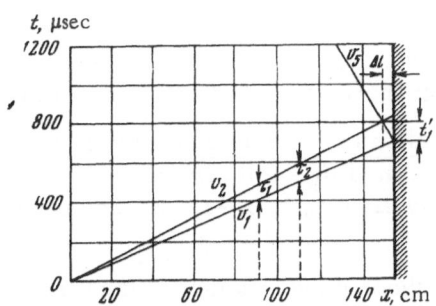

Fig. 16. Diagram of the propagation of the fronts of the incident and reflected shock waves in the 50% CO + 50% Ar mixture.

This peak is not observed (Fig. 14) behind the reflected shock wave for lateral observation. The luminescence oscillogram was obtained in undecomposed light. However, since the main contribution to the luminescence is made by the Swan bands under these conditions, it can be assumed that the oscillograms given characterize the luminescence of the $C_2$ molecule. Observation was carried out through a narrow slit (0.5 mm in width) cut in the light-shielding envelope of the tube in the direction normal to the motion of the shock wave front at a distance of 3 mm from the end. A small aperture 0.02 mm in diameter was located at the distance of 20 cm from this slit, and a photomultiplier was placed a certain distance away from it. The triggering of the system was achieved from the signal produced by another photomultiplier situated at the end of the tube or to the side of the tube, which was operated by the incident shock wave. For such a geometry of the experiment the resolving power of the equipment relative to the front of a reflected shock wave moving at the velocity $v_5 = 0.65 \cdot 10^5$ cm/sec was $\tau = 1 \mu$sec. The luminescence pulse, as is evident from the oscillogram, increases slowly during 10 $\mu$sec. After this the luminescence brightness remains almost constant during 150-180 $\mu$sec. The linear growth of the luminescence pulse during 10 $\mu$sec can be explained by two causes: 1) the process of establishing equilibrium concentration of $C_2$ molecules, which corresponds to an increase in gas temperature behind the front of the reflected shock wave; 2) the inhomogeneity of the gas layer located near the tube walls behind the front of the reflected shock wave.

The luminescence oscillogram (Fig. 15) obtained for each sequence of bands for observation from the end of the tube revealed a linear radiation growth which corresponds to the growth of the thickness of the radiating gas behind the reflected shock wave.

The growth time changes somewhat from experiment to experiment, but on the average it is equal to $\sim 50 \mu$sec. The calculated time interval from the instant of reflection of the shock wave front to the instant of its encounter with the contact surface is equal to $\sim 120 \mu$sec. Figure 16 shows the diagram of the propagation of the fronts of the incident and reflected shock waves. The measured velocity of the incident wave $v_1 = 2.30$ km/sec (t = 5 · 16.8 = 84 $\mu$sec; the base L = 19.3 cm). The velocity of the contact surface is calculated according to the equation $\rho_1 v_1 = \rho_2 v_2'$. Since $v_2' = (\rho'/\rho_2)v_1$, it follows that having determined the ratio $\rho_1/\rho_2$ we find $v_2'$, $\rho_1 = 9.72 \cdot 10^{-6}$ g/cm³ for a 50% CO + 50% Ar mixture and a pressure $P_1 = 5$ mm Hg, $\rho_2 = 5.55 \cdot 10^{-5}$ g/cm³ from the data of the thermodynamic calculation; from this we obtain $v_2' = 0.40$ km/sec. Thus, the velocity of the contact surface in the coordinate system associated with the tube will be $v_2 = 2.30 - 0.40 = 1.90$ km/sec. In this connection the length of the light pulses on the photomultipliers (2) and (3) must be equal to $\tau_1 = 80 \mu$sec and $\tau_2 = 100 \mu$sec. The experimental values (see Fig. 11) $\tau_1 = 45$ and $\tau_2 = 55 \mu$sec differ approximately by a factor of 2.

The maximum thickness $\Delta l$ of the luminescence layer, according to the diagram in Fig. 16, must be equal to 70-75 mm. The experimental thickness, which is determined from photographs taken by the streak camera (see Fig. 13), is equal to 33 mm. This value is in good agreement with the velocity of the reflected shock ($v_5 = 0.66$ km/sec) which was obtained from calculated data, and with the actual linear rise time which is equal to 50 $\mu$sec; actually 0.66 × 50 = 33 mm.

The linear growth of the luminescence of the $C_2$ bands during 40-60 $\mu$sec (depending on the experimental conditions) is evidence of the fact that there is no reabsorption of radiation.

b. The Experimental Values of the Square of the Matrix Element. The measurements of the oscillator force of the electronic $d^3\Pi_g - a^3\Pi_u$ transition of the $C_2$ molecule was carried out in radiation for five sequences $\Delta v = 2, 1, 0, -1, -2$. The calculation was carried out according to Eqs. (15) and (16).

The temperature of the gas behind the reflected shock wave was determined from the velocity of the incident wave on the basis of a gasdynamic calculation. The values of molar con-

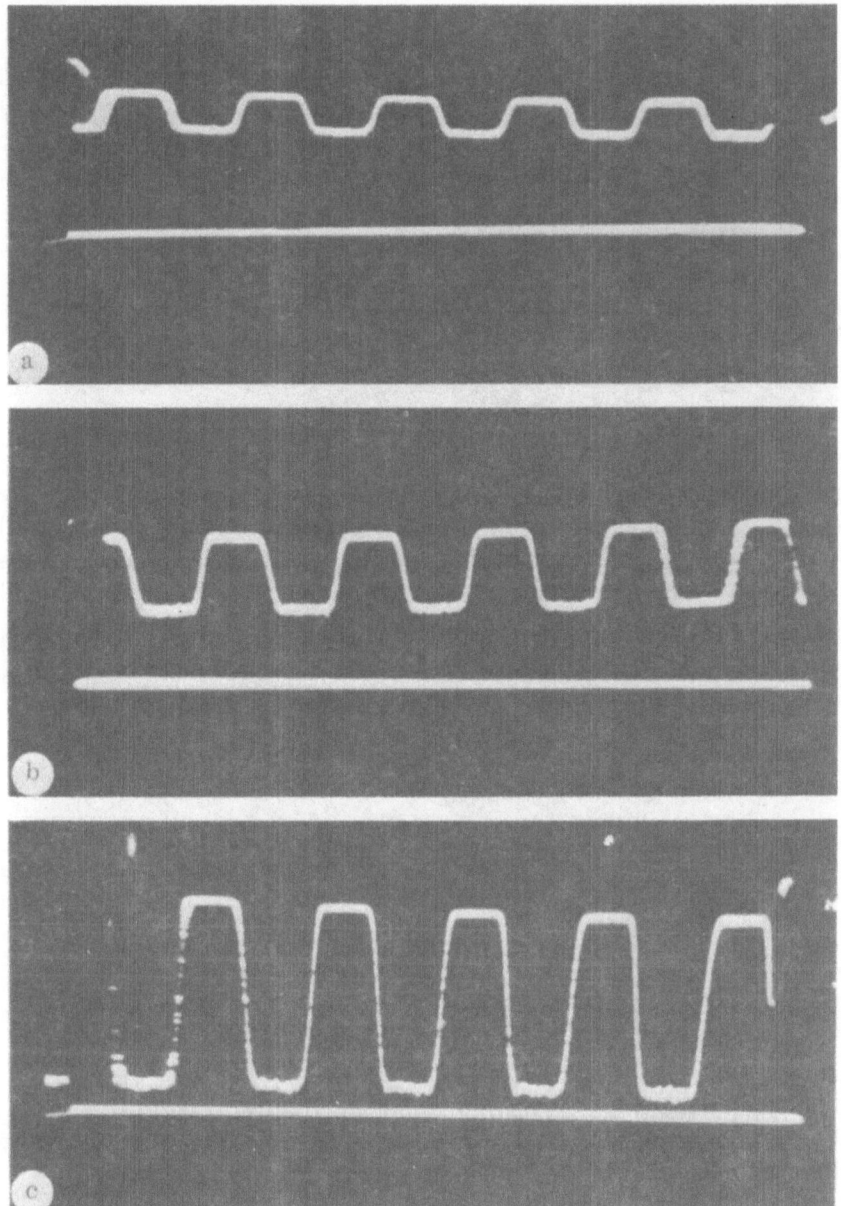

Fig. 17. Calibration pulses of luminescence brightness which were obtained with the temperature lamp for the spectral range 5168–4737 Å. a) $S_{0.65} = 1873°K$; b) $S_{0.65} = 1973°K$; c) $S_{0.65} = 2133°K$.

centration of $C_2$ were taken from this same calculation, and the concentration of $C_2$ molecules was determined according to Eq. (54). Then the total number of molecules along the optical path was found by multiplying $N_{C_2}$ by the thickness of the radiating layer $l$ which was determined experimentally. The resultant energy of radiation from the sequence of $C_2$ bands was found by comparing the maximum residue on the linear region of the luminescence oscillogram with the magnitude of the rectangular pulses used to calibrate the equipment with respect to the temperature lamp (Fig. 17). The energy of the temperature lamp radiation (in erg/sec·$sec^{-1}·cm^2·sr$)

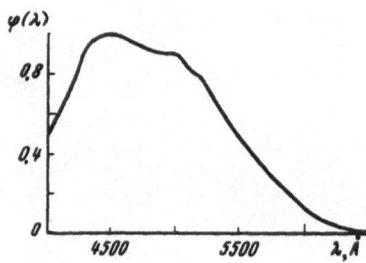

Fig. 18. Curve for the spectral
sensitivity of the FEU-25 photomultiplier (in relative units).

was calculated according to the Wien formula

$$\mathscr{I} = \frac{8\pi h \nu^3}{c^2} \exp\left(-\frac{h\nu_{av}}{kS_\lambda}\right) \Delta\nu, \tag{72}$$

where h, c, and k are known constants; $S_\lambda$ is the brightness temperature of the lamp for $\lambda_{av} = c/\nu_{av}$ ; $\Delta\nu$ is the spectral interval in sec⁻¹ which is transmitted by the output slit of the spectrograph; $\nu_{av}$ is the average frequency of this spectral interval. The width of the input slit of the spectrograph was chosen in the range 0.01-0.05 mm.

A comparison of the values of the radiation power of the lamp and the $C_2$ bands was carried out with allowance for the fact that not all of the radiation determined according to Eq. (10) entered the solid angle $\Delta\Omega$ from the shock tube, but only a part equal to

$$\frac{\Delta\Omega}{4\pi} \int I_\nu d\nu. \tag{73}$$

The radiation power of the temperature lamp in the solid angle (the same angle as the one used in measuring the radiation power of $C_2$ luminescence) was calculated according to the Wien formula for the brightness $B_\nu^0$ :

$$B_\nu^0 \Delta\nu\Delta\Omega = \frac{2h\nu^3}{c^2} \exp\left(-\frac{h\nu_{av}}{kS_\lambda}\right) \Delta\nu\Delta\Omega. \tag{74}$$

In practice it was necessary to consider the change in the sensitivity of the photomultiplier over the spectral interval (Fig. 18) and the difference between the spectral radiation of the lamp and the spectral radiation of the bands in analyzing the oscillograms.

Assume $n_1$ (mm) is the deflection of the oscilloscope beam corresponding to the radiation of a temperature lamp having a known brightness temperature $S_\lambda$, while $n_2$ (mm) is the deflection corresponding to radiation of the sequence of bands. Then

$$n_1 = \beta \frac{\int_{\nu_1}^{\nu_2} B_\nu^0 \, \varphi(\nu) \, d\nu,}{\int_{\nu_1}^{\nu_2} \varphi(\nu) \, d\nu} = \beta \, B_\nu' \Delta\nu, \tag{75}$$

where $\beta$ is a coefficient which considers the spectral and geometric characteristics of the instrument (dispersion, height, and width of the input and output slits, etc.) and the gain of the rf circuits; $\varphi(\nu)$ is the spectral sensitivity of the photomultiplier; $B_\nu'$ is the effective brightness of

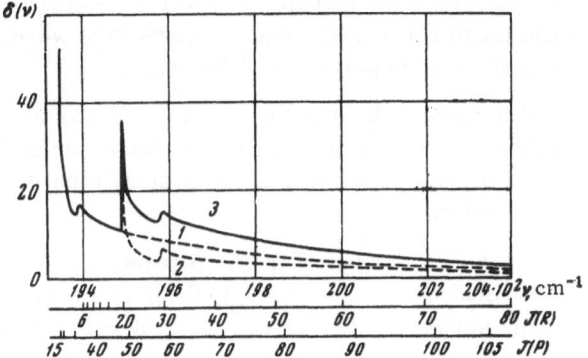

Fig. 19. Graphs of the function $\delta(v) = A \exp\left[-\dfrac{G_0(v)}{kT}\right] q_{v'v''} \sum\limits_{J_{\min}}^{J_{\max}} S_{J'J''} \exp\left[-\dfrac{F(J)}{kT}\right] n(v)$

for the R- and P-branches of the (0,0) and (1,1) vibrational bands of the electronic transition $d^3\Pi_g - a^3\Pi_u$ of the $C_2$ molecule at T = 7000°K. A is a constant; $n(v)$ is the relative density of the spectral lines per the spectral interval $\Delta v$. The rotational quantum numbers J for the R- and P-branches corresponding to the wave numbers $v$ are laid off along the axis of abscissas. Curves 1 and 2 replicate the spectral distribution of the radiation energy (or of the integrated absorption factor) with an accuracy of up to a factor of $v^4$ or $v$ for the (0,0) band or the (1,1) band, respectively. Curve 3 is the resultant distribution for the (0,0) and (1,1) bands.

TABLE 10. The Results of Determining $|R_e|^2$ of the Swan Bands by the Radiation Method

| $\Delta\lambda$, Å | $\Delta v$ | $n$ | $|R_e|^2$ |
|---|---|---|---|
| 6200—6013 | −2 | 18 | 3.01 |
| 5770—5312 | −1 | 13 | 2.18 |
| 5168—4737 | 0 | 11 | 2.64 |
| 4737—4482 | 1 | 13 | 2.00 |
| 4390—4253 | 2 | 20 | 2.26 |

the lamp, and $\Delta v = v_2 - v_1$:

$$n_2 = \beta \frac{\int_{v_1}^{v_2} B_v \varphi(v)\,dv}{\int_{v_1}^{v_2} \varphi(v)\,dv} = \beta \frac{B_{av}\int_{v_1}^{v_2} \delta(v)\varphi(v)\,dv}{\int_{v_1}^{v_2} \varphi(v)\,dv} = \beta B_{av}\Phi(v)\,dv, \quad (76)$$

where $B_v$ is the spectral brightness of the bands studied and can be represented in the form of the product of the brightness $B_{av}$ and the coefficient $\delta(v)$ which is the relative change of the brightness of the bands over the sequence; i.e.,

$$B_v = B_{av}\delta(v). \tag{77}$$

Under these conditions the variation of $\delta(v)$ for the sequence can either be calculated approximately (Fig. 19) or can be determined from additional spectrograms. The quantities

$$B_v' = \int B_v^0 \varphi(v)\,dv / \int \varphi(v)\,dv, \quad \Phi(v) = \int \delta(v)\varphi(v)\,dv / \int \varphi(v)\,dv$$

can easily be calculated by the method of numerical integration. In order to simplify the calculation of $B_v'$ a table of brightness temperatures $S_\lambda$ was calculated (see Appendix I). Thus, comparing the values of $n_1$ and $n_2$ obtained experimentally, it is easy to obtain the quantity which is numerically equal to the radiation energy of the sequence of bands (i.e., $\Sigma\Sigma \int I_v\,dv$).

The losses accompanying reflection from the mirror (~11-13%) and the losses which accompany the passage of the beam through the end windows (8%) were mutually compensating. The results of the experiments are displayed in Table 10.

The first column of this table indicates the spectral sector extracted onto the photomultiplier, while the second indicates the sequence of bands characterized by $\Delta v = v' - v''$; the next column indicates the number of experiments, and the last column indicates the mean values of the squares of the matrix elements.

The arithmetic mean value is extracted from 75 values of the squares of the matrix elements. This mean value is equal to $|R_e|^2 = (2.44 \pm 0.83) \cdot 10^{-36}$ or, in atomic units, $|R_e|^2 = 0.385 \pm 0.131$. Such a value of the square of the matrix element corresponds to an oscillator force $f_e = 0.022 \pm 0.008$.

CHAPTER IV

THE ABSORPTION METHOD

## §1. Description of the Apparatus and the Experimental Procedure

The absorption spectra were obtained by means of a large shock tube operating in essentially different regimes. The general appearance of the apparatus is shown in Fig. 20. The inside diameter of the tube (1) is 91 mm, the length of the high pressure chamber is 1 m, and the length of the low pressure chamber is 3.4 m. The inside surface of the tube is chrome-plated and polished. Two pairs of observation windows are built into the very end of the tube. The gas investigated was a mixture of carbon monoxide and argon (50% CO + 50% Ar) which was prepared in advance in a separate container. The accelerating gas was hydrogen. Aluminum diaphragms having a thickness of 4 mm were used. The range of pressures at which the diaphragms ruptured was equal to 90-120 atm, the range of incident-wave velocities constituting 2.40-3.16 km/sec.

The velocity of the incident wave was measured by means of ionization pickups $(P_0 - P_5)$ placed along the tube in the immediate vicinity of the end of the tubes. The signals from the ionization pickups were applied to one input of a two-beam OK-17M oscilloscope. A sinusoidal voltage from a quartz stabilized oscillator was applied to the other input of the oscilloscope.

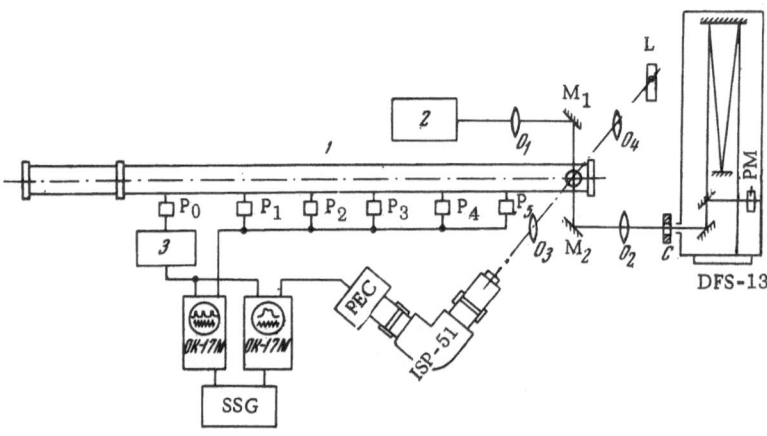

Fig. 20. Overall view of the apparatus.

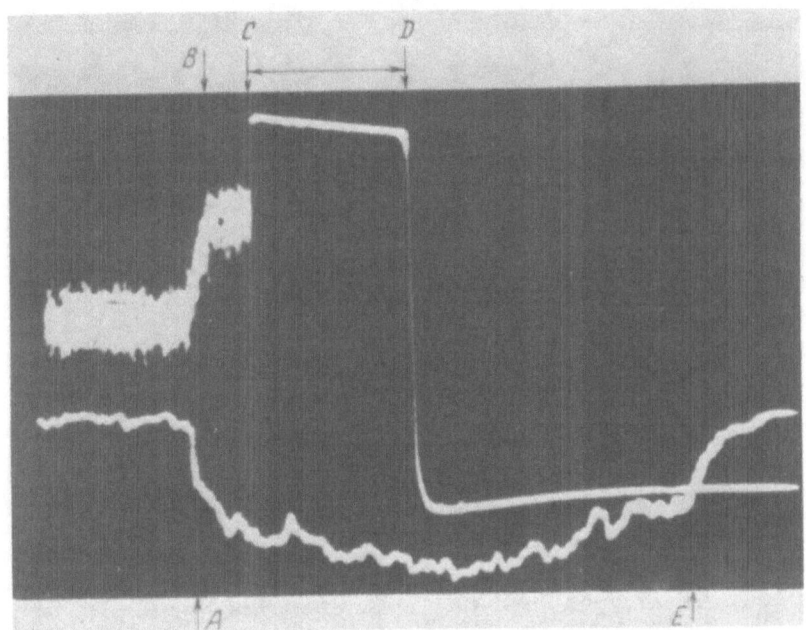

Fig. 21. Oscillogram for monitoring the operation of the shutter
in the synchronization circuit. A is the front of the reflected wave;
B is the luminescence of the gas; C is the beginning of the spectrum exposure; E is the contact surface. From the oscillogram
it is evident that the slit opened before the arrival of the reflected
shock wave and closed before the arrival of the contact surface.
The time interval between the beginning of the flash of the transilluminating source and the time of shutter closure is the exposure time of the absorption spectrum (CD).

Under these conditions the accuracy with which the velocity was determined was estimated to
be 2-2.5%.

In order to obtain the absorption spectra a pulsed source 2 of a continuous spectrum [94]
was used. In the range investigated it radiated as an absolutely black body with a temperature
39,000°K. Such a temperature was reached in the source during the discharge of an artificial
line through an aperture having a diameter of 2 mm in a textolite plate 11 mm thick. The discharge voltage of the artificial line was equal to 3 kV. A luminescence pulse having a table-
top shape was produced by the discharge. The duration of the pulse is 190 $\mu$sec.

A special shutter [95] was used in photographing the spectra. This was necessary to ensure that the exposure began when the temperature behind the wave front had already been
established, and to guarantee that the plasma radiation or the radiation from the transilluminating source was not additionally superimposed on the absorption spectrum. The shutter operated as follows. An adapter with a traveling slit and a wire pusher is fitted over the spectrograph slit. The width of the traveling slit is 0.1 mm. The pusher is made of nichrome wire
0.9 mm in diameter. Before the experiment the traveling slit was pressed against the wire in
such a way that one of its sides covered the spectrograph slit. At the required time a bank of
capacitors was discharged through the wire to ground. The wire lengthened and pushed the
traveling slit. By varying the discharge voltage of the capacitors one could regulate the speed
of the traveling slit and thus the exposure time. In our experiments the voltage was varied from
1500 to 2500 V, which corresponded to a range of exposure-time variation from 150 to 40 $\mu$sec.

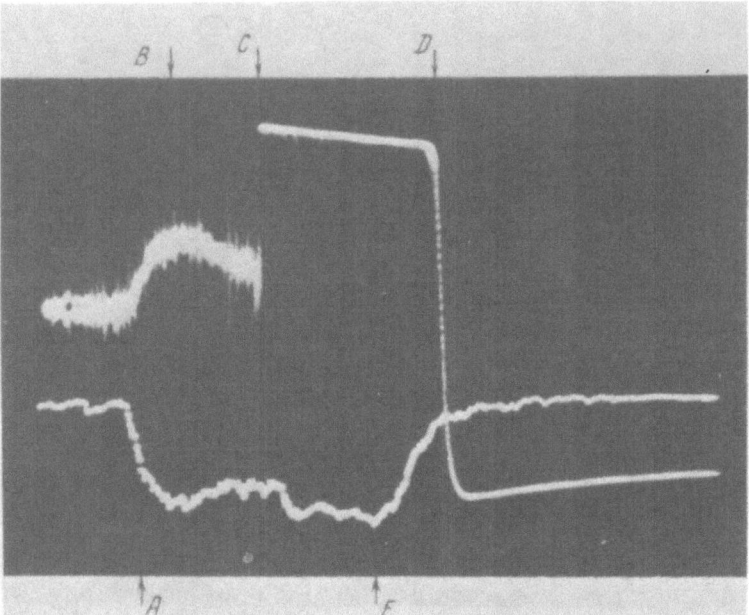

Fig. 22. Oscillogram for monitoring the operation of the
shutter and the synchronization circuit. A is the front of the
reflected wave; B is the luminescence of the gas; C is the
beginning of the flash of the light source; D is the time of
shutter closure; E is the contact surface. From the oscillogram it is evident that the slit opened before the arrival
of the reflected shock wave and closed after passage of the
contact surface. In this case the time interval between the
beginning of the flash of the transilluminating source and
the time of shutter closure is not the exposure time of the
absorption spectrum.

The material and shape of the wire pusher and the construction of the traveling slit were chosen
in such a way that the exposure time was replicated from experiment to experiment with an
accuracy up to 5 $\mu$sec. The duration of the exposure was monitored by means of a photoelectric circuit. A small mirror was mounted inside the spectrograph, which deflected an insignificant portion of the undecomposed light onto a photomultiplier. The signal from the load of the
photomultiplier was applied to the input of a two-beam pulse oscilloscope. The signal from the
photomultiplier that recorded the luminescence of the reflected shock wave was applied to the
other oscilloscope input. Thus, one could see the time at which the reflected shock wave began
to make a contribution to the exposure of the absorption spectrum (Figs. 21 and 22).

Synchronization of the recording of all the fast processes was achieved by means of device 3. At the required time this device triggered the oscilloscope sweep (the additional power
supply circuit for the high pressure xenon lamp serving to measure the temperature) and the
electrical circuit of the shutter; it also stipulated the time at which the pulsed source of the
continuous spectrum operated. The device itself was triggered from an ionization pickup on
the shock tube, which was situated at a distance of 1 m from the end of the tube.

The spectra were obtained on a four-meter DFS-13 diffraction spectrograph having a
grating with 600 lines/mm (dispersion 4 Å/mm) for a slit width of 0.02-0.03 mm.

Fig. 23.  Absorption spectrum of the (0,0) and (1,1) Swan bands of the $C_2$ molecule.

## §2. Analysis of the Spectrograms

Out of the large number of spectrograms obtained, 11 were chosen which were distinguished by a high spectrum quality and good blackening marks. The lines of the rotational structure of the R- and P-branches of the (0,0) and (1,1) Swan bands are clearly distinguishable right down to the small values of J (Fig. 23). The spectrograms were measured photometrically on a microphotometer at intervals of 0.01 mm and were traced on millimeter graph paper and then manipulated by the photographic photometry method. This is very painstaking work, and therefore a portion of the spectrograms was recorded mechanically. Since for quantitative measurements of our spectrograms the conventional circuit of the MF-4 recording microphotometer turned out to be unsuitable, we modified it. The photocell was replaced by an FEU-25 photomultiplier whose load voltage was applied to an electronic automatic recording potentiometer. Transfer characteristics which provided the dependence of the photocurrent of the photomultiplier on the light flux for various values of the high voltage across the photomultiplier (Fig. 24) were plotted for choosing the correct operating mode of the photomultiplier during the recording. The wavelength scale of the recording was expanded considerably. The length of one spectrum recorded on the paper was equal to 20-25 m, whereas on film its length was equal to 10.5 cm.

Then the spectrum, which was recorded on the paper tape for convenience in determining $\int k_\nu \, d\nu$, was traced in log $I/I_0$ units by means of the blackening marks, where $I/I_0$ is the transmissivity of the photometrized sector on the film. The planimetry of the spectral lines in calculating $\int k_\nu d\nu$ was carried out in the conventional manner. In order to reduce the errors which occur in the determination of $\Delta \nu$ as a consequence of small changes in the spectrum recording speed, constant monitoring of the recording speed was carried out. The individual spectrograms were analyzed several times each.

The pattern of the spectrum of the (0,0) and (1,1) bands is such that several lines — triplets — are completely or partially superimposed on the lines of other bands. In order to provide the possibility of analyzing not only resolved lines but also unresolved lines, the contributions from the individual lines were calculated. For this purpose the theoretical curves were plotted for the intensity distribution over the rotational lines in the (0, 0) and (1, 1) bands for the P- and R-branches at temperatures of 6000, 6500, and 7000°K (Fig. 25). If we know the resultant magni-

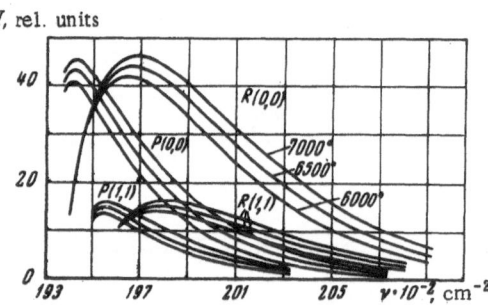

Fig. 25. The relative contribution of the individual rotational lines of the R- and P-branches of the (0,0) and (1,1) bands to the integrated absorption coefficient $\int k_\nu \, d\nu$ at temperatures of 6000, 6500, and 7000°K. The quantity

$$I = q_{v'v''} S_{J'J''} \exp\left[ \frac{-T_0'' + G_0''(v) + F_v''(J)}{kT} \right]$$

is laid off along the axis of ordinates.

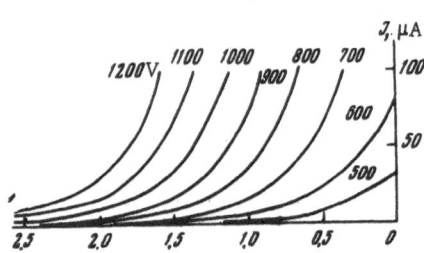

Fig. 24. Dependence of the photocurrent J of the FEU-25 on the blackening values S of the photometrized spectrogram for stipulated values of the high voltage across the photomultiplier.

tude of the integrated absorption factor $\int k_\nu d\nu$ for the group of lines in the vicinity of some frequency $\nu$, then the graph in Fig. 25 can be used to determine the contribution of each line to this quantity.

## § 3. Determination of the Gas Temperature behind the Shock Wave

At one time we carried out a detailed examination of the method of inversion of spectral lines and suggested a generalized version of it [96] based on measuring the radiation and absorption intensities of the spectral lines. In order to study the time variation of the temperatures of fast pressures having a duration of several tens of microseconds a photoelectric version of the inversion method was developed [97]; the circuit and equipment parts of the versions were used without change in the work described here.

A xenon DKSSh-1000 lamp with pulsed additional feed was used as the inverting source; its effective brightness temperature in the green region of the spectrum (near 5100 Å) was varied from 6800 to 9000°K as a function of the additional-feed voltage. The temperature measurements were carried out by the inversion method over various sectors of the $\Delta v = 0$ sequence of of the Swan bands. The accuracy of temperature measurement by this method was 3-4%. The calculated values of temperature varied from 6300 to 7200°K. For a large series of experiments carried out for the most varied regimes it was established that the experimentally observed temperatures are in agreement, within the limits of the experimental errors, with the temperatures calculated from the velocity of the incident shock wave.

If the rotational structure of the bands is well resolved, as is the case on our photographs, then it turns out to be possible to determine the rotational temperature of the gas [44]. Having solved the relationship (23) for

$$\exp\left[-\frac{F_v''(J)}{kT}\right],$$

we obtain

$$\exp\left[-\frac{F_v''(J)}{kT}\right] = \frac{3hc}{8\pi^3}\frac{Z\exp\left[\dfrac{T_0'' + G_0(v)}{kT}\right]}{Nlq_{v'v''}}\frac{\int k_\nu d\nu}{\nu S_{J'J''}}. \qquad (78)$$

The first two factors do not depend on the rotational quantum numbers J at a constant temperature. Taking account of this and taking the natural logarithm of (78), we obtain

$$-\frac{F_v''(J)}{kT} = \ln C + \ln\frac{\int k_\nu d\nu}{\nu S_{J'J''}}. \qquad (79)$$

If the distribution of the molecules over the rotational levels is determined by the single temperature $T_{rt}$, then the graph for the dependence of the quantity $\ln\left(\int k_\nu d\nu / \nu S_{J'J}\right)$ on $F_v'(J)$ will be a straight line. From the slope of the straight line relative to the axis of abscissas it is possible to calculate the rotational temperature. If its equilibrium has been established with respect to the rotational, vibrational, and translational degrees of freedom in the absorbing gas, then the rotational temperature obtained is the true gas temperature. The rotational temperature in each experiment was determined as the average over four branches. An example of a graph for determining the rotational temperature is shown in Fig. 26. The straight lines were drawn for each band using the method of least squares.

Beforehand we used a series of 20 "runs" carried out in the most varied regimes to establish good coincidence between the rotational temperature and the temperature calculated

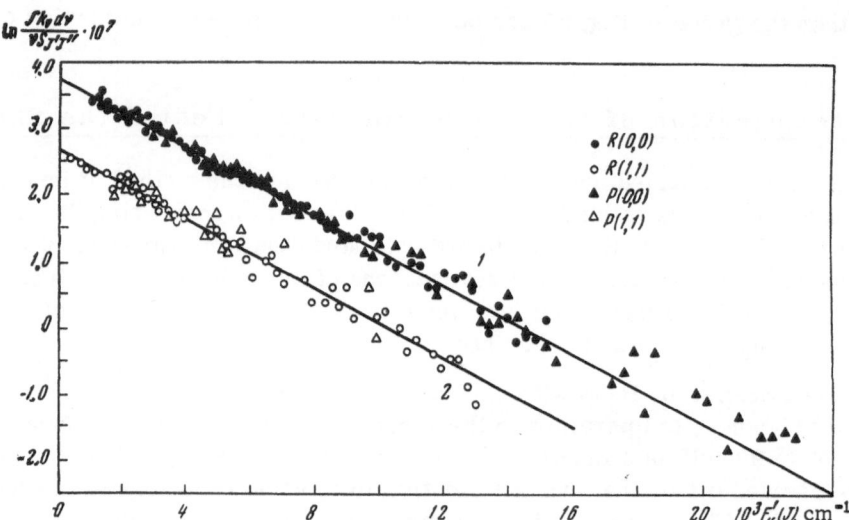

Fig. 26. Graph for determining the gas temperature behind the front of the reflected shock wave from the change of the integrated absorption coefficient of the rotational lines of the (0,0) and (1,1) bands of the $C_2$ molecule. The velocity $v_1 = 2.68$ km/sec, which corresponds to the calculated value $T = 5600°K$ ($P_1 = 50$ mm Hg CO). The temperature corresponding to the slope of the straight line 1 on the graph is equal to 5680°K while the slope of the straight line 2 is equal to 5500°K.

TABLE 11. Temperature behind the Shock Wave in the 50% CO + 50% Ar Mixture for an Initial Pressure $P_1 = 50$ mm Hg

| $v_1$, km/sec | $T_s$ | $T(0,0)$ | $T(1,1)$ | $T_{av}$ | $\Delta T$ |
|---|---|---|---|---|---|
| 2.45 | 6520 | 6600 | 6450 | 6525 | −5 |
| 2.45 | 6520 | 6630 | 6800 | 6745 | −225 |
| 2.46 | 6540 | 6400 | 6550 | 6475 | 65 |
| 2.49 | 6590 | 7030 | 6200 | 6615 | −25 |
| 2.51 | 6630 | 6620 | 6400 | 6510 | 120 |
| 2.53 | 6630 | 6300 | 6840 | 6570 | 110 |
| 2.53 | 6680 | 6550 | 7000 | 6675 | 5 |
| 2.53 | 6680 | 6800 | | | |
| 2.54 | 6690 | 5970 | 6650 | 6310 | 380 |
| 2.55 | 6720 | 6500 | 6260 | 6380 | 340 |
| 2.56 | 6740 | 6900 | 7200 | 7050 | −310 |
| 2.58 | 6780 | 6360 | 6900 | 6630 | 150 |
| 2.58 | 6780 | 7100 | 7000 | 7050 | −330 |
| 2.58 | 6780 | 6800 | 7090 | 6945 | −165 |
| 2.59 | 6800 | 6970 | 6350 | 6660 | 140 |
| 2.61 | 6830 | 6420 | 6450 | 6435 | 395 |
| 2.61 | 6830 | 6900 | 7180 | 7040 | −210 |
| 2.62 | 6850 | 6450 | 7160 | 6805 | 45 |
| 2.63 | 6870 | 6720 | 6900 | 6810 | 60 |
| 2.65 | 6900 | 6450 | 7210 | 6830 | 70 |

$$\overline{\Delta T} = \pm 170°\ K$$

TABLE 12. The Results of Determining $|R_e|^2$ of the Swan Bands of
the $C_2$ Molecule by the Absorption Method

| Test number | $P_1$, mm Hg | $v_1$, km/sec | $T_5°$, K | $P_5$, atm | $lN_{C_2}$ $10^{-17}$ | $Z$ $10^4$ | The (0,0) band | | The (1,1) band | |
|---|---|---|---|---|---|---|---|---|---|---|
| | | | | | | | $n$ | $|R_e|^2$ | $n$ | $|R_e|^2$ |
| 1 | 10 | 3.16 | 7280 | 11.4 | 1.05 | 4.73 | 20 | 0.53 | | |
| 2 | 25 | 2.62 | 6650 | 17.3 | 1.06 | 3.87 | 23 | 0.48 | | |
| 3 | 50 | 2.37 | 6350 | 23.5 | 1.28 | 3.49 | 15 | 0.33 | 59 | 0.41 |
| 4 | 50 | 2.45 | 6520 | 30.3 | 1.71 | 3.67 | 69 | 0.42 | 59 | 0.41 |
| 5 | 50 | 2.50 | 6600 | 31.5 | 1.90 | 3.80 | 144 | 0.41 | 123 | 0.41 |
| 6 | 50 | 2.50 | 6600 | 31.5 | 1.90 | 3.80 | 102 | 0.45 | 53 | 0.43 |
| 7 | 50 | 2.51 | 6630 | 31.8 | 1.95 | 3.86 | 36 | 0.47 | 17 | 0.39 |
| 8 | 50 | 2.53 | 6630 | 32.4 | 2.10 | 3.90 | 16 | 0.42 | — | — |
| 9 | 50 | 2.59 | 6300 | 34.0 | 2.45 | 4.07 | 134 | 0.45 | 66 | 0.35 |
| 10 | 50 | 2.71 | 7000 | 37.0 | 3.23 | 4.36 | 31 | 0.53 | 26 | 0.51 |
| 11 | 50 | 2.71 | 7000 | 37.0 | 3.23 | 4.36 | 111 | 0.45 | 55 | 0.50 |

according to the velocity of the shock wave. Table 11 shows the results of these tests. The
first column of the table contains values of the velocity $v_1$ of the incident wave; the second contains the corresponding values of the temperature $T_5$ behind the shock wave. The next two
columns give the values of the rotational temperatures obtained from the (0,0) and (1,1) bands,
respectively. Then the average values of the rotational temperatures are given, and the last
column of the table provides the difference between the calculated temperature and the average
temperature. On an average these deviations do not exceed 170°K. Therefore henceforth in
determining $|R_e|^2$ we use the calculated values of the temperature obtained from the measured
velocity of the incident wave. The experimental "runs" in which the values of these temperatures differed noticeably from the values obtained from the change in the integrated absorption
coefficient of the rotational lines (by more than 300°K) were discarded.

§ 4. Results of the Absorption Experiments

The experimental conditions and results for each of the 11 "runs" are displayed in Table
12. The second and third columns of the table contain the values of the pressure $P_1$ of the investigated mixture in the low-pressure chamber and the values of the velocity of the incident
wave $v_1$ determined experimentally; the fourth and fifth columns given the temperature $T_5$ and
pressure $P_5$ behind the reflected shock wave, as determined from the velocity $v_1$ by means of
the graphs in Fig. 7. The sixth column contains the values of the concentration of $C_2$ molecules,
which were calculated according to the Eq. (34); the seventh contains the sums over the states
(see Fig. 8). The eighth through eleventh columns give the number n of analyzed lines and the
mean value $|R_e|^2$ for the (0,0) and (1,1) bands.

The arithmetic mean value of the results of analyzing 1100 rotational lines in 11 runs was
equal to $|R_e|^2 = 0.44 \pm 0.08$ atomic units.

§ 5. Estimating the Oscillator Force for Fox — Herzberg

Bands

The $C_2$ molecule radiates Fox — Herzberg bands corresponding to the $f^3\Pi_g - a^3\Pi_u$ transition, which were first discovered in an electrical discharge in benzene vapors [16], besides
Swan bands. The calculated values [79] of the oscillator forces for these bands are 17 times as
large as they are for the Swan bands. Therefore, it can be expected that the contribution of
these bands to the radiative capacity of heated gas will be large. The lower electronic state
for these bands is the same as it is for the Swan bands. The rotational structure of the Fox —
Herzberg bands is the same as it is in the Swan bands. There are slight quantitative differ-

ences: for example, the triplet splitting is larger than in the Swan bands. The bands are shaded toward longer wavelengths.

The (0,3), (0,4), (0,5), and (0,6) bands are the most intense [98]. They are situated in the ultraviolet range, and the wavelengths of the edges of these bands are respectively equal to 2855, 2987, 3129, and 3283 Å.

We investigated the spectral range 2800–3200 Å in which the most intense Fox – Herzberg bands might be situated, for the most varied operating modes of the shock tubes both in the radiation of the gas situated behind the reflected shock wave and in the absorption. The lines of the impurity atoms were registered. Besides atomic lines, no attributes of the molecular construction were detected on the spectrograms in this range. Additional photographs of the arc discharge in the CO + Ar atmosphere that were taken by means of a spectrograph having an intermediate dispersion similarly failed to reveal these bands.

The experimental observation of the Fox – Herzberg bands is more difficult to achieve than observation of the Swan bands for several reasons: 1) the change in the equilibrium internuclear distance in this transition is very large (t = 2.32), whereas for the Swan bands [98] it is small (t = 0.414); therefore, the intensity is dispersed over a large number of bands; 2) the Fox – Herzberg bands do not have sharp edges as do the Swan bands; 3) it is very possible that for thermal excitation under equilibrium conditions a free dissociation of the upper $f^3\Pi_g$ level takes place which can lower the observable radiation intensity considerably (true, with an increase in pressure the degree of predissociation decreases strongly); 4) the maximum of the Franck – Condon parabola for the Fox – Herzberg bands lies in the ultraviolet range in approximately the same place where the continuous dissociation spectrum is located. When the radiation spectrum is recorded by means of a spectrograph having a small or intermediate dispersion the first two coordinates are insignificant. The same also applies to the recording of the absorption spectrum by an instrument having good dispersion. The estimates which we carried out with allowance for the concentration of $C_2$ molecules and the temperatures $T_5$ on the basis of spectrograms obtained with an instrument having good dispersion led us to the conclusion that the oscillator force for the Fox – Herzberg bands is less than 0.011; i.e., it is almost two orders of magnitude lower than that predicted by theory.

This fact is difficult to explain without resorting to additional assumptions. In fact, the Swan bands are caused by transitions which are similar to the sub-Rydberg series, and the value of the oscillator forces for them is comparatively small. On the contrary, the electronic transition for the Fox – Herzberg bands includes the transition of an electron from the $\Pi_u$-orbital to the $\Pi_g$-orbital. In accordance with Mulliken this form of electronic transition always has a very high intensity, and Schumann – Runge bands [14] serve as an example of this.

We do not know of a single case in which the Fox – Herzberg bands could be observed under equilibrium conditions. Several years ago [99] Herzberg was unable to observe them in the spectrum of a King furnace.

§6.  Comparison of the Methods of Investigation

Let us recall that in Chapter III we used the radiation method. Each of the methods which we used to determine $|R_e|^2$ for the Swan bands has its own merits and shortcomings. The absorption method, according to general opinion, is more reliable, since first it is not connected with calibration of the comparison source in absolute units, and second it does not require special care in reducing or considering the effect of reabsorption; however, it is not free from shortcomings. One of the basic shortcomings of this method lies in the fact that in the majority of cases it is not known exactly where the initial reference lines should be taken from which the absorption in the lines should be measured (i.e., $k_\nu$) in analyzing the spectrograms.

The position of this line, which characterizes the initial brightness of the transilluminating source, varies smoothly on the photometer tracing of the absorption spectrum with the variation of the spectral brightness of the source. The determination of its location from the shape of the wing of the spectral line is possible only for the case in which these lines are situated far from one another, as is the case for large J for the R-branches of the (0,0) Swan bands. However, in the general case of a "palisade" of rotational lines, as is observed in most molecular spectra, such a problem becomes unsolvable without additional investigations of the shape of the spectral lines or additional spectrograms characterizing radiation of the transilluminating source. A certain criterion of correct (or incorrect) establishment of the initial line can be the determination of $\int k_\nu d\nu$ for single and superimposed line (as is described in §2 of this chapter). Actually, the error in determining the initial line has less influence on the magnitude of $\int k_\nu d\nu$, obtained from the resultant magnitude of $\int k_\nu d\nu$ for several superimposed lines than on the magnitude obtained for a single line.

The radiation method (in the form in which it was realized in our tests) is not connected with a photographic plate and allows observation of the time variation of the radiation brightness. By recording the radiation over a wide spectral interval using a narrow spectrogram input slit we can ignore the apparatus function of the instrument. The use of comparatively wide spectral intervals does not introduce a substantial error if the variation of the photomultiplier sensitivity with wavelength and the spectral distribution of the comparison source radiation are correctly considered, and the approximate brightness distribution over the spectrum is known for the irradiated object. The most essential factor is the exclusion or consideration of the effect of reabsorption. In our radiation experiments we evidently were able to avoid reabsorption as a result of using those tube operating modes for which the total number of $C_2$ molecules along the optical path was 50-100 times as small as in the absorption experiments.

It should be noted that the experimental circuit which we used for simultaneous recording of the rise of the luminescence pulse and the uniform increase in the thickness of the luminescent gas layer allows automatic monitoring of the absence of reabsorption of radiation.

In a number of cases both methods can be recommended as complementing one another for a more reliable determination of the oscillator forces for the molecules.

CHAPTER V

STUDY OF THE SHAPE AND WIDTH OF THE ROTATIONAL LINES

## §1. Statement of the Problem

A very large number of both experimental and theoretical papers have been devoted to an investigation of the contours of atomic spectral lines. The results of these papers not only allow varied physical causes leading to broadening of spectral lines to be clarified, but also lead to the possibility of constructing a physical theory of the varied types of broadening of spectral lines. Based on modern theoretical concepts and data on atomic constants, we can at present not only solve the direct problem (i.e., calculate the contour of a spectral line) but also the inverse problem: for example, the luminescence of broadening collisions, and the temperature and concentration of charged particles in a plasma can be found from the contour of a spectral line. Thus, the plasma temperature [100] is determined from the Doppler broadening of spectral lines, while the temperature and concentration of charged particles [101-103] are determined from Stark broadening. Recently [104] a method was advanced by means of which one can determine the effective cross sections of elastic and inelastic collisions of electrons with excited atoms from the broadening cross sections.

The situation is considerably worse with respect to the study of the contours of spectral lines of diatomic molecules. Whereas a considerable number of papers has been devoted to the study of the contours of purely rotational or vibrational — rotational lines, there are no theoretical papers on the study of electron-vibrational — rotational lines at all, while we know of only three experimental papers [105-107]. The cause of this small number of papers on the study of the contours of electron-vibrational — rotational lines lies in the exceptionally unfavorable experimental conditions. Usually, the bands of molecular spectra contain a considerable number of superimposed rotational lines and it is often not possible to isolate an individual line from a large number of spectral lines.

In [105] a method was developed for determining the width of a line in absorption with allowance for the distorting effects of the apparatus function and possible superpositions of neighboring lines. The method worked out was used to study the contour of one rotational line of the system of Schumann—Runge bands of the oxygen molecule, and the broadening collision cross sections $\sigma(O_2, O_2)$ and $\sigma(O_2, O)$ were determined from the values of the line widths which were found. The measurements were carried out in absorption over the contour of two comparatively thoroughly overlapping lines $R_{27}$ and $R_{23}$ of the (0,14) band of the Schumann—Runge system of the oxygen molecule. It was established that for collisions of two oxygen molecules the effective broadening cross section is equal to $85 \cdot 10^{-16}$ cm$^2$, while for collision of an oxygen molecule with an oxygen atom it differs insignificantly from the value given above and is equal to $72 \cdot 10^{-16}$ cm$^2$.

In [106] the contour of the $P_{45}$ line of the (0,1) band of the violet system of CN was studied. Introducing the assumption that the cross sections which broaden the rotational lines of the CN molecule during its collision with $N_2$ and CO are equal, the authors found the broadening cross section to be equal to $42 \cdot 10^{-16}$ cm$^2$.

The present chapter is devoted to the study of the width and shape of individual rotational lines of the (0,0) electron-vibrational band of the Swan system of the $C_2$ molecule as a function of the excitation conditions in the shock tube. A more thorough treatment, which considers the difference between the light intensity $I_0$ which is incident on the investigated gas volume and the light intensity which is transmitted through the gas layer at the wavelength corresponding to the absorption minimum near the line investigated, allowed the value of $|R_e|^2$ to be refined for the Swan bands of the $C_2$ molecule by comparison with the value obtained in chapter IV.

## §2. Basic Information from the Theory of the Broadening of Spectral Lines

Five fundamental processes exist [108] which lead to broadening of spectral lines: natural broadening as a consequence of the finite lifetime of the excited state; Doppler broadening as a consequence of the thermal motion of the molecules; Lorentz broadening, which is caused by collisions with atoms and molecules of spurious gases; resonance broadening as a consequence of collisions with atoms and molecules of the same gas, and broadening caused by the Stark effect as a result of collisions with electrons and ions. Of these processes we shall consider only Doppler broadening and Lorentz broadening, since the remaining processes do not play any noticeable role under our conditions. The width of the spectral line shall be called the width of the contour at the value of the ordinate which is equal to one half the maximum value: $I = \frac{1}{2} I_0$.

In the case of Doppler broadening the shape of the spectral line can be described by an exponential Gaussian equation having the width (in cm$^{-1}$)

$$\Delta \nu_D = \frac{2\nu_0}{c} \sqrt{\frac{2RT \ln 2}{\mu}}, \tag{80}$$

where c is the velocity of light; R is the gas constant; T is temperature; $\mu$ is the molecular weight. The fall-off of the intensity on the wings of the Doppler contour is noticeably steeper than it is for the natural contour.

The intensity distribution in the case of Lorentz broadening, just as in the case of natural broadening, can be described by the dispersion formula

$$I_\nu = \frac{(1/\tau_0)^2}{4\pi\,(\nu - \nu_0)^2 + (1/\tau_0)^2}\,, \tag{81}$$

where $\tau_0$ is the mean free time which depends on the conditions to which the gas is subjected. The width of the spectral line for the Lorentz contour is equal to

$$\Delta\nu_L = \frac{1}{\pi\tau_0}\,(\text{sec}^{-1}) = \frac{1}{\pi c\tau_0}\,(\text{cm}^{-1}). \tag{82}$$

If impact broadening of the spectral line of a molecule occurs due to collisions with n species of particles, then the total number $\zeta$ of broadening collisions of the molecule with these particles will be

$$\zeta = \sum \zeta_i = \sum N_i\sigma_i v_i, \tag{83}$$

where $N_i$ is the number of particles of species i in cm$^3$; $\sigma_i$ is the effective cross section of the broadening collision of the investigated molecule with a particle of species i, and $v_i$ is the mean relative velocity of the molecule investigated and the particle of the i-th species,

$$v_i = 2\sqrt{\frac{2RT}{\pi}\left(\frac{1}{\mu_i} + \frac{1}{\mu}\right)}\,, \tag{84}$$

where $\mu_i$ and $\mu$ are respectively the molecular weights of the particle of species i and the molecule whose line broadening is investigated. The time interval $\tau_0$ between two collisions of the molecule investigated and any of the molecules of species i is equal to

$$\tau_0 = \frac{1}{\zeta} = \frac{1}{\sum \zeta_i} = \frac{1}{\sum N_i\sigma_i v_i}\,. \tag{85}$$

Substituting (85) into (82), we obtain

$$\Delta\nu_L = \frac{1}{\pi c} \sum N_i\sigma_i v_i. \tag{86}$$

From this it follows that the resultant Lorentz broadening can be represented in the form of the sum of individual Lorentz broadenings

$$\Delta\nu_L = \sum_i (\Delta\nu_L)_i. \tag{87}$$

Since the conditions of our experiment were such that a high temperature and a high particle concentration existed, it follows that the lines broadened simultaneously due to the presence of the Doppler effect and collisions. In order to describe the intensity distribution in such a line one can use the Voigt function [109] in the form in which it is written in [108]:

$$I = I_0\,\frac{a}{\pi} \int_{-\infty}^{+\infty} \frac{e^{-y^2}\,dy}{a^2 + (\omega - y)^2}\,, \tag{88}$$

where $a$ is the Lorentz width in $\Delta\nu_D^*$ units,

$$a = \frac{\Delta\nu_L}{\Delta\nu_D^*} , \tag{89}$$

$$\Delta\nu_D^* = \frac{\Delta\nu_D}{\sqrt{2\ln 2}} , \qquad \omega = \frac{\nu - \nu_0}{\frac{1}{2}\Delta\nu_D^*} \tag{90}$$

is the distance from the center of the line $\nu_0$ in $(1/2)\Delta\nu_D^*$ units, while y is the integration variable and is equal to

$$y = \frac{2\sqrt{\ln 2}\,(\nu - \nu_0)}{\Delta\nu_D} . \tag{91}$$

## §3. Description of the Experimental Conditions

The absorption spectra of the Swan bands of $C_2$ were obtained by means of the shock tube described in Chapter III. The low-pressure chamber was filled either with carbon monoxide CO or with a mixture of carbon monoxide with argon. The degree of purity of the carbon monoxide was specially checked. According to exchange chemical analysis, carbon dioxide $CO_2$ was completely absent, and 0.1% $O_2$, 0.3% $H_2$, and 0.9% $N_2$ were present.

In the investigation described the resolving power and dispersion of the receiving diffraction instrument were doubled by replacing grating having 600 lines/mm by a grating having 1200 lines/mm; this allowed the half-width of the apparatus function to be cut almost in half. Thus, the dispersion of the spectrograph in the first order became equal to 2 Å/mm.

The spectra corresponding to various regimes were used for investigation. The parameters of the plasma (the temperature $T_5$, the pressure $P_5$, the molar fractions of the various components, etc.) behind the reflected shock wave were determined on the basis of a gasdynamic calculation from the velocity of the incident shock wave, just as was done in Chapter III. The temperature behind the reflected shock wave was varied from 5000 to 7600°K in the experiments, while the pressure in the plug was varied from 11.3 to 66.9 atm.

## §4. Determination of the Apparatus Function
### of the Spectrograph

The distribution of illumination on the spectrogram always differs from the true one, since any practical spectral instrument inroduces distortion into this distribution. The shape and dimensions of the distribution are determined by various causes; in particular, the finite drift of the spectrograph slit, diffraction by the irises of the optical system, scattering by the photosensitive layer of the photographic film, etc., can be included here.

Assume $E(\nu)$ is the function which stipulates the shape of the spectral line being studied, while $a(x)$ is the apparatus function (i.e., the intensity distribution which is created in the focal plane of the spectrograph by an infinitely narrow spectral line). Then the intensity distribution created by the line under study, as we know, can be represented in the form

$$E(x) = \int_{-\infty}^{+\infty} E(\nu)\,a(x-\nu)\,d\nu. \tag{92}$$

During the transformation of an ideal image into a practical one only a redistribution of the energy occurs in the image; however, the overall energy remains unchanged. Therefore, the apparatus function is usually normalized to unity:

$$\int\limits_{-\infty}^{+\infty} a\,(x - \nu)\,d\nu = 1. \tag{93}$$

For an experimental determination of the apparatus function of the spectral instrument a source is required which yields very narrow spectral lines. We used a low-pressure mercury lamp as such a source. The measurements were carried out on the mercury line $\lambda = 5789.69$ Å corresponding to the $6^3D - 6^1P_1$ transition. No hyperfine structure was revealed for this line. The spectra were obtained on photographic film having a high sensitivity with several exposures.

For a slit width equal to 0.02 mm, the width of the apparatus function was 0.057 Å; for a slit width of 0.03 mm, the apparatus function had a width of 0.064 Å, which corresponds to 0.24 and 0.26 cm$^{-1}$ for operation in the portion of the $C_2$ spectrum of interest to us. For narrower slit widths the width undergoes practically no reduction, and the blackenings on the spectrogram during exposure to the pulse source having the continuous spectrum turn out to be inadequate for photometry. Wider slits noticeably increase the width of the apparatus function. Therefore, for our work we chose a slit width of 0.02 mm.

The shape of the apparatus function is displayed in Fig. 27. For comparison purposes the dispersion and Gaussian contours are similarly plotted. From Fig. 27 it is evident that the apparatus function can be described sufficiently well by a Gauss function.

Strictly speaking, the apparatus function of the instrument should be determined under the same conditions as those under which photographing of the investigated lines of the spectrum is carried out. Actually, in line radiation spectra obtained by means of a spectrograph with a grating, so-called "ghosts" are observed at a certain distance from each spectral line. Therefore, in an absorption spectrum the centers of the lines increase their brightness due to the "ghosts" which develop from the neighboring sectors of the continuum. As a consequence of this the width of the apparatus function increases. The light which appears in the spectrograph as a consequence of scattering and reflection from the optical surfaces and walls of the spectrograph similarly degrades the apparatus function.

However, it is practically impossible to determine the apparatus function of an instrument for an individual line or group of lines in the absorption spectrum of a diatomic molecule. Therefore, one usually attempts to acquire a diffraction grating of good quality so as to be able to make use of the apparatus function obtained from the radiation lines.

Notwithstanding the fact that the diffraction grating in our spectrograph is of comparatively good quality (the relative intensity of the "ghosts" amounts to just 0.15% in the first order), we carried out additional

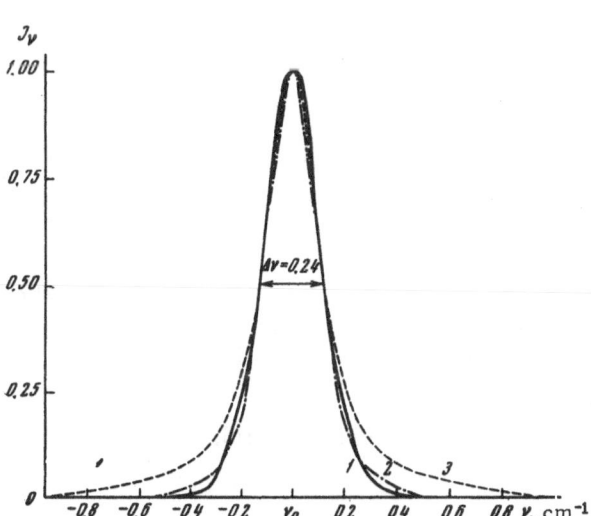

Fig. 27. Apparatus function of the spectrograph (1) for a slit width of 0.02 mm. For comparison purposes the Gaussian (2) and dispersion (3) curves have been plotted.

measurements of the equipment function in absorption. The tests were carried out with a low-pressure mercury lamp. The transillumination was carried out by a pulsed source having a continuous spectrum. The mercury line $\lambda = 5789.69$ Å was not observed in the absorption. Therefore, the measurements were carried out according to the central component of the $(7^3S - 6^3P_2)\lambda = 5460.74$ Å line whose intensity is 86.7% of the resultant intensity of the multiplet. No noticeable change in the absorption line contour was observed. The increase in the width of the apparatus function did not exceed 10%. The influence of the "ghosts" and scattered light in the spectrograph on the apparatus function during the photographing of the absorption spectra of the Swan bands of the $C_2$ molecule will therefore be still smaller, and thus it can be ignored.

## § 5. Procedure for Analyzing the Contours of Spectral Lines.

## Determination of the True Shape and Width of Spectral Lines

As is well known, the rotational lines of Swan bands have a triplet structure [9]. We succeeded in selecting three lines whose multiplet splitting was less than 0.01 cm$^{-1}$ from a large number of lines — namely, the lines $R_{70}$, $R_{69}$, $R_{68}$ in (0,0) bands whose wave numbers are respectively equal to 20190.52, 20171.95, and 20152.13 cm$^{-1}$. These lines are also free from superpositions of the lines of other branches and are isolated to a fair degree.

After photometry, the values of the blackenings of the line images were recalculated in values of $I_\nu/I_0$ and $k_\nu l$ by means of the characteristic photographic film curve taken with a nine-stage attenuator and a pulsed source of a continuous spectrum, and then the contours of the spectral lines were plotted.

The radiation intensity $I(\nu)$ incident on the spectrograph slit is related to the intensity $I'(x)$ at the output, which is recorded on the photographic film, by the relationship

$$I'(x) = \frac{\int_{-\infty}^{+\infty} I(\nu) \, a(x-\nu) \, dx}{\int_{-\infty}^{+\infty} a(x) \, dx}, \tag{94}$$

where x is the coordinate along the spectrum in the focal plane of the spectrograph. If the intensity $I_0$ of the transilluminating source is constant over the width of the line, then the true and observed distributions of the intensity will have the form

$$I(\nu) = I_0 \exp(-k_\nu l), \tag{95}$$

$$I'(\nu) = I_0 \exp(-k'_\nu l), \tag{96}$$

where $k_\nu$ and $k'_\nu$ are the true and observed absorption coefficients, and $l$ is the length of the optical path.

For the spectral absorption coefficient the following relationship is valid:

$$1 - \exp(-k'_\nu l) = \frac{\int_{-\infty}^{+\infty} (1 - \exp(-k_\nu l)) \, a(x-\nu) \, dx}{\int_{-\infty}^{+\infty} a(x) \, dx}. \tag{97}$$

In order to determine the true shape and width of the absorption spectral line from the observed value it is necessary to choose those values of $k_\nu l$ for which Eq. (97) would be fulfilled. In choosing these values one can use the auxilliary condition [105]

$$\int_{-\infty}^{+\infty} (1 - \exp(-k_\nu' l))\, dx = \int_{-\infty}^{+\infty} (1 - \exp(-k_\nu l))\, d\nu, \tag{98}$$

which is a corollary of Eq. (97) and expresses the fact that the total absorption does not depend on the distorting effect of the apparatus function.

The contour of the spectral line under the conditions of our experiment ($T_5 = 5000$-$7600°K$; $P_5 = 11.9$-$66.9$ atm) is best described by the Voigt curve (88) which can be represented in the following form for the absorption case:

$$k_\nu = k_0 \frac{a}{\pi} \int_{-\infty}^{+\infty} \frac{e^{-y^2} dy}{a^2 + (\omega - y)^2}, \tag{99}$$

where $k_0$ is the absorption factor at the center of the line (i.e., for $\nu = \nu_0$), and $a$, $\omega$, and $y$ are determined by Eqs. (89), (90), and (91).

By changing the values of $k_0$ and $a$ in (99), a better coincidence between the observed and true contours of the absorption spectral lines can be achieved (i.e., the solution of Eq. (97) is selected). In order to determine the quantity $a$ it is necessary to know $\Delta\nu_L$ and $\Delta\nu_D$. The Doppler broadening $\Delta\nu_D$ was calculated according to Eq. (80) from the known temperature $T_5$ behind the reflected shock wave. Measuring $\Delta\nu_{obs}$ and knowing $\Delta\nu_L$, we found the approximate value of $\Delta\nu_L$ by simple subtraction. Since the amount by which the apparatus function broadens $\Delta\nu_{tru}$ is not known a priori, several rough estimates were made for $\Delta\nu_L$ (and consequently for $a$ also). The value of $a$ was chosen in such a way that the following equation was best fulfilled:

$$\int_{\nu_1}^{\nu_2} \left(1 - \exp\left\{-\left[k_0 \frac{a}{\pi} \int_{-\infty}^{+\infty} \frac{e^{-y^2} dy}{a^2 + (\omega - y)^2}\right] l\right\}\right) d\nu = \int_{\nu_1}^{\nu_2} (1 - \exp(-k_\nu' l))\, dx. \tag{100}$$

The contour chosen in this way was convoluted with the apparatus function; i.e., the quantity

$$\frac{\int_{\nu_1}^{\nu_2} \left(1 - \exp\left\{-\left[k_0 \frac{a}{\pi} \int_{-\infty}^{+\infty} \frac{e^{-y^2} dy}{a^2 + (\omega - y)^2}\right] l\right\}\right) a\,(x - \nu)\, dx}{\int_{-\infty}^{+\infty} a\,(x - \nu)\, d\nu} \tag{101}$$

was determined.

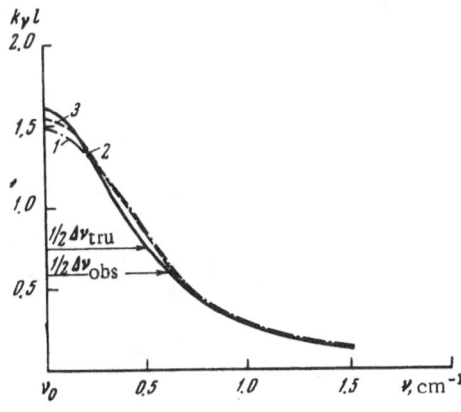

Fig. 28. Contour of the $R_{70}$ line (run No. 54). 1) Observed contour, $\Delta\nu_{obs} = 1.12$ cm$^{-1}$; 2) true contour, $\Delta\nu_{tru} = 0.96$ cm$^{-1}$; 3) contour after analysis of the apparatus function.

The calculation of (101) was carried out graphically. If the contour obtained after analysis of the apparatus function coincided with the observed contour in shape and width, then the problem turned out to be solved, and the contour was finally represented in values of $k_\nu l$; if not, then new values of $a$ and $k_0$ were chosen, and the calculation was repeated.

Figure 28 shows three contours of the $R_{70}$ line in $k_\nu l$ values: the contour 1, which was obtained as the result of photometry of the spectrogram; the supposed true contour 2, and the contour 3 after broadening of contour 2 by the apparatus function. In the figure it is evident that the contour described by the Voigt function coincides fairly well with the experimental contour after transformation of the apparatus function. The slight difference in the center is obviously a consequence of photographic effects.

Having determined $a$ and knowing $\Delta\nu_D$, one can determine $\Delta\nu_L$ from Eq. (89). The absorption coefficient $k_\nu'$ is determined from the relationship

$$k_\nu' = \frac{1}{l} \ln \frac{I_0}{I(x)} .\tag{102}$$

Therefore, in determining line widths (and the integrated absorption coefficient) from absorption spectra the position of the initial line corresponding to the value $I_0$ plays a very substantial role; in the majority of cases the position of the line is chosen in a manner which is not fully specified, and as a result an error arises.

Figure 30 reproduces the tracing of a sector of the absorption spectrum of the $C_2$ molecule on a recording microphotometer. The axis of abscissas corresponds to the direction of motion of the tape, and the axis of ordinates corresponds to the pen deflections of the automatic recorder. These deflections can be graduated in relative intensity units using the characteristic photolayer curve obtained by means of the blackening marks. The straight line $I_0$ yields the average value of the background intensity, which is determined from two spectrograms of the radiation from the pulsed light source which were obtained before and after exposure of the absorption spectrum. Preliminary tests were used to establish the fact that the luminescence intensity of the source decreases insignificantly from flash to flash. This rule was similarly noted in [94]. Therefore, for each lamp flash the intensity level of the background can be represented well as the average of two successive flashes. From Fig. 29 it is evident that even at a considerable distance from the line the minimum absorption $I_\nu^{min}$ in the absorption spectrum does not reach the $I_0'$ level.

The question arises as to what value of $I_0$ must be substituted into Eqs. (90) and (91) to determine $k_\nu$: the averaged value $I_0$, the minimum value $I_\nu^{min}$ between the lines, or some value intermediate between $I_\nu^{min}$ and $I_0$?

The point is that the difference of $I_\nu^{min}$ from $I_0$, which was always observed experimentally, may be caused by: the wing of the line under study, the wings of more distant lines,

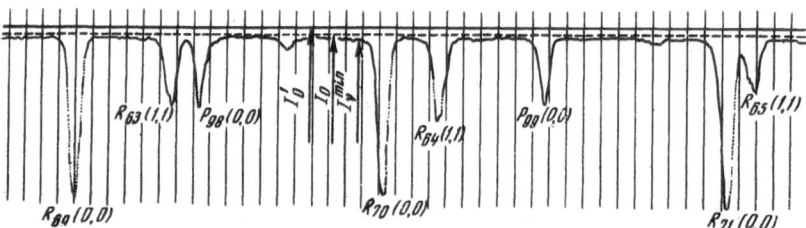

Fig. 29. Sector of the $C_2$ molecule spectrum recorded on the microphotometer.

TABLE 13

| No. | Filler | $P_5$ | $T°, K$ | $\Delta\nu_L,$ cm$^{-1}$ | $\dfrac{I_0' - I_\nu^{min}}{I_0'}$ | $\dfrac{I_0' - I_0}{I_0'}$ |
|---|---|---|---|---|---|---|
| 1 | CO + Ar | 53.6 | 6200 | 1.43 | 0.19 | 0.15 |
| 2 | CO | 50.0 | 5740 | 1.10 | 0.19 | 0.14 |
| 3 | CO + Ar | 13.3 | 7500 | 0.87 | 0.18 | 0.13 |
| 4 | CO | 13.0 | 6730 | 0 77 | 0.15 | 0.08 |

weak uninterpreted lines of other branches, or of $C_2$ bands, or of other molecules and atoms, and by a continuous absorption (or scattering) spectrum of unknown origin.

We cannot immediately say which of the causes produces the difference between $I_\nu^{min}$ and $I_0$. If this difference is produced by the first cause, then one should take $I_0$ in the calculation. If it is produced by the second or third causes, then in analyzing a certain line it is necessary to consider the entire group of lines interacting with it. The assumption that weak uninterpreted lines and a continuous absorption spectrum do not alter the shape of the line investigated allows a complicated examination to be avoided. Numerous attempts at matching the experimental contours of rotational lines of the $C_2$ molecule with the Voigt contour have shown that one should take a certain intermediate value between $I_0$ and $I_\nu^{min}$ as the zero line which determines $I_0$. This fact means that the difference between $I_0$ and $I_\nu^{min}$ is produced by all four causes. In practice the choice of $I_0$ in the given work was achieved according to the wings of the Voigt contour; this contour coincided with the experimental contour after transformation of the apparatus function.

As an example, Table 13 displays the values of the relative difference between $I_0'$ and $I_\nu^{min}$ or between $I_0'$ and $I_0$ for several operating modes of the shock tube. In addition the table indicates: the type of gas in the low-pressure chamber, the pressure $P_5$ and the temperature $T_5$ behind the reflected shock wave, and the width $\Delta\nu_L$ of the rotational line due to collisions in cm$^{-1}$.

The data displayed in the tables show that the position of the $I_\nu^{min}$ lines, which characterizes the minimum absorption in the spectrum, and the position of the zero line $I_0'$ do not remain constant relative to the line $I_0$ representing the averaged brightness level of the transilluminating source. The relative position of these lines changes as a function of the regime of each experiment.

It is natural that any arbitrariness in the choice of the initial reference line corresponding to $I_0$ (and the errors which are permitted under these conditions) can lead to substantial errors in determining the width of the line and in determining $\int k_\nu \, d\nu$ when the conventional procedure is used for analyzing the spectrograms. According to our estimates, these errors can reach 15 and 30%, respectively.

## §6. Results of the Investigation

a. Determination of the Effective Cross Sections of Broadening Collisions of $C_2$ Molecules with CO Molecules and Ar, C, and O Atoms. The results of the tests are displayed in Table 14. Columns 2-4 show the experimental conditions: the composition of the mixture in the low-pressure chamber, the initial pressure $P_1$ in mm Hg, and the velocity of the incident shock wave in km/sec. Columns 5-14 contain the characteristics of the gas behind the reflected shock wave: the temperature $T_5$ in °K, the pressure $P_5$, and the concentration of molecules and atoms behind the reflected shock wave. Column 15 contains the values of broadening of the rotational lines due to the Doppler effect, column 16 contains the width of the contour observed on the photometer tracing, column 17 con-

TABLE 14. Results of Tests on Determining the Effective Cross Sections of the Broadening Collisions of $C_2$ Molecules with CO Molecules and Ar, C, and O Atoms

| Exper. Run No. | Gas composition | $P_1$, mm Hg | $v_1$, km/sec | $T_s$, K | $P_s$, atm | $N_{Ar} \cdot 10^{-16}$ | $N_{CO} \cdot 10^{-16}$ | $N_G \cdot 10^{-17}$ | $N_O \cdot 10^{-17}$ | $N_{C_2} \cdot 10^{-16}$ | $N_{CO_2} \cdot 10^{-16}$ | $N_e \cdot 10^{-16}$ |
|---|---|---|---|---|---|---|---|---|---|---|---|---|
| 13 | CO + Ar | 100 | 2.25 | 6200 | 53.6 | 3.26 | 3.06 | 8.20 | 8.20 | 1.77 | 17.7 | 5.0 |
| 14 | CO + Ar | 100 | 2.30 | 6350 | 55.5 | 3.20 | 2.85 | 8.75 | 8.75 | 1.98 | 19.5 | 5.60 |
| 45 | CO + Ar | 10 | 3.40 | 7640 | 14.5 | 0.59 | 0.386 | 20.4 | 20.4 | 1.73 | 0.60 | 13.3 |
| 54 | CO + Ar | 10 | 3.30 | 7500 | 13.3 | 0.56 | 0.397 | 16.9 | 16.9 | 1.51 | 0.77 | 10.1 |
| 57 | CO + Ar | 10 | 3.23 | 7330 | 12.3 | 0.53 | 0.392 | 14.9 | 14.9 | 1.33 | 0.58 | 8.35 |
| 62 | CO + Ar | 10 | 3.15 | 7260 | 11.3 | 0.51 | 0.385 | 12.0 | 12.0 | 1.31 | 0.65 | 7.32 |
| 63 | CO + Ar | 10 | 3.15 | 7260 | 11.3 | 0.51 | 0.385 | 12.0 | 12.0 | 1.31 | 0.65 | 7.32 |
| 119 | CO | 100 | 2.39 | 5030 | 59.4 | — | 8.55 | 1.36 | 1.16 | 0.78 | 36.7 | 0.09 |
| 120 | CO | 100 | 2.50 | 5390 | 66.3 | — | 8.82 | 3.00 | 2.90 | 1.57 | 44.0 | 0.22 |
| 121 | CO | 100 | 2.51 | 5420 | 66.9 | — | 8.92 | 3.24 | 3.11 | 1.67 | 45.2 | 0.23 |
| 78 | CO | 50 | 2.68 | 5850 | 38.6 | — | 4.70 | 5.35 | 5.35 | 1.43 | 20.3 | 0.83 |
| 79 | CO | 50 | 2.68 | 5850 | 38.6 | — | 4.70 | 5.35 | 5.35 | 1.43 | 20.3 | 0.83 |
| 80 | CO | 50 | 2.62 | 5740 | 37.0 | — | 4.60 | 4.10 | 4.10 | 1.31 | 18.7 | 0.56 |
| 106 | CO | 10 | 3.34 | 6640 | 12.4 | — | 1.13 | 10.3 | 10.3 | 1.18 | 2.89 | 2.37 |
| 107 | CO | 10 | 3.40 | 6730 | 13.0 | — | 1.16 | 11.0 | 11.0 | 1.30 | 3.14 | 2.80 |
| 108 | CO | 10 | 3.38 | 6700 | 12.8 | — | 1.15 | 10.7 | 10.7 | 1.24 | 3.08 | 2.64 |

| Exper. Run No. | Gas composition | $N_{C^+} \cdot 10^{-16}$ | $\Delta\nu_D$, cm$^{-1}$ | $\Delta\nu'_{obs}$, cm$^{-1}$ | $\Delta\nu_{tru}$, cm$^{-1}$ | $\delta$, cm$^{-1}$ | $a$ | $\Delta\nu_e$, cm$^{-1}$ | $l/k_\nu d_\nu$ | $|R_{el}|^2 \cdot 10^{36}$ | $|R_{el}|^2$, a.u. |
|---|---|---|---|---|---|---|---|---|---|---|---|
| 13 | CO + Ar | 3.60 | 0.232 | 1.63 | 1.50 | 0.13 | 5.1 | 1.42 | 3.44 | 3.78 | 0.596 |
| 14 | CO + Ar | 4.00 | 0.234 | 1.60 | 1.41 | 0.19 | 4.7 | 1.32 | 3.24 | 3.27 | 0.516 |
| 45 | CO + Ar | 9.50 | 0.257 | 1.18 | 1.01 | 0.17 | 3.1 | 0.96 | 2.69 | 3.60 | 0.568 |
| 54 | CO + Ar | 7.36 | 0.255 | 1.12 | 0.96 | 0.16 | 2.9 | 0.89 | 2.40 | 3.45 | 0.545 |
| 57 | CO + Ar | 5.90 | 0.252 | 1.11 | 0.94 | 0.15 | 3.0 | 0.91 | 2.44 | 4.00 | 0.630 |
| 62 | CO + Ar | 5.24 | 0.251 | 0.93 | 0.78 | 0.14 | 2.3 | 0.69 | 2.40 | 3.64 | 0.574 |
| 63 | CO + Ar | 5.24 | 0.251 | 0.93 | 0.79 | 0.14 | 2.3 | 0.69 | 2.48 | 3.77 | 0.595 |
| 119 | CO | 0.05 | 0.208 | 1.43 | 1.37 | 0.06 | 5.2 | 1.30 | 1.17 | 2.94 | 0.464 |
| 120 | CO | 0.02 | 0.216 | 2.03 | 1.87 | 0.16 | 7.0 | 1.82 | 1.93 | 2.39 | 0.377 |
| 121 | CO | 0.02 | 0.216 | 1.90 | 1.85 | 0.05 | 7.0 | 1.82 | 2.29 | 2.66 | 0.420 |
| 78 | CO | 0.12 | 0.225 | 1.26 | 1.06 | 0.17 | 4.0 | 1.07 | 2.37 | 3.23 | 0.510 |
| 79 | CO | 0.12 | 0.225 | 1.24 | 1.07 | 0.17 | 3.8 | 1.03 | 2.48 | 3.38 | 0.533 |
| 80 | CO | 0.06 | 0.223 | 1.32 | 1.16 | 0.16 | 4.1 | 1.10 | 2.27 | 3.40 | 0.536 |
| 106 | CO | 1.80 | 0.240 | 1.01 | 0.82 | 0.19 | 2.6 | 0.75 | 2.28 | 3.94 | 0.620 |
| 107 | CO | 2.22 | 0.241 | 1.02 | 0.84 | 0.18 | 2.7 | 0.77 | 2.46 | 3.80 | 0.600 |
| 108 | CO | 2.00 | 0.241 | 1.01 | 0.83 | 0.18 | 2.6 | 0.75 | 2.05 | 3.37 | 0.530 |

tains the width of the Voigt contour which coincides with the observed contour after the apparatus function has been analyzed, column 18 contains the magnitude of the broadening of the true lines due to the apparatus function, columns 19-20 contain the values of the parameter $a$ and the value found for the broadening of the line due to collisions. The data in column 18 show that the observed widths of the rotational lines increase due to the apparatus function by an average of 0.17 cm$^{-1}$ compared with the true widths. The results of analyzing other lines (for example, $R_{69}$ and $R_{68}$) substantiate this conclusion. Thus, for certain approximate estimates of the widths of individual rotational lines one can avoid carrying out painstaking work when the position of the zero line is well known from additional spectrograms and the correction obtained for broadening due to the apparatus function can be used to obtain the true half-width.

Knowing the width $\Delta\nu_L$ of the line caused by collisions, one can use Eqs. (84), (86), (87) to determine the effective cross sections of the broadening collisions. For this purpose we shall begin by considering the group of runs 119, 120, 121 or the group of runs 78, 79, 80. In the first approximation it can be assumed that the broadening of the $C_2$ lines in these runs occurs only due to collisions with CO molecules, since the concentration of all the remaining particles is smaller by at least two orders of magnitude. Having determined $\sigma$ ($C_2$, CO) approximately, one can then, for example, consider runs 13 and 14, where the basic broadening particles are CO and Ar atoms, and find the cross section $\sigma$ ($C_2$, Ar).

In the next approximation one can consider the broadening effects of C and O atoms whose concentration is just two orders of magnitude smaller than that of Ar and CO. Since in all the tests the concentrations of C and O are equal, one can find only the resultant broadening cross section for ($C_2$, O; $C_2$, C) which is caused by the collision of $C_2$ molecules with C and O atoms. For this purpose one can consider, for example, runs 106, 107, 108 or the group of runs 45, 54, 57, 62, and 63. The values obtained in this manner were further refined by the method of successive approximations while considering the broadening effect of other molecules. Finally, the result

$$\sigma\,(C_2,\,Ar) = (63.0 \pm 5.0)\cdot 10^{-16}\ cm^2,$$

$$\sigma\,(C_2,\,CO) = (57.0 \pm 6.0)\cdot 10^{-16}\ cm^2,$$

$$\sigma\,(C_2,\,C;\,C_2,\,O) = (920 \pm 76)\cdot 10^{-16}\ cm^2.$$

was obtained. The large value of the collision cross section $\sigma(C_2,\,C;\,C_2,\,O)$ attracts attention. Possibly, this is caused by the high chemical activity manifested during collisions of a $C_2$ molecule with C and O atoms.

In our calculations we neglected the broadening effect of $C_2$, $CO_2$, $C^+$, and electrons. Whereas this is admissible in the series of runs 13, 14; 119-121, and 78-80, it is no longer admissible in the series 45, 54, 57, 62, and 63, for example, since the electron concentration is just two orders of magnitude smaller than the concentrations $N_{Ar}$ and $N_{CO}$. However, if we consider the fact that the velocity of the electrons is two orders of magnitude higher than the velocity of the molecules and atoms, it follows that for the cross sections of collisions of electrons and $C_2$, which have an order of magnitude equal to the cross sections of collisions of $C_2$ with Ar, and CO, the broadening of the $C_2$ line due to collisions with electrons will be of the same order of magnitude as it is for collisions with Ar and CO. An attempt was made at determining the cross section of the collisions of $C_2$ with electrons, but it did not lead to satisfactory correlation of the results over all of the runs.

    b. Determination of $l\int k_\nu d\nu$ and Calculation of the Matrix Element of the Dipole Moments of the Electronic Transition of the System of Swan Bands for the $C_2$ Molecule. One of the columns of Table 14 displays the values of the integrated absorption coefficient $l\int k_\nu d\nu$, which were determined from the true contour

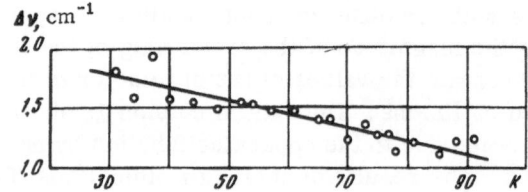

Fig. 30. Dependence of the half-width of the rotational lines of the R-branch of the (0,0) band on the quantum number K.

of the spectral lines. The next columns contain the values of the matrix element of the electronic transition $a^3\Pi_u - d^3\Pi_g$ for the (0,0) band. The values of $|R_e|^2$ were calculated according to Eq. (10) given in Chapter II. On the basis of 16 runs the average value was $|R_e|^2 = 0.54 \pm 0.03$ atomic units (a.u.).

c. Dependence of the Width of a Rotational Line on the Quantum Number K. In order to obtain more complete characteristics of the radiation and absorption of the $C_2$ molecule it is necessary to know how the width of the rotational line varies as a function of the quantum number K. Due to difficulties of a fundamental character (for example, superposition of lines of different bands or branches), experimental data on the majority of the investigated molecules are not available. There are no theoretical calculations at all.

We measured the widths of the rotational lines which were free of superpositions for the R (0,0), R (1,1), and T (0,0) bands. Figure 30 displays the data for the analysis of the lines of the R-branch of the (0,0) band for run No. 79. Since the half-width of the lines varies weakly as a function of K, it can be assumed that the apparatus function introduces an identical increment which is approximately equal to 0.17 cm$^{-1}$. As is evident from Fig. 30, the observed widths of the lines decrease from 1.6 to 1.0 cm$^{-1}$ if K varies from 30 to 90. In analyzing the results displayed in Fig. 30, one should consider [17] that the distance between the components of the triplet of each rotational line varies within limits of 0.6 cm$^{-1}$ when K varies from 30 to 90.

Thus, on this basis it may be assumed that the broadening of the rotational lines of the R-branch of the (0,0) band does not depend on the quantum number K; if it does occur, then its magnitude does not exceed 0.1 cm$^{-1}$ (the accuracy of our experimental measurements). Analogous results were also obtained for the lines of the R- and P-branches of the (1,1) and (0,0) bands.

CONCLUSION

Using the radiation method, we obtained for $|R_e|^2$ the value 0.38±0.14 a.u. while with the absorption method we obtained the values 0.44±0.08 and 0.54±0.03 a.u. Averaging the results obtained with a statistical weight which is the reciprocal of the measurement error, we obtain the value $|R_e|^2 = 0.50 \pm 0.04$ a.u. which should be considered to be the most reliable. This value corresponds to an oscillator force $f_e = 0.028 \pm 0.002$.

It should be noted that the values given for $|R_e|^2$ and $f_e$ can deviate from the real values by ±30% as a consequence of the fact that the calculation of the plasma composition requires knowledge of the equilibrium constants $K_e$, which have been determined with insufficient accuracy.

Table 15 displays the values of the oscillator force for the electronic transition $a^3\Pi_u - d^3\Pi_g$ of the $C_2$ molecule as obtained by various calculated and experimental methods, for comparison purposes.

The data obtained by Hagan [81] ($f_e = 0.005$), and Jeunehomme and Schwenker [82] ($f_e = 0.0043$) evidently refer to one band (0,0) rather than to the electronic transitions; otherwise one cannot explain such a pronounced difference between their data and ours. If this is so, then in

TABLE 15. Summary Table of the Values of the Oscillator Forces for the Electronic Transitions of the Swan Bands of the $C_2$ Molecule

| Authors | Year | Method | $[f_e]$ |
|---|---|---|---|
| Lyddane, Rogers, Roach [72] .......... | 1941 | Calculation | 0.024 |
| Shull [75] ..................... | 1950 | " | 0.13 |
| Stevenson ..................... | 1951 | " | 0.029 |
| Coulson, Lester [78] ............... | 1955 | " | 0.24 |
| Clementi [79] ................... | 1960 | " | 0.0485 |
| Hicks [80] ..................... | 1957 | Radiation, King furnace ? | 0.034 |
| | | | (0.005± 0.003) |
| Hagan [81] ..................... | 1963 | Radiation, absorption, King furnace ? | 0.021 |
| | | | (0.0043± 0.00012) |
| Jeunehomme, Schwenker [82] ......... | 1965 | From the lifetime | 0.018 |
| Sviridov, Sobolev, Sutovskii [111]....... | 1965 | Radiation, shock tube | 0.022± 0.008 |
| Sviridov, Sobolev, Novgorodov [112] ...... | 1965 | Absorption, shock tube | 0.025± 0.005 |
| Sviridov, Sobolev, Novgorodov, Arutyunova [113]..................... | 1965 | " " " | 0.031± 0.002 |
| Fairbairn [110]................... | 1966 | " " " | 0.033± 0.012 |
| Harrington, Modica, Libby [114] ........ | 1966 | " " " | 0.017± 0.005 |
| | | | 0.018± 0.009 |

order to obtain the value of the oscillator force $f_e$ for the electronic transition the values which they give should be multiplied by the coefficient $g/q_{00}$, where g is the degeneration of the original level and $q_{0,0}$ is the Franck—Condon factor. These values are given in the table without parentheses.

An analysis of the table showed that the calculated data obtained by various methods differ considerably from each other, sometimes by one order of magnitude. Thus, the experimental data obtained by us and by other investigators for the $C_2$ molecule can at present serve as a good base for perfecting various theoretical methods of calculation.

After completion of the present investigation, two more papers appeared which were devoted to the determination of the electronic oscillator force of the Swan bands. In the first of them [110] the absolute luminescence brightness of a specific fraction of the (0,0), (1,1), and (2,2) bands was measured behind the incident shock wave in mixtures of $C_2H_2$, $C_2N_2$, and CO with argon. Since the measurements were carried out behind the incident shock wave and there was no automatic monitoring of radiation reabsorption, the author was compelled additionally to calculate the growth curves. The value $f_e = 0.033 ± 0.012$ published by the author is in amazingly good agreement with our average value. In the second paper [114] a method was used similar to ours, but with the difference that $C_2ClF_3$, $CF_4$, or $C_2F_4$ gas was fed into the shock tube. Since the dissociation energy of $CF_2$ and CF is not known exactly, the extremal values known from literature sources were used. In this connection two values of the oscillator force were obtained.

Thus, in our work we did the following.

I. A method of determining the matrix element $|R_e|^2$ of the electronic transition $d^3\Pi_g - a^3\Pi_u$ (the system of Swan bands) of the $C_2$ molecule from radiation was worked out and tested. The luminescence spectrum was excited in a glass shock tube in comparatively low operating modes in a 50% Ar + 50% CO mixture. The initial pressure was $P_1 = 5$ mm Hg; the pressure $P_4$ at which the diaphragm ruptured was varied within the range from 10 to 45 atm. The working gas was hydrogen. The luminescence brightness of a whole sequence of bands was measured photoelectrically. A linear growth of the radiation brightness with the thickness of the layer of luminescent gas behind the front of the reflected shock wave was observed. The maximum thickness of the luminescent gas was determined from photographs of the propagation of the reflected shock wave which were obtained using a streak camera. The calibration of the sensitivity of the equipment was carried out by means of a temperature lamp having a tungsten ribbon. The gas temperature and the concentration of $C_2$ molecules were determined by calculating the state of the gas behind the reflected shock wave from the velocity of the incident wave which was determined experimentally. The temperature $T_5$ of the gas behind the reflected shock wave was varied from 5000 to 6000°K; the pressure $P_5$ was varied from 2 to 4.5 atm under these conditions, and the number of $C_2$ molecules on the optical path was varied from $0.5 \cdot 10^{15}$ to $4.7 \cdot 10^{15}$. From the results of analyzing 75 runs on five sequences of bands with $\Delta v$ = -2, -1, 0, 1 and 2 the value $|R_e|^2 = (2.44 \pm 0.83) \cdot 10^{-36}$ was obtained (or, in atomic units, $|R_e|^2 = 0.385 \pm 0.131$), which corresponds to an oscillator force $f_e = 0.022 \pm 0.008$. The method described is suitable for determining matrix elements of the electronic transitions of a number of other molecules. The merit of the method suggested lies in the automatic monitoring of the absence of radiation reabsorption whose characteristic attribute is the linear rise of the radiation brightness with the thickness of the layer of luminescent gas.

II. The integrated absorption coefficient was measured for a large number of rotational lines of the (0,0) and (1,1) bands of the Swan system for the $C_2$ molecule. Gas samples having a known temperature $T_5$ (6300-7300°K) and pressure $P_5$ (11-37 atm) were obtained behind the reflected shock wave in a CO + Ar mixture. The concentration of $C_2$ was determined by calculation from the measured velocity of the incident shock wave. The number of $C_2$ molecules along the optical path varied from $10^{17}$ to $3.2 \cdot 10^{17}$ in the runs. The temperature of the heated gas in each test was determined by two independent methods: 1) from the velocity of the incident shock wave; 2) from the absorption in the rotational structure of the Swan bands. The coincidence of the indicated temperatures with the temperature obtained by the generalized method of spectral line inversion was established in advance. The value $|R_e|^2 = 0.43 \pm 0.08$ a.u. was obtained from the measured integrated absorption coefficients of the rotational lines of $C_2$. An attempt was made to obtain the absorption spectrum of the Fox – Herzberg bands under equilibrium conditions. The upper bound $f_e < 0.011$ of the oscillator force of the electronic transition of these bands was determined.

III.

1. The absorption spectra were obtained behind the reflected shock wave in a CO + Ar mixture and in pure carbon monoxide CO at temperatures from 5000 to 7600°K and pressures from 11.3 to 66.9 atm.

2. The true values of the broadening of spectral lines due to collisions were determined from the observed contours of the individual rotational lines $R_{78}$, $R_{69}$, and $R_{68}$ of the (0,0) band of the $C_2$ molecule; this allowed calculation of the effective cross sections of broadening collisions of $C_2$ molecules with CO molecules and Ar, C, and O atoms, whose values turned out to be equal to

$$\sigma(C_2, Ar) = (68 \pm 5) \cdot 10^{-16} \text{ cm}^2,$$

$$\sigma(C_2, CO) = (57 \pm 6.0) \cdot 10^{-16} \text{ cm}^2,$$

$$\sigma(C_2, C; C_2, O) = (920 \pm 76) \cdot 10^{-16} \text{ cm}^2.$$

Thus, the cross sections $\sigma(C_2, Ar)$ and $\sigma(C_2, CO)$ of the broadening collisions are of the order of the gaskinetic cross section, whereas the cross section $\sigma(C_2, C; C_2, O)$ is one order of magnitude larger than the gaskinetic cross section; this is evidently connected with the chemical activity of C and O.

3. Measurements of the half-widths of rotational lines as functions of the quantum number K showed that the observed broadening does not depend on K.

4. The values of $\int k_\nu$ were determined from the true contours of the spectral lines, and on the basis of these values more reliable values of $|R_e|^2$ were obtained for the electronic transitions $a^3\Pi_u - d^3\Pi_g$ of the $C_2$ molecule. $|R_e|^2 = 0.50 + 0.04$ a.u. and $f_e = 0.028 \pm 0.002$ should be assumed to be the most reliable values.

5. An analysis of the conventional methods of analyzing the absorption spectra showed that the errors in determining the integrated absorption coefficient $\int k_\nu \, d\nu$ can reach 30% due to the incorrect determination of the position of the initial reference line which determines $I_0$.

In conclusion the author expresses his deep appreciation to the scientific supervisor of the present work, Professor N. N. Sobolev for his constant attention and repeated fruitful discussions of the results which were obtained. The author warmly thanks V. M. Sutovskii, M. Z. Novgorodov, and G. A. Arutyunov, who provided much help in performing the experiments. The author thanks senior scientific associate of the Physics Institute, Academy of Sciences of the USSR, L. N. Turnitskii for his help in discussing the results obtained.

The work presented could not have been carried out without the great help provided by A. A. Sapronov, N. F. Lunyakov, I. M. Kholinov, as well as A. T. Matachun and I. N. Berestovaya, to whom the author expresses his deep thanks.

APPENDIX

TABLES OF THE BRIGHTNESS TEMPERATURES

OF THE TUNGSTEN RIBBON-FILAMENT LAMP

The temperature lamp with a tungsten ribbon-filament is at present one of the most convenient and reliable standard sources. Calibration of the lamp with respect to current is carried out either at the Teller temperature or at the brightness temperature in a specified spectral sector, usually at $\lambda = 0.65\,\mu$.

The brightness of the lamp in various sectors of the spectrum is determined by the true temperature of the tungsten ribbon and by its spectral radiative capacity. The radiation brightness of the ribbon is uniquely characterized by its brightness temperature. Before 1954 the majority of metrological laboratories (the Mendeleev All-Union Scientific-Research Institute of Metrology, The National Bureau of Standards of the USA, the Physico-Engineering Bureau in the German Democratic Republic, etc.) used the values of the radiative capacity of tungsten obtained by Ornstein and associates [115, 116]. The most complete investigation of the radiative capacity of tungsten, which covered a broad spectral range from 0.23 to 2.70$\mu$ and a wide temperature range from 1600 to 2800°K was carried out by DeVos [117]. The radiative capacity of tungsten was determined by comparing the spectral radiation brightness of the wall with the radiation brightness of the aperture of a tubular blackbody coiled from a tungsten ribbon 50$\mu$ thick. The length of the tube was 16 cm, and the cross section was an equilateral triangle with sides of 5 mm. Five apertures having a diameter of 0.3 mm were made in the center of one of the tube walls for observation purposes. A blackbody with such geometric dimensions is close to an ideal black radiator on the basis of calculations. Even under unfavorable conditions (high reflectivity and low temperature) the deviation from an ideal black radiator remains less than 0.5%.

More thorough investigations of the radiative capacity of tungsten having a high purity (99.99%) were carried out by Larrabee [118] in the spectral range from 0.31 to 0.80$\mu$ at temperatures ranging from 1600 to 2400°K. The investigations were similarly carried out by

TABLE A.1. The Radiative Capacity of Tungsten at $\lambda = 0.65\mu$

| $T$, °K / Authors | 1000 | 1200 | 1400 | 1600 | 1800 | 2000 | 2200 | 2400 | 2600 | 2800 | 3000 |
|---|---|---|---|---|---|---|---|---|---|---|---|
| Ornstein [115] | 0.446 | 0.444 | 0.442 | 0.440 | 0.433 | 0.436 | 0.434 | 0.432 | 0.430 | 0.423 | 0.426 |
| De-Vos [116] | | | | 0.450 | 0.446 | 0.442 | 0.438 | 0 431 | 0.430 | 0.427 | |
| Larrabee [118] | | | | 0.441 | 0.437 | 0.433 | 0.429 | 0 425 | | | |

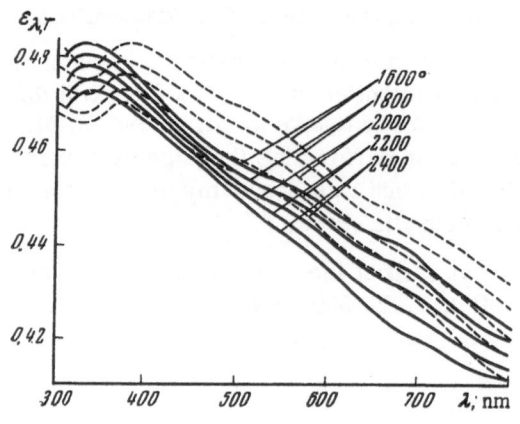

Fig. 31. Spectral radiative capacity of tungsten. The solid lines represent the Larrabee data, while the dashed lines represent the DeVos data.

means of a thin-walled tungsten furnace. Larrabee was able to consider the error caused by the influence of the additional radiation from the edges of the aperture, which distorts the blackbody radiation. The values of radiative capacity 1.5-2.0% which he obtained were lower than the data obtained by DeVos and Ornstein in the visible range of the spectrum (see Table A.1 in which the values of the radiative capacity of tungsten for $\lambda = 0.65\mu$ are given as a sample for comparison).

Notwithstanding the substantial difference between the absolute values obtained by De Vos and Larrabee, the dependence of the radiative capacity on wavelength is almost identical in the spectral range from 400 to 800 nm (Fig. 31). Therefore, the spectral brightness distribution and the brightness temperatures were found to be identical, whether we used the DeVos or the Larrabee data in the recalculation, provided that we used the brightness temperature (for example, $S_{\lambda=0.65}$) as the original temperature. The calculated true temperatures, as determined according to the stipulated brightness temperature, will naturally differ from each other. This difference reaches several degrees.

We calculated the values of brightness temperature $S_\lambda$ in the range from 0.60 to 0.35$\mu$ according to the stipulated brightness temperature for $\lambda = 0.65\mu$. The calculation was carried out according to the equation

$$\frac{1}{S_\lambda} - \frac{1}{S_{0.65}} = \frac{\lambda_{0.65} - \lambda}{C_2} \ln \frac{\varepsilon_{0.65T}}{\varepsilon_{\lambda, T}} \tag{A.1}$$

for $C_2 = 1.438$ cm · deg, where $\varepsilon_{\lambda, T}$ is the radiative capacity of tungsten at the stipulated wavelength $\lambda$ and the stipulated temperature T. Under these conditions the Hamaker values of $\varepsilon_{\lambda, T}$ were used up to a temperature of 1600°K, since no other investigations of the radiative capacity $\varepsilon_{\lambda, T}$ are available; the DeVos data were used at temperatures from 1600-2800°K, and at temperatures above 2800°K the values of $\varepsilon_{\lambda, T}$ found by extrapolating the DeVos data were used.

TABLE A. 2. The Brightness $S_\lambda$, Color ($T_c$), and True (T) Temperatures for Tungsten

| [118] | T [115,116] | [117] | 6500 | 6300 | 6100 | 5900 | 5700 | 5500 | 5300 | 5100 | 4900 | 4700 | 4500 | 4300 | 4100 | 4000 | 3900 | 3800 | 3700 | 3600 | 3500 | $T_c$ |
|---|---|---|---|---|---|---|---|---|---|---|---|---|---|---|---|---|---|---|---|---|---|---|
| | 1002.0 | | 963.0 | 963.6 | 965.5 | 966.4 | 968.1 | 969.4 | 970.6 | 971.9 | 973.0 | 974.2 | 975.5 | 976.5 | 977.8 | 978.5 | 979.0 | 979.5 | 980.0 | 980.4 | 980.9 | |
| | 1052.0 | | 1008.6 | 1009.8 | 1012.0 | 1012.9 | 1014.8 | 1016.2 | 1017.5 | 1018.9 | 1020.2 | 1021.5 | 1022.8 | 1024.2 | 1025.6 | 1026.2 | 1026.8 | 1027.2 | 1027.8 | 1028.4 | 1028.6 | |
| | 1102.0 | | 1054.4 | 1055.9 | 1058.3 | 1059.4 | 1061.3 | 1062.9 | 1064.1 | 1065.9 | 1067.3 | 1068.7 | 1070.2 | 1071.7 | 1073.1 | 1073.8 | 1074.5 | 1075.0 | 1075.7 | 1076.2 | 1076.8 | |
| | 1152.0 | | 1100.2 | 1101.8 | 1104.5 | 1105.5 | 1107.7 | 1109.4 | 1111.0 | 1112.6 | 1114.2 | 1115.8 | 1117.3 | 1119.0 | 1120.6 | 1121.4 | 1122.1 | 1122.7 | 1123.4 | 1124.0 | 1124.6 | |
| | 1201.9 | | 1145.9 | 1147.5 | 1150.4 | 1151.6 | 1153.9 | 1155.8 | 1157.6 | 1159.2 | 1161.0 | 1162.7 | 1164.4 | 1166.2 | 1168.0 | 1168.8 | 1169.6 | 1170.3 | 1171.0 | 1171.7 | 1172.3 | |
| | 1251.9 | | 1191.2 | 1193.0 | 1196.2 | 1197.4 | 1199.9 | 1201.9 | 1203.8 | 1205.7 | 1207.6 | 1209.5 | 1211.3 | 1213.3 | 1215.2 | 1216.1 | 1216.9 | 1217.7 | 1218.5 | 1219.3 | 1219.9 | |
| | 1301.9 | | 1236.4 | 1238.4 | 1241.8 | 1243.2 | 1245.8 | 1248.0 | 1250.0 | 1252.0 | 1254.1 | 1256.1 | 1258.1 | 1260.2 | 1262.3 | 1263.3 | 1264.2 | 1265.0 | 1265.9 | 1266.7 | 1267.5 | |
| | 1351.9 | | 1281.5 | 1283.6 | 1287.3 | 1288.6 | 1291.5 | 1293.8 | 1296.0 | 1298.2 | 1300.4 | 1302.6 | 1304.8 | 1307.1 | 1309.3 | 1310.3 | 1311.3 | 1312.3 | 1313.3 | 1314.1 | 1314.9 | |
| | 1401.9 | | 1326.3 | 1328.6 | 1332.5 | 1334.0 | 1337.0 | 1339.6 | 1341.9 | 1344.3 | 1346.7 | 1349.0 | 1351.3 | 1353.8 | 1356.2 | 1357.3 | 1358.4 | 1359.4 | 1360.4 | 1361.4 | 1362.4 | |
| | 1452.1 | | 1371.4 | 1373.5 | 1377.6 | 1379.1 | 1382.4 | 1385.0 | 1387.6 | 1390.0 | 1392.7 | 1395.2 | 1397.8 | 1400.4 | 1402.9 | 1404.2 | 1405.3 | 1406.4 | 1407.5 | 1408.5 | 1409.4 | |
| | 1502.3 | | 1415.4 | 1418.0 | 1422.6 | 1424.2 | 1427.6 | 1430.5 | 1433.2 | 1435.9 | 1438.6 | 1441.3 | 1444.0 | 1446.8 | 1449.6 | 1450.9 | 1452.1 | 1453.4 | 1454.4 | 1455.6 | 1456.6 | |
| | 1552.4 | | 1459.8 | 1462.5 | 1467.3 | 1469.2 | 1472.6 | 1475.7 | 1478.6 | 1481.5 | 1484.4 | 1487.3 | 1490.1 | 1493.1 | 1496.1 | 1497.5 | 1498.8 | 1500.1 | 1501.3 | 1502.5 | 1503.6 | |
| 1602.5 | 1602.6 | 1600 | 1503.8 | 1506.7 | 1511.8 | 1513.8 | 1517.5 | 1520.8 | 1523.9 | 1526.9 | 1530.0 | 1533.1 | 1536.1 | 1539.3 | 1542.4 | 1543.9 | 1545.4 | 1546.8 | 1548.2 | 1549.4 | 1550.5 | 1618 |
| 1652.6 | 1652.5 | 1650 | 1547.7 | 1550.8 | 1556.2 | 1558.2 | 1562.2 | 1565.7 | 1569.0 | 1572.2 | 1575.5 | 1578.8 | 1582.0 | 1585.4 | 1588.6 | 1590.3 | 1591.8 | 1593.4 | 1594.8 | 1596.1 | 1597.3 | 1663 |
| 1702.6 | 1702.3 | 1700 | 1591.4 | 1594.7 | 1600.3 | 1602.7 | 1606.7 | 1610.5 | 1614.0 | 1617.4 | 1620.9 | 1624.3 | 1627.7 | 1631.3 | 1634.9 | 1636.6 | 1638.2 | 1639.9 | 1641.3 | 1642.8 | 1644.0 | 1718 |
| 1752.7 | 1752.1 | 1750 | 1635.7 | 1638.4 | 1644.3 | 1646.8 | 1651.1 | 1655.0 | 1658.8 | 1662.4 | 1666.1 | 1669.7 | 1673.4 | 1677.2 | 1680.9 | 1682.7 | 1684.5 | 1686.2 | 1687.7 | 1689.3 | 1690.7 | 1771 |
| 1802.7 | 1801.8 | 1800 | 1677.9 | 1681.7 | 1687.0 | 1690.5 | 1695.0 | 1699.2 | 1703.2 | 1706.9 | 1710.8 | 1714.7 | 1718.5 | 1722.5 | 1726.8 | 1728.4 | 1730.3 | 1732.2 | 1733.8 | 1735.4 | 1736.8 | 1821 |
| 1853.0 | 1851.6 | 1850 | 1721.3 | 1725.3 | 1731.1 | 1734.4 | 1739.3 | 1743.7 | 1748.0 | 1751.5 | 1756.0 | 1760.1 | 1764.2 | 1768.4 | 1772.6 | 1774.5 | 1776.6 | 1778.6 | 1780.3 | 1782.0 | 1783.8 | 1871 |
| 1903.3 | 1901.7 | 1900 | 1764.3 | 1768.6 | 1775.2 | 1778.2 | 1783.1 | 1787.8 | 1792.3 | 1796.5 | 1800.9 | 1805.2 | 1809.4 | 1813.9 | 1818.3 | 1820.4 | 1822.5 | 1824.6 | 1826.5 | 1828.3 | 1829.8 | 1921 |
| 1953.4 | 1951.7 | 1950 | 1806.4 | 1811.6 | 1818.5 | 1821.7 | 1826.8 | 1831.7 | 1836.5 | 1841.0 | 1845.5 | 1850.1 | 1854.5 | 1859.3 | 1863.9 | 1866.2 | 1868.3 | 1870.5 | 1872.5 | 1874.4 | 1876.4 | 1972 |
| 2003.7 | 2001.6 | 2000 | 1849.6 | 1854.4 | 1861.6 | 1865.0 | 1870.3 | 1875.5 | 1880.5 | 1885.3 | 1890.1 | 1894.8 | 1899.5 | 1904.5 | 1909.4 | 1911.7 | 1914.0 | 1916.4 | 1918.5 | 1920.4 | 1922.4 | 2020 |
| 2054.0 | 2051.8 | 2050 | 1891.9 | 1896.9 | 1904.5 | 1908.1 | 1913.7 | 1919.1 | 1924.3 | 1929.4 | 1934.6 | 1939.4 | 1944.3 | 1949.5 | 1954.7 | 1957.1 | 1959.6 | 1962.0 | 1964.3 | 1966.3 | 1968.1 | 2073 |
| 2104.2 | 2102.0 | 2100 | 1933.9 | 1939.2 | 1947.3 | 1950.9 | 1957.0 | 1962.6 | 1968.0 | 1973.3 | 1978.5 | 1983.8 | 1989.0 | 1994.4 | 1999.8 | 2002.1 | 2005.0 | 2007.6 | 2009.9 | 2012.0 | 2014.1 | 2125 |
| 2154.5 | 2152.0 | 2150 | 1975.9 | 1981.5 | 1990.7 | 1993.7 | 2000.1 | 2005.0 | 2011.5 | 2017.1 | 2022.6 | 2028.1 | 2033.5 | 2039.2 | 2044.9 | 2047.6 | 2050.3 | 2053.0 | 2055.5 | 2057.8 | 2059.8 | 2177 |
| 2204.8 | 2202.1 | 2200 | 2017.8 | 2023.5 | 2032.5 | 2036.4 | 2043.0 | 2049.2 | 2054.9 | 2060.7 | 2066.5 | 2072.2 | 2077.9 | 2083.9 | 2089.8 | 2092.6 | 2095.5 | 2098.4 | 2100.9 | 2103.4 | 2105.5 | 2228 |
| 2255.0 | 2252.1 | 2250 | 2059.4 | 2065.3 | 2074.1 | 2078.8 | 2085.7 | 2092.2 | 2098.2 | 2104.2 | 2110.2 | 2116.2 | 2122.2 | 2128.5 | 2134.7 | 2137.6 | 2140.5 | 2143.7 | 2146.3 | 2148.8 | 2151.0 | 2280 |
| 2305.4 | 2301.9 | 2300 | 2100.4 | 2106.8 | 2115.4 | 2120.9 | 2128.3 | 2135.1 | 2141.3 | 2147.6 | 2153.9 | 2160.2 | 2166.4 | 2172.9 | 2179.4 | 2182.5 | 2185.6 | 2188.8 | 2191.5 | 2194.1 | 2196.5 | 2330 |
| 2355.5 | 2351.6 | 2350 | 2141.8 | 2148.3 | 2158.4 | 2163.2 | 2170.8 | 2177.8 | 2184.8 | 2190.8 | 2197.5 | 2204.0 | 2210.5 | 2217.3 | 2224.1 | 2227.2 | 2230.4 | 2233.9 | 2236.7 | 2239.2 | 2241.8 | 2383 |
| 2405.6 | 2401.4 | 2400 | 2182.5 | 2189.5 | 2200.2 | 2204.9 | 2213.1 | 2220.4 | 2227.2 | 2234.0 | 2240.8 | 2247.0 | 2254.4 | 2261.5 | 2268.5 | 2271.9 | 2275.2 | 2278.8 | 2281.8 | 2284.4 | 2287.1 | 2436 |
| | 2451.0 | 2450 | 2223.2 | 2230.6 | 2241.8 | 2246.5 | 2255.1 | 2262.7 | 2269.8 | 2276.8 | 2283.9 | 2291.0 | 2298.2 | 2305.5 | 2312.9 | 2316.4 | 2319.9 | 2323.6 | 2326.6 | 2329.4 | 2332.3 | 2488 |
| | 2500.5 | 2500 | 2263.9 | 2271.7 | 2283.2 | 2288.2 | 2296.8 | 2304.8 | 2312.3 | 2319.5 | 2326.9 | 2334.3 | 2341.9 | 2349.5 | 2357.2 | 2360.8 | 2364.4 | 2368.2 | 2371.4 | 2374.3 | 2377.3 | 2540 |
| | 2550.2 | 2550 | 2304.0 | 2312.5 | 2324.5 | 2329.3 | 2337.5 | 2346.8 | 2354.7 | 2362.1 | 2369.5 | 2377.0 | 2385.4 | 2393.3 | 2401.3 | 2405.1 | 2408.9 | 2412.7 | 2416.1 | 2419.0 | 2422.3 | 2592 |
| | 2600.0 | 2600 | 2344.0 | 2353.1 | 2365.5 | 2370.2 | 2380.0 | 2388.6 | 2396.9 | 2404.5 | 2412.5 | 2420.4 | 2428.8 | 2436.9 | 2445.3 | 2449.3 | 2452.9 | 2457.1 | 2460.6 | 2463.9 | 2467.4 | 2643 |
| | 2650.0 | 2650 | 2384.1 | 2393.4 | 2405.3 | 2411.4 | 2421.4 | 2430.3 | 2438.9 | 2446.5 | 2455.1 | 2463.4 | 2472.1 | 2480.5 | 2489.2 | 2493.3 | 2497.0 | 2501.4 | 2505.1 | 2508.5 | 2511.8 | 2698 |
| | 2699.8 | 2700 | 2426.4 | 2433.5 | 2446.9 | 2452.5 | 2462.6 | 2471.9 | 2480.7 | 2489.0 | 2497.6 | 2506.2 | 2515.2 | 2523.9 | 2532.9 | 2537.2 | 2541.1 | 2545.6 | 2549.6 | 2552.9 | 2556.4 | 2749 |
| | 2749.7 | 2750 | 2467.5 | 2473.4 | 2487.3 | 2493.9 | 2503.8 | 2513.4 | 2522.3 | 2531.1 | 2540.0 | 2549.0 | 2558.2 | 2567.2 | 2576.7 | 2581.0 | 2585.4 | 2589.8 | 2593.7 | 2597.4 | 2601.0 | 2802 |
| | 2799.6 | 2800 | 2504.1 | 2513.1 | 2527.5 | 2534.2 | 2544.8 | 2554.6 | 2563.7 | 2573.0 | 2582.2 | 2591.4 | 2601.1 | 2610.4 | 2620.2 | 2624.7 | 2629.2 | 2633.7 | 2637.8 | 2641.6 | 2645.4 | 2852 |
| | 2849.3 | 2850 | 2543.1 | 2552.8 | 2567.5 | 2574.2 | 2585.4 | 2595.7 | 2605.1 | 2614.6 | 2624.3 | 2633.7 | 2643.8 | 2653.5 | 2663.6 | 2668.3 | 2673.0 | 2676.2 | 2681.9 | 2685.7 | 2689.7 | 2905 |
| | 2899.0 | 2900 | 2582.6 | 2592.4 | 2607.4 | 2614.8 | 2625.9 | 2636.5 | 2646.2 | 2656.3 | 2666.2 | 2676.0 | 2686.4 | 2696.5 | 2706.8 | 2711.7 | 2716.6 | 2721.4 | 2725.9 | 2729.9 | 2733.8 | 2956 |
| | 2948.5 | 2950 | 2621.4 | 2631.9 | 2647.4 | 2654.9 | 2666.1 | 2677.5 | 2687.0 | 2697.3 | 2708.0 | 2718.2 | 2728.8 | 2739.2 | 2750.0 | 2755.0 | 2760.0 | 2765.1 | 2769.6 | 2773.8 | 2778.0 | 3009 |
| | 2998.9 | 3000 | 2660.9 | 2670.9 | 2686.9 | 2694.9 | 2706.2 | 2717.5 | 2728.0 | 2738.2 | 2749.6 | 2760.1 | 2771.2 | 2781.9 | 2793.3 | 2798.2 | 2803.4 | 2808.5 | 2813.3 | 2817.6 | 2821.9 | 3060 |
| | | 3050 | 2698.7 | 2709.9 | 2726.5 | 2731.1 | 2746.2 | 2758.0 | 2768.7 | 2780.1 | 2791.1 | 2802.0 | 2813.6 | 2824.5 | 2835.9 | 2841.3 | 2846.7 | 2852.0 | 2856.9 | 2861.3 | 2865.9 | |
| | | 3100 | 2741.5 | 2748.7 | 2765.7 | 2773.8 | 2785.9 | 2798.2 | 2809.3 | 2820.5 | 2832.5 | 2843.6 | 2855.5 | 2866.8 | 2878.7 | 2884.0 | 2889.8 | 2895.3 | 2900.6 | 2904.9 | 2909.5 | |
| | | 3150 | 2775.5 | 2787.4 | 2804.8 | 2813.0 | 2825.0 | 2838.4 | 2849.8 | 2861.5 | 2873.7 | 2886.0 | 2897.2 | 2909.0 | 2921.6 | 2927.0 | 2932.8 | 2938.4 | 2943.7 | 2948.4 | 2953.2 | |
| | | 3200 | 2814.0 | 2826.0 | 2843.8 | 2852.5 | 2865.0 | 2878.4 | 2890.2 | 2902.1 | 2914.7 | 2926.1 | 2939.1 | 2951.3 | 2963.9 | 2969.8 | 2975.7 | 2981.6 | 2986.9 | 2991.9 | 2996.7 | |

The results of the calculation are given in Table A.2. Besides the brightness temperatures $S_\lambda$, the table gives the true temperatures T. The relationship between the brightness temperature $S_\lambda$ and the true temperature T is determined by the following equation [119]:

$$\frac{1}{T} - \frac{1}{S_\lambda} = \frac{\lambda}{C_2} \ln (\tau \varepsilon_{\lambda,T}), \tag{A.2}$$

where $\tau = 0.92$ is the transmission coefficient of the window of the lamp envelope. The first column of Table A.2 contains temperatures which correspond to the value $\varepsilon_{\lambda=0.65}$ obtained by Larrabee, the second column contains the corresponding data obtained by Hamaker and Ornstein, while the third contains the DeVos data. The last column of this table also gives the color temperatures $T_c$ which were calculated according to the equation

$$\frac{1}{T} = \frac{1}{T_c} - \frac{k}{C_2}, \tag{A.3}$$

the constant k being determined graphically by the best approximation of the curve $\varepsilon_{\lambda,T} = \beta e^{\frac{k}{\lambda}}$, where $\beta$ and k are quantities which are independent of $\lambda$.

## REFERENCES

1. TASS Communique, Pravda, No. 292 (17974), October 19 (1967).
2. E. T. Antropov, E. F. Gippius, A. P. Dronov, N. N. Krindach, N. N. Sobolev, E. M. Kudryavtseva, A. N. Pechenov, A. G. Sviridov, L. N. Tunitskii, F. S. Faizullov, and V. P. Cheremisinov, Transactions of the Spectroscopy Commission, Vol. 1 (1963), pp. 64–81
3. E. T. Antropov, Tr. Fiz. Inst. Akad. Nauk SSSR, 35:3 (1966).
4. E. M. Kudryavtsev, Tr. Fiz. Inst. Akad. Nauk SSSR, 35:74 (1966).
5. L. Herman, J. Grumberg-Akriche, and Ch. Granat, J. Quant. Spectr. Radiative Transfer, 5:4 (1965).
6. A. P. Dronov, A. G. Sviridov, and N. D. Sobolev, Opt. i Spektr., 10:3 (1961).
7. V. N. Alyamovskii, A. P. Dronov, V. F. Kitaeva, A. G. Sviridov, and N. N. Sobolev, Problems of Magnetohydrodynamics and Plasmadynamics, Riga (1962), pp. 377–386.
8. A. P. Dronov, A. G. Sviridov, N. N. Sobolev, Opt. i Spektr., 12:6 (1962).
9. N. N. Sobolev, V. N. Aljmovsky [Alyamovskii], A. P. Dronov, F. S. Faizullov, V. F. Kitaeva, and A. G. Sviridov, Proceedings of the Fifth International Conference on Ionization Phenomena in Gases, Amsterdam, Vol. 2 (1962), pp. 2122–2137.
10. A. P. Dronov, A. G. Sviridov, and N. N. Sobolev, Physical Problems of Spectroscopy, Vol. 1 (1962), p. 152
11. E. A. Ballik and D. A. Ramsay, Chem. Phys., 31:4 (1969).
12. W. Jevons, Band Spectra of Diatomic Molecules, London (1932).
13. W. E. Pretty, Proc. Phys. Soc., 40:71 (1927).
14. G. Herzberg, Spectra and Structure of Diatomic Molecules [Russian translation], IL (1949).
15. R. S. Johnson, Philos. Trans. Roy. Soc., 226:157 (1926/27).
16. J. A. Fox and G. Herzberg, Phys. Rev., 52:638 (1937).
17. J. D. Shea, Phys. Rev., 30:825 (1927).
18. J. Budo, Z. Phys., 105:579 (1937).
19. J. A. Phillips, Astrophys. J., 108:434 (1948).
20. G. Herzberg, Phys. Rev., 70:762 (1946).
21. R. S. Johnson and P. K. Assundi, Proc. Roy. Soc., 124:668 (1929).
22. G. A. Hornbeck and R. S. Herman, J. Chem. Phys., 18:763 (1950).

23. G. A. Hornbeck and R. S. Herman, Industrial and Engineering Chemistry, 43:2739 (1959).
24. J. A. Phillips, Astrophys. J., 110:73 (1949).
25. R. A. Durie, Proc. Roy. Soc., A221:110 (1952).
26. N. H. Kiess and M. Bass, J. Chem. Phys., 22:569 (1954).
27. A. Monfils and B. Rosen, Nature, 164:713 (1949).
28. R. Clusius and A. E. Douglas, Canad. J. Phys., 32:319 (1954).
29. A. E. Douglas, Astrophys. J., 114:466 (1951).
30. B. Rosen and P. Swings, Ann. Astrophys., 16:82 (1953).
31. R. Goupil and R. Herman, Ann. Astrophys., 16:441 (1963).
32. W. R. Garton, Fuel, 32:519 (1953).
33. P. Swings, Monthly Notices Roy. Astron. Soc., 103:92 (1943).
34. R. Johnson and N. Tawde, Proc. Roy. Soc., A137:575 (1932).
35. B. B. Laud, Ph. D. Thesis, Bombay Univ. (1951).
36. J. M. Patel, Ph. D. Thesis, Bombay Univ. (1947).
37. N. R. Patel and J. M. Patel, Ph.D. Thesis, Bombay Univ., VI, Part II (1937), p. 29.
38. N. Tawde and B. B. Laud, Proc. Nat. Inst. Sci. Ind., 20A:259 (1964).
39. R. King, Astrophys. J., 108:429 (1948).
40. J. A. Phillips, Mem. Soc. Roy. Sci. Liège, 14:111 (1954).
41. J. A. Phillips, Astrophys. J., 125:153 (1967).
42. V. I. Kondrat'ev, Structure of the Atoms of Molecules, Fizmatgiz (1959).
43. A. Gaydon, Dissociation Energy and Spectra of Diatomic Molecules [Russian translation], IL (1949).
44. S. S. Penner, Quantitative Molecular Spectroscopy and Radiative Capacity of Gases [Russian translation], IL (1959).
45. V. N. Kolesnikov and L. V. Leskov, Usp. Fiz. Nauk, 65:3 (1958).
46. F. S. Ortenberg, Usp. Fiz. Nauk, 90:2 (1966).
47. E. Hutchison, Phys. Rev., 37:45 (1931).
48. K. Wurm, Z. Astrophys., 5:620 (1932).
49. A. McKellar and W. Buscomb, Publ. Dom. Astrophys. Obs. Vict., 7:361 (1949).
50. N. Tawde and J. Patel, Astrophys. J., 112:210 (1950)
51. A. McKellar and N. R. Tawde, Astrophys. J., 113:440 (1951).
52. M. E. Pillow, Mem. Soc. Roy. Sci. Liège, 13:145 (1953).
53. M. E. Pillow, Proc. Phys. Soc., 64A:772 (1951).
54. A. G. Gaydon and R. W. Pearce, Proc. Roy. Soc., 173A:37 (1939).
55. Ta-jou Wu, Proc. Phys. Soc., 65A:965 (1952).
56. G. Wentzel, Z. Phys., 38:518 (1926).
57. H. A. Kramers, Z. Phys., 39:828 (1926).
58. L. Brillouin, J. Chim. Radium., 7:353 (1926).
59. A. A. Wyler, Mem. Soc. Roy. Sci. Liège, Vol. A137 (1953).
60. A. A. Wyler, Astrophys. J., 127:763 (1953).
61. H. M. Hulbert and J. D. Hirschfeld, J. Chem. Phys., 35:1961 (1961).
62. N. R. Tawde and B. B. Laud, Proc. Nat. Inst. Sci. Ind., 200A:259 (1954).
63. W. Jarmain, P. Fraser, and R. Nichols, Astrophys. J., 119:286 (1954).
64. F. S. Ortenberg, Opt. i Spektr., 16:729 (1964).
65. W. R. Jarmain, Canad. J. Phys., 38:217 (1960).
66. N. L. Singh and D. C. Jain, Canad. J. Phys., 40:520 (1960).
67. S. M. Read and J. T. Wanderslice, J. Chem. Phys., 36:2366 (1962).
68. R. W. Nichols, P. A. Fraser, and W. R. Jarmain, Combustion and Flames, 3:13 (1959).
69. P. A. Fraser, Canad. J. Phys., 32:515 (1954).
70. D. C. Jain, J. Quant. Spectr. Radiative Transfer, 4:3 (1964).
71. J. E. Mentall and R. W. Nichols, Proc. Phys. Soc., 86:552 (1965).
72. R. H. Lyddane, F. T. Rogers, and F. Roach, Phys. Rev., 60:281 (1941).

73. J. M. White, J. Chem. Phys., 8:79 (1940).
74. A. Stevenson, Proc. Phys. Soc., 64A:99 (1951).
75. H. Shull, Astrophys. J., 112:352 (1950).
76. H. Shull, Astrophys. J., 114:546 (1956).
77. H. Shull, J. Chem. Phys., 20:1095 (1952).
78. G. A. Coulson and R. G. Lester, Trans. Faraday Soc., 51:1605 (1955).
79. E. Clementi, Astrophys. J., 132:898 (1960).
80. W. T. Hicks, UCRL Report, 3696 (1957).
81. L. G. Hagan, Ph.D. Thesis, UCRL 10620 (1963).
82. M. Jeunehomme and R. P. Schwenker, J. Chem. Phys., 42(7):2406 (1965).
83. M. A. El'yashevich, Atomic and Molecular Spectroscopy, Fizmatgiz (1962).
84. A. S. Predvoditelev, Physical Gasdynamics, Izd. AN SSSR (1959).
85. P. P. Lazarev, Tr. Fiz. Inst. Akad. Nauk SSSR, 30:221 (1964).
86. Thermodynamic Properties of the Components of Combustion Products (Handbook of the Institute of Mineral Fuels, Academy of Sciences of the USSR, Institute of Applied Chemistry and Gamma-Ray Core Sampling, Council of Ministers of the USSR on Chemistry, 1962).
87. L. Brewer, W. T. Hicks, and O. H. Krikorian, J. Chem. Phys., 63:182 (1962).
88. M. A. Leontovich, Statistical Physics, OGIZ (1944).
89. F. Burhorn and R. Wienecke, Z. Chem., 212:105 (1959).
90. J. Hunaerts, Ann. Astrophys., 10:237 (1957).
91. E. Clementi, Astrophys. J., 133:303 (1961).
92. E. Hill and J. H. Van Vleck, Phys. Rev., 32:250 (1928).
93. J. A. Clouston and A. G. Gaydon, Spectr. Chim. Acta, 14:56 (1959).
94. N. M. Ogurtsova and I. V. Podmoshenskii, Opt. i Spektr., 4:439 (1959).
95. J. C. Camm, Rev. Sci. Instr., 31:270 (1960).
96. A. G. Sviridov and N. N. Sobolev, Zh. Éksperim. i Teor. Phys., 24:93 (1953).
97. F. S. Faizullov, N. N. Sobolev, and E. M. Kudryavtsev, Opt. i Spektr., 5:585 (1960).
98. R. W. Nichols, P. A. Fraser, and W. A. Jarmain, Combustion and Flame, 3:13 (1959).
99. G. Herzberg, private communication.
100. M. D. Raizer, A. G. Frank, and V. F. Kitaeva, Zh. Tekh. Fiz., 33:1011 (1963).
101. A. P. Dronov, A. G. Sviridov, and N. N. Sobolev, Opt. i Spektr., 10:312 (1961).
102. V. F. Kitaeva and N. N. Sobolev, Dokl. Akad. Nauk SSSR, 137:1091 (1961).
103. V. F. Kitaeva, V. V. Obukhov-Denisov, and N. N. Sobolev, Opt. i Spektr., 12:178 (1962).
104. I. I. Sobel'man, Zh. Éksperim. i Teor. Fiz., 48:965 (1965).
105. N. I. Krindach, L. N. Tunitskii, and N. N. Sobolev, Opt. i Spektr., 15:298 (1963).
106. E. M. Kudryavtsev, E. F. Gippius, S. S. Derbeneva, A. N. Pechenov, and N. N. Sobolev, Teplofiz. Vys. Temp., 1:376 (1963).
107. A. Treanor and W. Wuster, J. Chem. Phys., 32:758 (1960).
108. S. É. Frish, Optical Spectra of Atoms, Fizmatgiz (1963).
109. W. Voigt and S. B. Bayer, Akad. Wiss., 3:603 (1912).
110. A. R. Fairbairn, J. Quant. Spectr. Radiative Transfer, 6(3):325 (1966).
111. A. G. Sviridov, N. N. Sobolev, and V. M. Sutovskii, J. Quant. Spectr. Radiative Transfer, 5:525 (1965).
112. A. G. Sviridov, N. N. Sobolev, and M. Z. Novgorodov, J. Quant. Spectr. Radiative Transfer, 6:337 (1965).
113. A. G. Sviridov, N. N. Sobolev, M. Z. Novgorodov, and G. A. Arutyunova, J. Quant. Spectr. Radiative Transfer, 7:875 (1965).
114. J. A. Harrington, A. P. Modica, and D. R. Libby, J. Chem. Phys., 44:3380 (1966).
115 L. S. Ornstein, Physica, 3:561 (1936).
116. H. C. Hamaker, Thesis, Utrecht (1934).
117. J. C. DeVos, Physica, 20(10):690 (1964).
118. R. D. Larrabee, JOSA, 49:619 (1959).
119. G. Ribo, Optical Pyrometry, Gostekhizdat (1934).